VOLUME 1

Advances in Cell Biology

Series Editors:
David M. Prescott
Lester Goldstein
Edwin McConkey
Department of Molecular, Cellular and Developmental Biology, University of Colorado, Boulder, Colorado

VOLUME 1

Advances in Cell Biology

APPLETON-CENTURY-CROFTS
EDUCATIONAL DIVISION MEREDITH CORPORATION
NEW YORK

Softcover reprint of the hardcover 1st edition 1970

ISBN 978-1-4684-8481-6 ISBN 978-1-4684-8479-3 (eBook)
DOI 10.1007/978-1-4684-8479-3

6119-1

Library of Congress Catalog Card Number: 70-85896

390-71690-1

Contributors

B. R. BRINKLEY

The M.D. Anderson Hospital and Tumor Institute, The University of Texas, Texas Medical Center, Houston, Texas 77025

LESTER GOLDSTEIN

Department of Molecular, Cellular, and Developmental Biology, University of Colorado, Boulder, Colorado 80302

RICHARD J. GOSS

Department of Biology, Brown University, Providence, Rhode Island 02912

PETER L. KUEMPEL

Department of Molecular, Cellular, and Developmental Biology, University of Colorado, Boulder, Colorado 80302

DAVID M. PRESCOTT

Department of Molecular, Cellular, and Developmental Biology, University of Colorado, Boulder, Colorado 80302

HAROLD P. RUSCH

McArdle Laboratory for Cancer Research, Medical Center, University of Wisconsin, Madison, Wisconsin 53706

ELTON STUBBLEFIELD

The M.D. Anderson Hospital and Tumor Institute, The University of Texas, Texas Medical Center, Houston, Texas 77025

ANDERS ZETTERBERG

Institute for Cell Research, Karolinska Institutet, Stockholm, Sweden

Preface

Advances in Cell Biology has been initiated as a continuing, multi-volume series to report on the progress of a wide spectrum of problems of cell structure and cell function. In arranging these volumes individual contributors are asked not only to review the major new information, but especially to present the state of a given problem or area by discussing the current central issues, speculations, concepts, hypotheses, and technical problems. We intend, in addition, that these volumes will *not* be concerned with comprehensive reviews of the recent literature but will consist rather of presentations of an interpretative and integrative nature, based on selection of major research advances.

It is our aim that these volumes should provide the means whereby cell biologists may keep themselves reasonably well informed about the current progress in research areas in cell biology in which they are not immediately or directly involved themselves. The articles, nevertheless, are expected to bring into focus the experimental objectives of the specialists in a given research area.

D. M. P.

L. G.

E. M.

Contents

VOLUME 1

Advances in Cell Biology

Peter L. Kuempel

Department of Molecular, Cellular, and Developmental Biology, University of Colorado, Boulder, Colorado

1

Bacterial Chromosome Replication

INTRODUCTION

Studies of bacterial chromosome replication have progressed far enough that the broad outlines are now fairly well understood. Bacterial chromosomes are extremely long double helices of DNA which are circular in many if not all organisms. They are replicated in a sequential, nonrandom fashion from a definite point on the chromosome, and the initiation of replication is controlled by specific proteins. The time in the cell cycle when initiation occurs, the time required for the replication of a chromosome, and the number of growing chromosomes per cell are all definable properties of the cell. These properties vary, however, with the growth rate. At the end of a cell cycle, the chromosomes are segregated into the daughter cells. This segregation is probably effected by an attachment between the chromosome and the cell membrane. Very little is known concerning the nature of the attachment site(s) except that replication occurs at the cell membrane.

The purpose of this article is to review the experiments which have elucidated these and other properties of bacterial chromosome replication and which form the basis for future investigations in this field. This is an appropriate time for such a review, since the field is now entering a more molecular analysis of chromosome replica-

Written during the tenure of Grant GM 15905 from the National Institutes of Health.

tion. Other recent articles on this topic are by K. Lark (1966a, 1966b), Sueoka (1966, 1967), Maaløe and Kjeldgaard (1966), Cairns and Davern (1967), and K. Lark et al. (1967).

STRUCTURE OF THE BACTERIAL CHROMOSOME

Genetic and Physical Studies

The first indication that bacterial chromosomes are circular came from genetic studies. As a result of crosses between various *Escherichia coli* K12 Hfr and F^- strains, it was established that *E. coli* possesses a single, circular linkage group (Jacob and Wollman, 1961; Taylor and Trotter, 1967). This suggested that at least F^- bacteria have a circular chromosome. Subsequent matings between Hfr bacteria and F^- phenocopies made from Hfr bacteria demonstrated that there was close linkage between the origin and terminal loci of the Hfr chromosome (Taylor and Adelberg, 1961). Thus, the Hfr chromosome could also be circular, at least in stationary phase cells. Other bacteria for which circular genetic maps have been obtained are *Salmonella typhimurium* (Sanderson, 1967), which is closely related to *E. coli,* and the mycelial bacterium *Streptomyces coelicolor* (Hopwood, 1967).

Genetic analyses of *Bacillus subtilis* are not as complete as those of *E. coli* and *S. typhimurium,* and a circular genetic map has not yet been obtained for this organism. Most of the loci which have been studied have been placed in four linkage groups (Dubnau et al., 1967). These linkage groups could represent more than one chromosome. However, the genes in the linkage groups can be arranged in a single linear sequence by density-transfer experiments (Dubnau et al., 1967; O'Sullivan and Sueoka, 1967). When heavy (^{2}H-DNA), completed chromosomes were synchronously replicated in light medium, the sequence in which the genes appeared in the hybrid and light DNA indicated that the linkage groups were replicated sequentially. This suggests that the linkage groups were contained in one as yet incompletely mapped, possibly circular chromosome.

Genetic evidence alone is not sufficient to establish the structure of a chromosome, since structures other than a circle can give circular genetic maps, and vice versa (Stahl, 1967). That the *E. coli* chromosome is physically a circle, has been demonstrated by autoradiography. Cairns (1963b) labeled K12 Hfr 3000 cells with radioactive thymidine for two generations, released the chromosomes from the cells by means of lysozyme treatment in sucrose, and subjected the extended chromosomes to autoradiography. The resulting autoradiographs show that the chromosome is circular (Fig. 1). Sections A and B in Figure 1 are of equal length, and they are the growing daughter chromosomes. Sections Y to C and C to X are the unreplicated part of the parental chromosome. The length of an unreplicated chromosome would be about 1,100 μ (sections A or B plus Y to C and C to X).

An interesting feature of the replicating chromosome is that it contains two forks, X and Y, in the circular molecule. By means of the grain density in the autoradiograph, it can be deduced that one of these forks was the point at which replication was occurring (the replication fork), and that the other was the

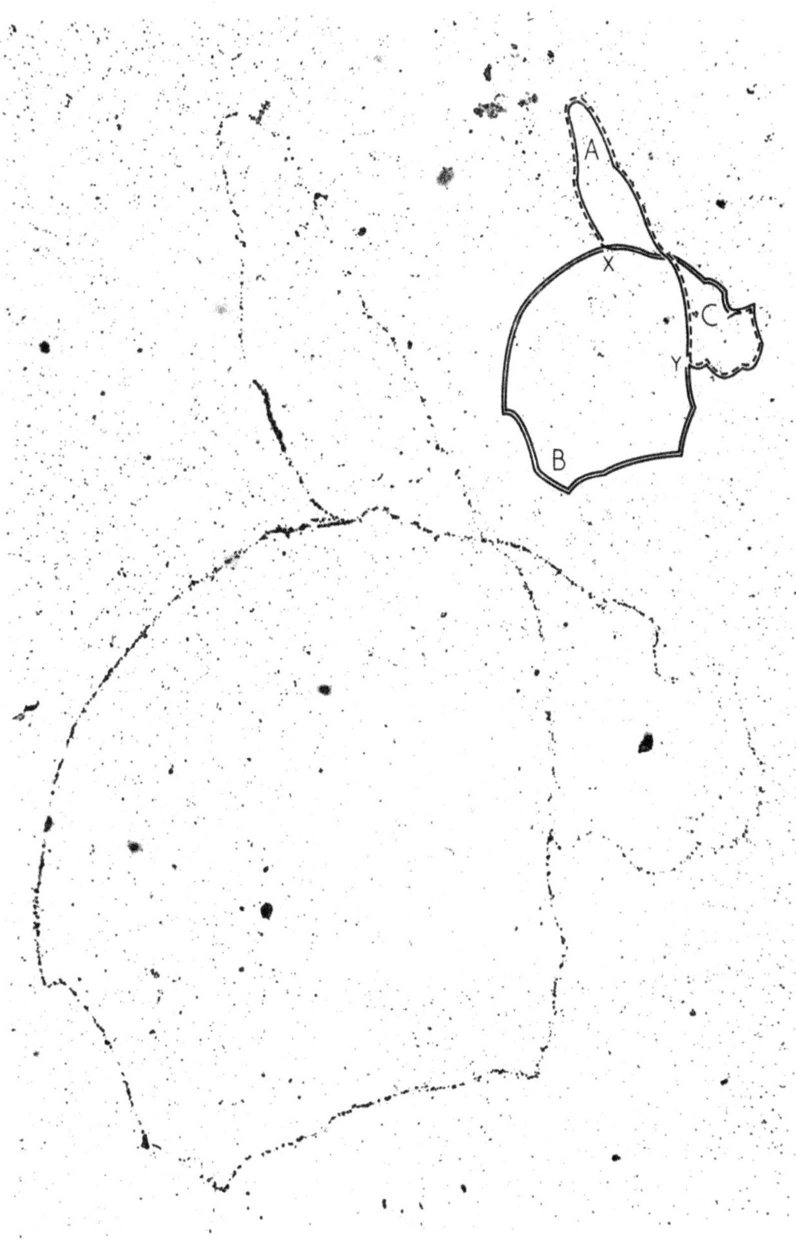

FIG. 1. Autoradiograph of the chromosome of *E. coli* K12 Hfr 3000, labeled with tritiated thymidine for two generations and extracted with lysozyme. Inset, the same structure is shown diagrammatically and divided into three sections (A, B, and C) that arise at the two forks, X and Y. Sections A and B are the growing daughter chromosomes, and section C is still unduplicated. Y is the replication fork and X is the replication origin and terminus. The labeled DNA is represented by solid lines. (From Cairns, 1963. *Cold Spring Harbor Symp. Quant. Biol.*, 28:43.)

point at which replication was initiated (the replication origin). This can be done as follows. One of the parental strands was labeled in only one section (C to X). The other parental strand was completely labeled. Consequently, the parental chromosome was a hybrid for most of its length, consisting of a labeled and an unlabeled strand. The replication of such a hybrid chromosome would produce a fork at which one fully labeled and two hybrid labeled sections were joined. The replication fork is thus at Y. The origin and terminus of replication is at X, since this fork is joined to the labeled end (C to X) of the largely unlabeled parental strand. This labeled end is the terminal part which was replicated immediately after the addition of the tritiated thymidine.

Cairns' autoradiographs do not demonstrate whether the circular chromosome contains only DNA, or whether other components, such as protein linkers (Freese, 1958) or swivels (Cairns, 1963a, 1963b) are present. The degree of resolution obtained in the autoradiographs would not show the presence of such non-DNA components. In addition, if these components are covalently linked to the DNA, the lysozyme extraction procedure would not break these bonds. In his earlier studies on chromosome structure, Cairns extracted the chromosomes by means of duponol in sucrose (Cairns, 1963a). *E. coli* strains B3 and K12 Hfr 3000 were used, and no intact, replicating, circular chromosomes were obtained. It could be argued that the duponol removed linkers or swivels from the chromosome. This implies, however, that the non-DNA units were not covalently linked to the chromosome. Cairns proposed a more likely alternative for this lack of complete replicating circles (Cairns, 1963b). Duponol probably caused the loss of basic proteins and polyamines which stabilized the chromosome against breakage by turbulence or tritium decay during the lengthy extraction procedure.

Electron microscopy has also been used to study the structure of bacterial chromosomes, and with this technique better resolution is attained than is possible with autoradiography. H. Bode and Morowitz (1967) have obtained electron micrographs of the *Mycoplasma hominis* chromosome, and their pictures demonstrate that its structure is similar to the *E. coli* structure observed by Cairns. In the non-replicating form, the average length of the chromosome is 262 μ. In the replicating form, there are two forks in the chromosome (Fig.2). The two forks, which are the replication fork and the replication origin, look alike, and the chromosome appears to have the same width at all points. If there is any material such as protein linkers or swivels inserted in the chromosome, it must have the same dimensions as a double helix. Also, since pronase was used in making the preparations, the material must be resistant to this enzyme or the chromosome would have been broken. This indicates that the *M. hominis* chromosome consists entirely of DNA. Similar looking replicating molecules have been observed in electron micrographs of mitochondrial DNA (Kirschner et al., 1968). This demonstrates that non-DNA swivels or linkers are not a necessary requisite for the replication of circular chromosomes.

Electron micrographs of the *Micrococcus lysodeikticus* chromosome suggest that it, too, might be circular (Kleinschmidt et al., 1961). In some of the pictures, no ends are observed on the chromosome. However, the preparations are not completely extended, and the ends could be hidden in a region of convergence. Electron

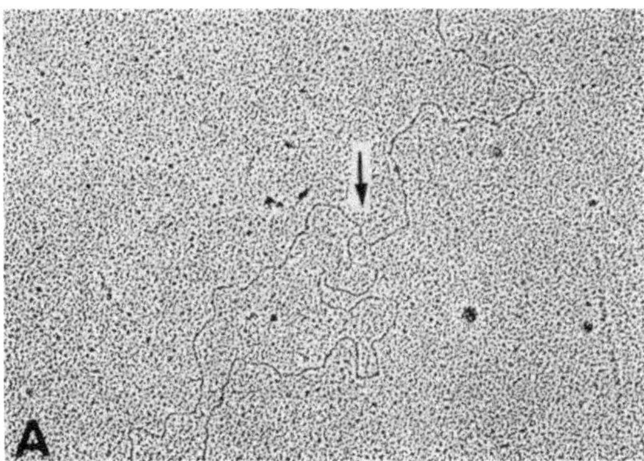

FIG. 2. Electron micrographs of forks in the *M. hominis* chromosome. (A) Fork (indicated by arrow) in a fragment of a DNA molecule. × 16,000. (B and C) Two overlapping micrographs of a double-fork fragment. × 17,000. The arrows indicate the two forks and the number the point of overlap. (From Bode and Morowitz, 1967. *J. Molec. Biol.*, 23:191.)

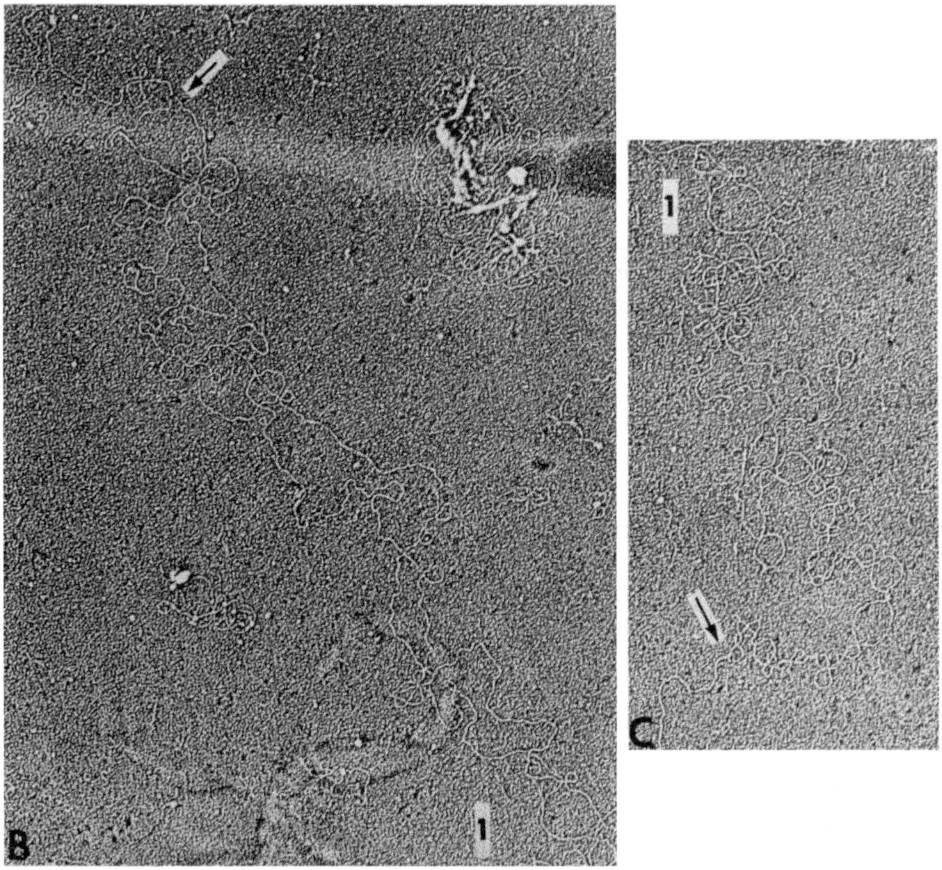

micrographs of the *Hemophilus influenzae* chromosome have revealed linear structures which are about 800 μ long (MacHattie et al., 1965). These molecules do not appear to be fragments of a larger chromosome, since all of the cellular DNA can be contained in one such structure. The linear form could be the normal structure of the chromosome, or it could be caused by the shearing of a circular chromosome. No replication forks can be seen in either the *M. lysodeikticus* or *H. influenzae* preparations. This is not surprising, since the cells were taken from the stationary growth phase.

The *B. subtilis* chromosome has been studied by both electron microscopy and autoradiography, and no intact, circular chromosomes have yet been observed. Earlier investigations showed linear molecules with lengths of 600 to 900 μ (Ganesan, 1963; Dennis and Wake, 1966). More recently, Dennis and Wake (1967) have obtained autoradiographs of structures with contour lengths of up to 1,360 μ. This length does not seem to represent the entire chromosome, however. The DNA content of a *B. subtilis* spore is 5.1 to 5.5 $\times 10^{-15}$ g (Fitz-James and Young, 1959; Dennis and Wake, 1966, 1967), and this is sufficient for 1,600 to 1,700 μ of DNA. Spores contain completed chromosomes (Yoshikawa et al., 1964), and if there is only one chromosome per spore, its length would obviously be about 1,700 μ. Autoradiographic determinations of the number of conserved DNA units present in spores indicate that there is only one chromosome (Ryter and Jacob, 1967; Dennis and Wake, 1967; Goldring and Wake, 1968).

Using the same procedure, however, Yoshikawa (1968) has obtained evidence that there are two chromosomes in spores. This could indicate that the chromosomes are less than 1,700 μ long. But Yoshikawa did not determine the DNA content of the spores he studied, and they possibly contained enough DNA for two chromosomes 1,700 μ long (about 10^{-14} g). Fitz-James and Young (1959) have observed that spores of some *Bacillus* species give multiples of the DNA content found in *B. subtilis,* and such differences might also occur in *B. subtilis* strains.

The autoradiographs of Dennis and Wake (1967) indicate that the *B. subtilis* chromosome is circular. No intact, replicating circles were observed, but some structures appear to be derived from replicating circles. These linear structures are divided for part of their length, and the grain density indicates the divided region is the growing daughter chromosomes. The replication fork and the replication origin would be the points at which the divided region converges. The length of these structures is considerably less than 1,700 μ, suggesting that part of the unreplicated region has been lost.

Yoshikawa (1967) has also obtained evidence that the *B. subtilis* chromosome is circular. Spores were germinated synchronously in medium containing 5-bromouracil and the DNA was then banded in alkaline CsCl gradients. This separated the strands and let them band at their corresponding densities. The initially synthesized DNA, which contained 5-bromouracil, banded at a density between the densities for thymine and 5-bromouracil strands. This intermediate density could be caused by a physical linkage between the strands at the terminus and the origin of replication, such as could be present in a circular chromosome. Such a linkage is difficult to prove by this type experiment, however. Strands with intermediate density could also result from a nonexchangeable pool of thymine derivatives, from uncompleted chromosomes in some of the spores, or from interstrand links in the DNA.

The question of whether all bacterial chromosomes are circular is obviously still unanswered. When circular chromosomes have been diligently sought, they have been found (Cairns, 1962, 1963a, 1963b). Studies of the *B. subtilis* chromosome seem to be leading to a similar conclusion. The major obstacle which must be overcome in analyses of chromosome structure is the extreme shear sensitivity of the molecules. Circular chromosomes have been observed by electron microscopy or sedimentation analyses in phage (Kleinschmidt et al., 1963; V. Bode and Kaiser, 1965; Marvin and Schaller, 1966; Mitra et al., 1967), sex factors (Hickson et al., 1967; Freifelder, 1968), colicinogenic factors (Roth and Helinski, 1967), resistance-transfer factors (Mickel et al., 1968), and animal cells (see Chapter 2). These numerous observations demonstrate that circularity may be a general property of chromosome structure. Even in preparations of small chromosomes, however, some linear structures were usually observed. These linear structures were most likely caused by shearing of circular molecules and probably do not represent a stage of the replication cycle. Since the shearing force exerted by a given gradient is expected to increase with the square of the molecular weight (Levinthal and Davison, 1961), the absence of circular molecules in prepared material is not yet sufficient proof of a linear structure for the chromosome. Until shear-free extraction procedures have been developed, apparently linear chromosomes must still be considered as potentially circular.

The Association of the Chromosome and Cell Membrane

An aspect of bacterial chromosome structure which is at present very poorly understood, is the nature of the association which exists between the chromosome and the cell surface. Such an association has been observed to exist in both gram-positive and gram-negative species (see Ryter, 1968, for a recent review). In gram-positive bacteria, the association frequently appears to be mediated by a mesosome. (A mesosome is a membranous structure consisting of an invagination of the cell membrane in which tubules are rolled up.) Figure 3 shows a direct connection between the chromosome and membrane at a mesosome in a protoplast of *B. subtilis*. The mesosome in this cell has been extruded from the cell due to the hypertonic medium. Mesosomes, however, are not essential for this association, since it can also be observed in cells which contain no mesosomes, such as germinating spores and reverting protoplasts. The association between the chromosome and cell surface is more difficult to observe in gram-negative bacteria. A direct contact between the nucleus and the cell membrane can be seen in spheroplasts of *E. coli,* however (Ryter and Jacob, 1966).

It is possible that this association involves an attachment factor which is inserted into the chromosome. If so, this factor might then be considered a part of the chromosome. The electron micrographs of the *M. hominis* chromosome argue against this possibility, however, at least for this organism. An alternate possibility is that this attachment occurs at the point at which replication occurs, the membrane-bound replication machinery acting as the attachment site. Evidence in support of this

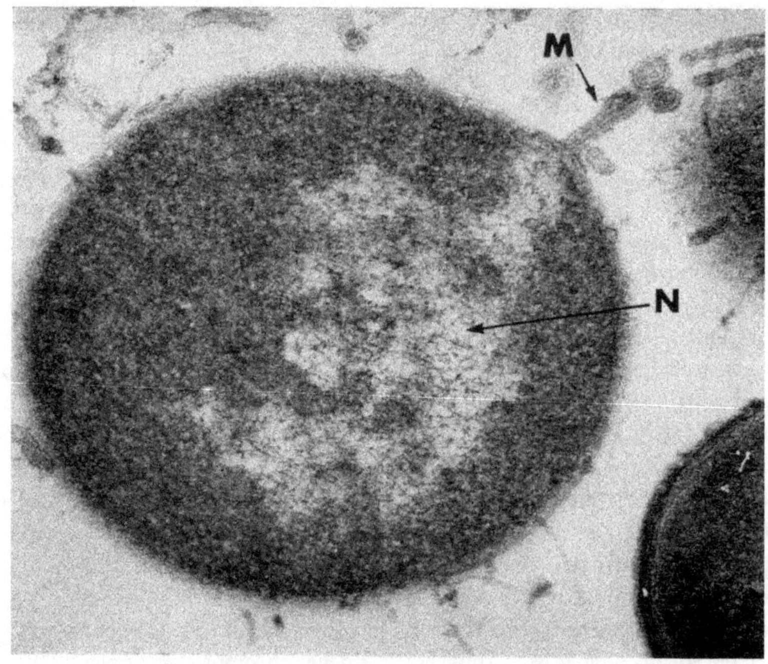

FIG. 3. Protoplast of *B. subtilis* showing association of the nucleus (N) with the membrane. A portion of a mesosomal tubule (M) is attached at the same place. × 52,500. (From Ryter and Jacob. 1966. *Ann. Inst. Pasteur,* 110:801.)

hypothesis comes from the experiments of Ganesan and Lederberg (1965) and Smith and Hanawalt (1967), who studied *B. subtilis* and *E. coli,* respectively. These investigators observed, by sedimentation analysis, that newly synthesized DNA was initially associated with the rapidly sedimenting membrane fraction. Shortly after synthesis, however, the DNA was found with the bulk of the DNA, which sediments more slowly.

The above observations do not preclude the possibility that the chromosome is attached to the cell surface at more than one point. As a result of an analysis of the chromosome transferring properties of *E. coli* K12 Hfr bacteria, Jacob et al. (1963) have proposed that the chromosome is attached to the cell surface at the site of insertion of the F factor. This attachment would be in addition to other sites of attachment. More recently, Sueoka and Quinn (1968) have tested the genetic identity of the DNA which is associated with the membrane in *B. subtilis.* They have observed that the membrane-bound DNA was enriched, relative to the membrane-free DNA, for the *ade* 16 and *met* markers. These markers are close to the origin and terminus of replication, respectively. These results, in conjunction with the evidence that replication occurs at the membrane, indicate that the *B. subtilis* chromosome is attached to the cell surface by at least two different sites. No direct evidence is available for *E. coli* concerning the attachment of the chromosome to the cell surface at a point other than the site of replication.

PROTEIN SYNTHESIS AND THE INITIATION OF CHROMOSOME REPLICATION

The initiation of a cycle of chromosome replication requires the synthesis of certain specific proteins. Once initiation has occurred, however, the cycle can be completed in the absence of further protein synthesis. That such a relationship exists between protein synthesis and intiation was first proposed by Maaløe and co-workers, who based this hypothesis on experiments concerned with thymineless death in *E. coli* strain 15 TAU, a strain that is auxotrophic for thymine, arginine, and uracil (Maaløe and Hanawalt, 1961; Hanawalt et al., 1961; Maaløe, 1961). It was observed that cells could be rendered immune to thymineless death if they were first incubated in the presence of thymine and the absence of arginine and uracil; subsequent removal of thymine did not kill the cells. However, if such immune cells were incubated in the absence of thymine and presence of arginine and uracil, they gradually became susceptible to thymineless death. Autoradiographs of cells from similar experiments indicated a strong correlation between the number of cells capable of synthesizing DNA and the susceptibility of the cells to thymineless death. On the basis of these observations, Maaløe and Hanawalt proposed that cells can enter thymineless death only if they are capable of replicating the chromosome when subjected to thymineless conditions, and that protein and/or RNA synthesis is necessary for the initiation, but not the completion, of a cycle of replication. According to this proposal, cells incubated in the presence of thymine and absence of arginine and uracil finish, but do not initiate cycles of replication, and are thus rendered immune to thymineless death. Incubation of such immune cells in the presence of arginine and uracil puts the cells in a state where a new round of synthesis is or can be initiated, and they lose their immunity.

Further indication of the requirement for protein and/or RNA synthesis for initiation, comes from the observation that DNA increased only about 40 to 50 percent in a culture of exponentially growing cells deprived of arginine and uracil. Maaløe and Hanawalt (1961) point out that this is close to the 39 percent increase expected if all of the chromosomes in an exponentially growing culture finish replication then in progress and no new cycles of replication are initiated.

K. Lark and co-workers have conducted a number of experiments which confirm and extend the above hypothesis. The experiments of Maaløe and co-workers did not allow a determination of whether the macromolecules required for initiation are protein or RNA, since both protein and RNA synthesis ceased as a result of the starvation procedures used. K. Lark et al. (1967) starved cells for required amino acids in the presence of high concentrations of chloramphenicol. Under these conditions, RNA synthesis continued in the absence of protein synthesis. No new rounds of replication were initiated, which demonstrates that proteins, and not just RNA, are required for initiation.

When DNA synthesis is reinitiated following a period of amino acid starvation, the replication occurs at a specific portion of the chromosome. This has been demonstrated by the use of density-transfer experiments (K. Lark et al., 1963; Pritchard and Lark, 1964). Exponentially growing cells of an *E. coli* $15T^-$ strain were starved for required amino acids until DNA synthesis had ceased, the cells were resupplemented with the amino acids, and a pulse of tritiated thymine was given as DNA synthesis resumed. This specifically labeled the portion of the chromosome which was replicated when initiation occurred (Fig. 4). The cells were then grown for several generations to randomize growth, and they were again starved for the required amino acids. Once DNA synthesis had ceased, the cells were transferred to complete medium which contained 5-bromouracil instead of thymine. Samples were removed after various periods of growth in this medium, and the amount of the tritiated thymine present in the hybrid density DNA was determined. If replication is initiated at the same portion of the chromosome following each period of amino acid deprivation, it would be expected that the tritiated thymine labeled DNA would become hybrid more rapidly than the bulk of the DNA. The results of the experiment demonstrated that a preferential replication occurred (curve A, Fig. 5). If the second synchronization of replication via amino acid deprivation was

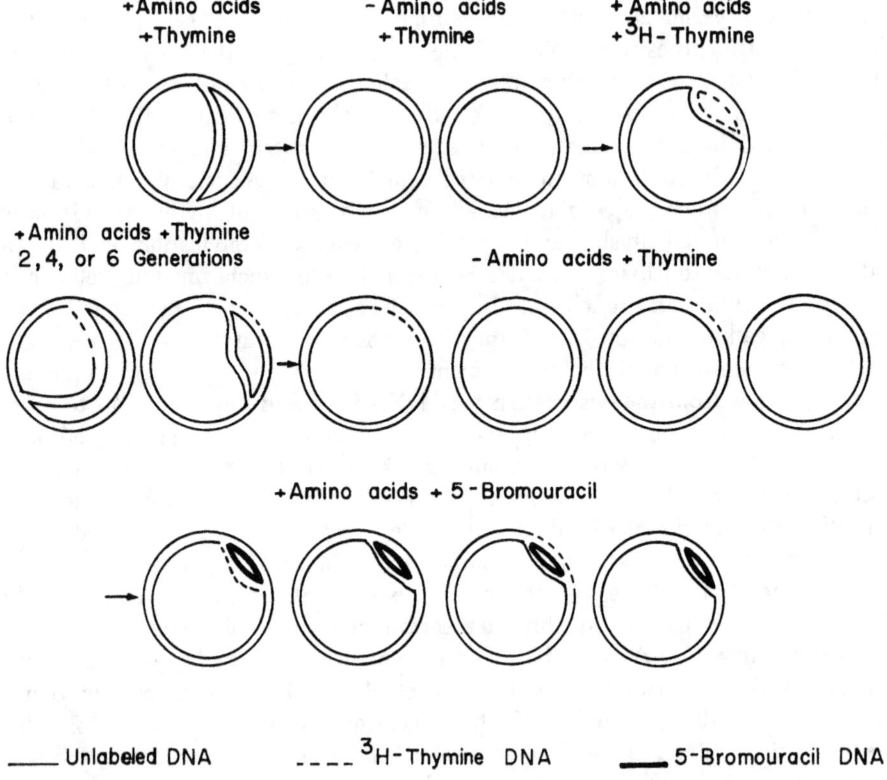

FIG. 4. Diagrammatic representation of what happens when amino acid deprivation is used to line up chromosomes for labeling the replication origin region first with tritiated thymine and subsequently with 5-bromouracil. Exponentially growing cells are derprived of required amino acids until DNA synthesis ceases. A pulse of ^{3}H-thymine is then used to label the DNA synthesized immediately after the addition of the amino acids. The cells are then grown for two, four, or six generations to randomize growth, and the amino acids are then removed from the medium. Upon the readdition of the amino acids, the cells are transferred to 5-bromouracil containing medium. (Adapted from K. Lark, 1966. *Bact. Rev.*, 30:3.)

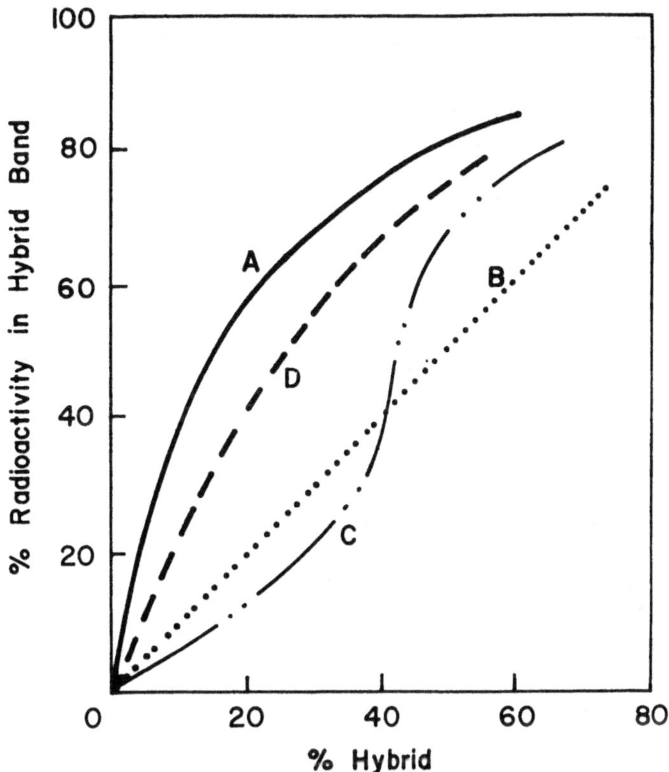

FIG. 5. (A) Preferential replication of the tritium labeled DNA in the experiment described in Figure 4 (Pritchard and Lark, 1964). (B) Replication of the tritium labeled DNA if the second period of amino acid deprivation is omitted (Pritchard and Lark, 1964). (C) An experiment of the type described in Figure 4, in which the tritium labeling occurred after replication of 10 percent of the chromosome in nonradioactive medium (Lark et al., 1963). (D) An experiment of the type described in Figure 4, in which 40 minutes of thymine starvation was substituted for the second period of amino acid deprivation (Pritchard and Lark, 1964). (From K. Lark, 1966. *Bact. Rev.*, 30:3.)

omitted, there was no preferential replication of the tritium labeled DNA (curve B, Fig. 5).

The part of the chromosome at which replication is initiated following a period or amino acid deprivation is called the replication origin. Replication is also initiated at this part of the chromosome following a number of other treatments. Pritchard and Lark (1964) observed that if exponentially growing *E. coli* 15T$^-$ cells are starved of thymine, the normal replication sequence is altered. When thymine is added back to the starved cells, the chromosomes are now replicated from two points, the preexisting replication fork and the replication origin. This phenomenon has been termed premature initiation. To demonstrate that premature initiation occurred at the replication origin, the origin was labeled with tritiated thymine as described before. The cells were then grown for several generations to randomize growth, and the cells were starved for thymine for 40 minutes. 5-bromouracil was then added to the medium, and the time of replication of the tritium labeled DNA was determined. There was a preferential shift of radioactivity to the hybrid density, indicating that replication had occurred at the replication origin (curve D, Fig. 5). The rate of this shift was not as great as that observed in experiments of the type described in the preceding paragraph, however. It was proposed that this occurs because DNA synthesis continues at the preexisting replication fork, and because premature initiation occurs at only one of the replication origins of the partially replicated chromosomes.

The preceding proposal predicts that cells which have undergone premature initiation should give a 110 percent increase in their DNA in the absence of protein synthesis. This is close to the result obtained by Pritchard and Lark. But similar results would have been obtained if half of the chromosomes did not undergo premature initiation, and the other half had premature initiation occurring at both replication origins. Experiments which test these possibilities have been reported by K. Lark and Bird (1965a). Autoradiographs were made of cells which were first incubated in the absence of thymine, and then in the presence of tritiated thymine in the absence of required amino acids. If premature initiation occurred at one of the replication origins of all of the chromosomes, a unimodal distribution for the number of labeled conserved units would be expected. A unimodal distribution was obtained, which is consistent with Pritchard and Lark's proposal.

A phenomenon similar to premature initiation is observed when *E. coli* K12 thymine-requiring strains are shifted to medium containing 5-bromouracil. Abe and Tomizawa (1967) have reported experiments which demonstrate that after such a shift, replication was initiated at the replication origin. This premature initiation was different, in some respects, from that observed in *E. coli* 15T$^-$. The 15T$^-$ strain, unlike the K12T$^-$ strains that were tested, did not exhibit premature initiation in 5-bromouracil medium unless growing cells were first starved for thymine (Pritchard and Lark, 1964). Also, Abe and Tomizawa propose that the premature initiation in K12T$^-$ strains occurs at both of the replication origins of a partially replicated chromosome, and that DNA synthesis at the preexisting growth point ceases after a shift to the 5-bromouracil medium. This cessation of synthesis does not occur immediately after the shift, however (see p. 20).

Replication is also initiated at the replication origin when DNA synthesis resumes following the removal of phenethyl alcohol. Treick and Konetzka (1964) have proposed that phenethyl alcohol inhibits the initiation of chromosome replication, but not the completion of rounds of replication already in progress. K. Lark and C. Lark (1966) tested the effect of phenethyl alcohol on chromosome replication, and they observed that, following the removal of phenethyl alcohol inhibition of DNA synthesis, chromosome replication was initiated at the same place on the chromosome as following amino acid deprivation. The design of these experiments was similar to that used in studying premature initiation.

In addition to studying the chromosome locus at which replication is initiated following amino acid deprivation, thymine deprivation, and phenethyl alcohol treatment, Lark and co-workers have also investigated the types of proteins which must be synthesized to allow initiation to occur after these treatments. As a result of these studies, Lark has proposed that more than one protein is required for the initiation of chromosome replication (C. Lark and K. Lark, 1964; K. Lark, 1966a; K. Lark et al., 1967). The pertinent observations are the following. Protein synthesis is required for the initiation of chromosome replication; this initiation does not occur in the absence of required amino acids. However, the synthesis of at least some of the proteins required for this initiation is resistant to low levels of chloramphenicol and 5-fluorouracil (C. Lark and K. Lark, 1964). The amount of initiation observed in the presence of these inhibitors is reduced, but it is significant. Protein synthesis is also required for the premature initiation that is caused by thymine starvation; this initiation is not observed if the cells are deprived of a required amino acid during the thymine starvation (Pritchard and Lark, 1964). Unlike initiation after amino acid deprivation, premature initiation is completely inhibited by low levels of chloramphenicol or 5-fluorouracil (C. Lark and K. Lark, 1964). Premature initiation is also inhibited by amino acid analogues, whereas the analogues do not inhibit initiation after amino acid deprivation (K. Lark et al., 1967). If these two types of initiation are related, these results suggest that at least two proteins are required for normal initiation. (For the postulated roles of these proteins in the initiation of replication, see Lark's model for the control of replication discussed below.)

Experiments involving the use of phenethyl alcohol also indicate that at least two proteins are involved in initiation (K. Lark and C. Lark, 1966). Following the treatment of cells with phenethyl alcohol, the removal of a required amino acid completely inhibits the initiation of chromosome replication. If low levels of chloramphenicol are used instead to inhibit protein synthesis once the phenethyl alcohol is removed, DNA synthesis resumes at a high rate. This demonstrates that the synthesis of a chloramphenicol insensitive protein is required for initiation. If low levels of chloramphenicol are used to inhibit protein synthesis during the phenethyl alcohol treatment, and if chloramphenicol is left in the culture after the phenethyl alcohol is removed, only a low amount of synthesis is observed. This suggests that a phenethyl alcohol resistant, chloramphenicol sensitive initiator protein accumulates during the phenethyl alcohol treatment, and that this protein functions in initiation. But this protein is not sufficient for initiation. Initiation also requires the synthesis of a chloramphenicol resistant, phenethyl alcohol sensitive initiator protein.

Chloramphenicol appears to act by inhibiting the growth of nascent polypeptide chains (Weber and DeMoss, 1966) and it seems unusual that protein synthesis should be differentially sensitive to this antibiotic. Proteins whose syntheses are not inhibited by low levels of chloramphenicol, are not unique to the initiation protein(s) of *E. coli,* however. The synthesis of progeny RF (replicative form) in S13 (Tessman, 1966) and ΦX174 (Levine and Sinsheimer, 1968) infected cells, requires a protein whose synthesis is resistant to chloramphenicol. The synthesis of the Lac repressor is also resistant to low levels of chloramphenicol and 5-methyltryptophan (Pardee and Prestidge, 1959). It is possible that this differential sensitivity is caused by a protein synthesis system that is relatively insensitive to chloramphenicol. A more likely possibility is that only very small amounts of these proteins are required for the expression of activity *in vivo,* and that sufficient amounts are still produced when protein synthesis is severely reduced (Levine and Sinsheimer, 1968).

Protein synthesis is required for the initiation of a cycle of chromosome replication in other bacteria besides *E. coli*. Yoshikawa (1965) has studied the relationship between protein synthesis and chromosome replication in cultures of synchronously germinating spores of *B. subtilis*. By adding chloramphenicol at different times during the growth of the culture, he observed that chloramphenicol inhibited the initiation of new rounds of replication, but not the completion of rounds of replication in progress. This inhibition is somewhat different from that observed in *E. coli*. Twenty μg/ml of chloramphenicol is sufficient to inhibit the initiation of replication in *B. subtilis,* whereas 150 μg/ml is usually used for complete inhibition in *E. coli* (Clark and Maaløe, 1967). *B. subtilis* is similar to *E. coli* in that premature initiation results if growing cells are deprived of thymine (Kallenbach and Ma, 1968). Whether this premature initiation involves the synthesis of a distinct initiator protein is unknown. *Lactobacillus acidophilus* also has properties similar to those observed in *E. coli*; protein synthesis is required for the initiation of a round of replication, and growth in the absence of DNA synthesis results in premature initiation (Soska and Lark, 1966). Only a small number of species have been investigated, but these properties are probably found in many kinds of bacteria.

SEQUENTIAL REPLICATION OF THE BACTERIAL CHROMOSOME

Escherichia coli

The simplest process for replicating a long, continuous DNA molecule such as the *E. coli* chromosome consists of a single replication fork which starts at a specific locus and then moves around the chromosome in a sequential, nonrandom fashion. The first experiments that suggested replication occurs in this fashion were the density-shift experiments of Meselson and Stahl (1958). One generation after the shift from heavy to light medium, virtually all of the DNA was of hybrid density. This demonstrates that once a point on the chromosome was replicated, that point was not replicated again until the rest of the chromosome was replicated. Such a nonrandom mode of replication is most readily explained by a single, sequentially moving replication fork. Meselson and Stahl's results also suggest that initiation occurred at a specific locus. If the point of initiation had shifted randomly in each replication cycle, one generation after the medium shift 11 percent heavy, 78 percent hybrid, and 11 percent light DNA would have been expected (Moran, 1963). This is near the limit of sensitivity of the procedure, however.

The experiments of Bonhoeffer and Gierer (1963) also indicate that *E. coli* chromosome replication is sequential and nonrandom. Cells were labeled with 5-bromouracil for various intervals, and the density and molecular weight of the 5-bromouracil-containing DNA were determined. Knowing these values, the rate at which the 5-bromouracil-containing fragments became completely hybrid, and consequently the rate of synthesis at the replication fork, could be determined. This rate was then compared to the rate at which 5-bromouracil was incorporated into a nucleus. Assuming that each nucleus contains one chromosome, the results indicate that a replicating chromosome contains one and at the most two replication forks. A single replication fork could most readily replicate a large chromosome by moving sequentially around it.

The most elegant demonstration of the sequential replication of the *E. coli* chromosome, the autoradiographs of Cairns (1963a, 1963b), has already been discussed above. Given the evidence that the circular chromosome contains two forks, the replication origin and the replication fork, it is virtually impossible to provide a model for replication which is other than sequential. The grain density in the autoradiographs is also often used as a demonstration that initiation always occurs at the same place on the chromosome. It should be noted that the grain density in Figure 1 lends itself to an alternate interpretation, however. The density near the origin of the half-labeled daughter chromosome is consistent with the proposal that the point of initiation has shifted in a clockwise direction since the partly labeled parental strand was synthesized. If replication was always initiated at the same point on the chromosome, the replication origin of the hybrid daughter chromosome should not have label in both strands. The excess grains near the replication origin are conceivably caused by tangling, but the published autoradiographs do not contradict the preceding proposal.

The first experiments that attempted to determine the genetic identity of the replication origin and the direction of replication were conducted by Nagata (1963). Synchronous cultures of *E. coli* K12 Hfr strains H and CS101 that contained either prophage λ or prophages λ and 424 were used (Fig. 6). At various times in the cell cycle, the prophage(s) in these strains were induced with ultraviolet light, and either the number of λ prophage per cell or the ratio of λ to 424 in the phage produced was determined. It was found that these values changed by a factor of two at specific times in the cell cycle (Fig. 7). Assuming that the abrupt twofold changes were caused by the replication of the parts of the chromosome containing the different prophages, and knowing the amounts of DNA synthesized between these changes and the positions of the prophages on the chromosome, the results indicate that chromosome replication in Hfr bacteria proceeds in an F factor to origin direction. Furthermore, if it is assumed that initiation occurs at the time of cell division, the site of initiation would be at the F factor. F^- bacteria containing prophage λ or prophages λ and 424 showed no abrupt changes during the cell cycle in the number of λ prophage per cell or the ratio of λ to 424 in the phage produced. Nagata interpreted this as indicating that the F^- bacteria have no specific locus at which a cycle of chromosome replication is initiated.

Since Nagata's studies, many other investigators have worked on this problem, and a number of conflicting observations have been made. Vielmetter et al. (1968) have conducted two different types of experiments which provide support for Nagata's proposals. In one type of experiment cells were labeled with ^{32}P and stored until ^{32}P decay had caused an average of nine lethal hits (about 10^{-4} survivors). The ^{32}P decay induced mutations, and about one percent of the survivors were auxotrophs of different kinds. Many of these mutants were heterozygotes; they produced sectored colonies. Vielmetter et al. propose that the closer a locus is to the replication origin, the greater the probability that heterozygous mutants will be produced. As one of several models for how this could occur, they propose that ^{32}P

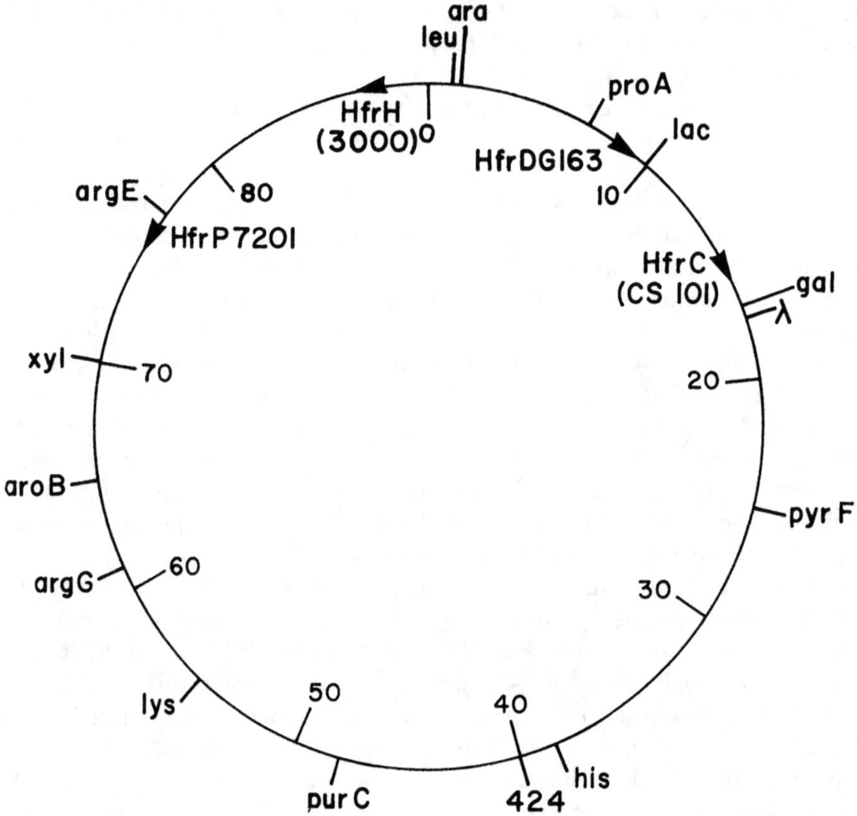

FIG. 6. Linkage map of *E. coli* showing pertinent genetic loci and the chromosome transfer properties of Hfr strains.

mutations are induced in single strands. All of the mutants would consequently be heterozygous if ^{32}P decay only induced mutagenesis. But ^{32}P also causes lethal hits which inactivate the chromosomes. A lethal hit in the unreplicated part of a chromosome inactivates the entire chromosome. A lethal hit in one of the replicated branches of a partially replicated chromosome kills the newly arising chromosome that derives from that branch. The newly arising chromosome that is derived from the undamaged branch is not inactivated (Koch, 1966). Consequently, if a mutation is in the unreplicated part of a chromosome which suffers a lethal hit in one of the replicated parts, and if the mutation is in the strand that goes into the active progeny chromosome, completion of the round of replication will produce a homozygous mutant. Mutations behind the replication fork will remain heterozygous. Since the chromosomes were at different stages of the replication cycle when stored to allow ^{32}P decay to occur, an increasing gradient for the proportion of heterozygous mutants would be expected for genes at increasing distances from the terminus of replication.

Vielmetter et al. determined the frequency of heterozygous mutants for different genes in several Hfr strains. They observed that the proportion of ^{32}P induced

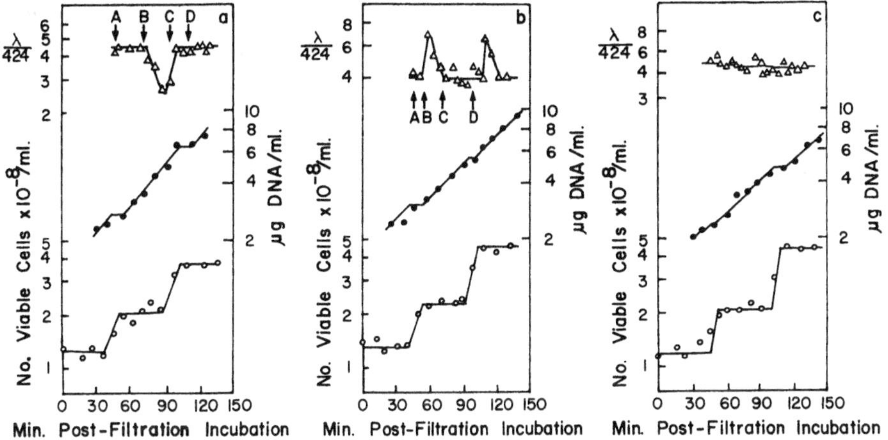

FIG. 7. Change in the relative number of prophages λ and 424 during one replication cycle of DNA in synchronized populations of *E. coli* K12. The relative number was expressed in terms of the ratio (△) of phage λ to phage 424 in the lysate obtained after induction of double lysogens with UV. (●) DNA amounts; (○) viable counts of bacteria. (a) Hfr H (λ, 424), (b) Hfr CS101 (λ, 424), (c) F-Z260 (λ, 424). (From Nagata, 1963. *Proc. Nat. Acad. Sci. U.S.A.*, 49:551.)

heterozygous/total mutants was higher for loci near the F factor as compared to loci near the conjugation origin. This gradient in the proportion of heterozygous mutants was not observed in F^- cells. These results are consistent with the proposals that replication in Hfr cells is initiated at the F factor end of the chromosome and proceeds towards the conjugation origin, and that F^- cells have no specific replication origin. Vielmetter et al. also conducted more complex experiments in which the segregation kinetics of mutations induced by nitrosoguanidine were studied. The results of these experiments are also consistent with Nagata's proposals.

Other investigators have obtained results which lead to a different proposal for the origin and direction of chromosome replication. Berg and Caro (1967) have studied the effect of the location and orientation of the F factor on chromosome replication, and they have concluded that the F factor has no effect. The rationale behind their experiments was similar to that used by Yoshikawa and Sueoka (1963a) in investigating chromosome replication in *B. subtilis*. Markers that are near the origin of chromosome replication are expected to be about twice as frequent in exponentially growing cells as markers which are near the terminus of replication. If the F factor affects replication of the chromosome, marker frequencies should be different in different Hfr's. Transduction was used to determine the frequency of markers in several isogenic Hfr strains with different origins and directions of conjugation. No significant differences were observed. This similarity of marker frequencies suggests that the different strains have a common origin and direction of replication. The results do not indicate, however, what the origin and direction are.

Cutler and Evans (1967) have obtained evidence that indicates replication occurs in a clockwise direction in *E. coli* K12 Hfr 3000 (λ, 424), a Hayes-type Hfr that contains prophages λ and 424. Synchronous cultures were labeled with 5-bromouracil at various times in the cell cycle, and the hybrid DNA molecules were isolated by means of density gradient centrifugation. This procedure isolated the segments of the chromosome that were replicated at different times during the cell cycle. The isolated segments were then hybridized with phage λ and 424 messenger RNA's, and the extent of the hybridization was determined. This allowed a determination of the times at which the λ and 424 containing segments of the chromosome were replicated. A clockwise direction of replication was indicated, but the results do not demonstrate the point at which initiation occurred. An interesting aspect of these experiments is that they indicate that 5-bromouracil induced initiation does not occur immediately after a shift to 5-bromouracil containing medium, and that replication at the preexisting replication fork continues for some time after such a shift. Little cross-annealing occurred between various DNA samples that were isolated after about 0.1 generation growth in 5-bromouracil. If initiation were induced immediately by the 5-bromouracil, it would be expected that the DNA samples taken at the different times in the cell cycle would all cross-anneal with each other.

Abe and Tomizawa (1967) also used 5-bromouracil to study chromosome replication. In these experiments, however, the incubation periods in 5-bromouracil medium were long enough to induce initiation at the replication origin. *E. coli* K12T$^-$ strains Hfr H, Hfr C, and F$^-$ 3110 were used (Fig. 6). Exponentially growing cells were transferred to medium containing 5-bromouracil, and at various times the cells were shifted back to medium containing thymine and infected with phage P1. Phage P1 is a generalized transducing phage. After the cells had lysed, the frequency of transducing phage for various markers was determined in the light, hybrid, and heavy phage bands in CsCl gradients. The results demonstrate that in all of the strains tested, the *lys* marker was among the first to exhibit extensive incorporation of 5-bromouracil, and the markers in a clockwise direction from *lys* incorporated decreasing amounts of the analogue in a given period of incorporation. Assuming that 5-bromouracil induced initiation did not produce an abnormal origin and direction of replication, the results demonstrate that the F factor has no effect on chromosome replication, that the replication origin is between the *lys* and *his* markers (between 39 and 55 minutes), and that replication occurs in a clockwise direction. An unusual result from these experiments is that *leu,* instead of *lys,* was the first marker to be found in the heavy phage. The reason for this is unknown, but the result could have been caused by a second initiation of replication near *leu* before the first round of 5-bromouracil induced replication was completed.

Wolf et al. (1968) used a somewhat different procedure to study the origin and direction of chromosome replication in two K12 F$^-$ and two Hfr strains (P7201 and DG163, Fig. 6). Either the "ends" or the "beginnings" of the chromosomes were labeled with 5-bromouracil, and the markers present in the 5-bromouracil-containing DNA were determined. The "ends" of the chromosomes were labeled by incubating the cells in the presence of 5-bromouracil in the absence of required amino acids. The "beginnings" were labeled by depriving the cells of required amino acids until DNA synthesis had ceased, and then incubating the cells in the presence of 5-bromouracil and required amino acids for one-half generation. Following the labeling of the "ends" or the "beginnings", the cells were transferred to medium containing thymine and infected with phage P1. When the cells had lysed, the progeny phage were banded in CsCl density gradients to separate the phage containing 5-bromouracil labeled DNA. These are the transducing phages that picked up a piece of the 5-bromouracil labeled "ends" or "beginnings" of the host cell chromosomes. The transduction frequencies of these phages were then determined, and the transduction frequencies from phage grown on exponential cells were used to normalize the data. Figure 8 shows the normalized transduction frequencies for F⁻ strain DG75. The results indicate that the two F⁻ strains and Hfr strain P7201 had the origin and terminus of replication in the region between *lys* and *xyl* (55 to 70 minutes). This replication origin is distinctly different from the location of the F factor in P7201. The direction of replication in these strains was clockwise, as indicated by the gradient in transduction frequencies from *ara* to *lys* in the "ends" experiments. The other Hfr strain they studied had the origin and terminus

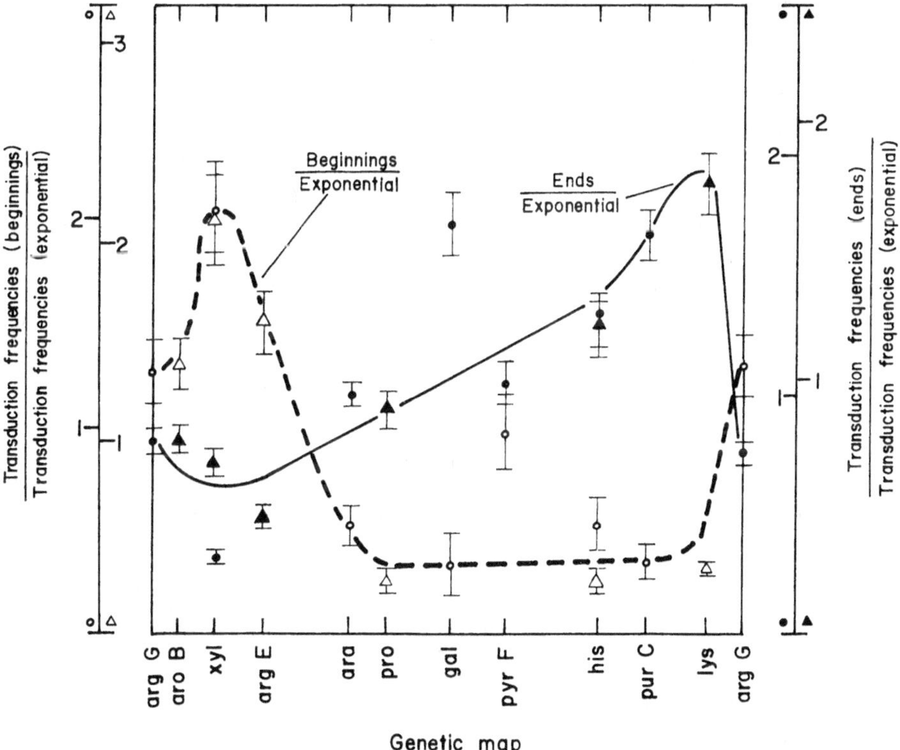

FIG. 8. Normalized transduction frequencies for F-strain DG75. Errors shown are standard deviations: ○, ● assayed on recipient DG90; △, ▲ on DG111. Beginnings ○, △. Ends ●, ▲. Beginnings and ends transduction frequencies were divided by transduction frequencies obtained from exponential phage lysate P1-DG75 to obtain the normalized values. The abscissa is the circular linkage map of Figure 6 opened at *arg* G. (From Wolf et al., 1968. *J. Molec. Biol.*, 32:611.)

in the region between the *pro* and *gal* markers (7 to 17 minutes), and the direction of replication was counterclockwise. An interesting feature of this strain is that it appears to have another replication origin between *lys* and *xyl*; markers from both the *pro* to *gal* and the *lys* to *xyl* regions were found in the transducing phage when the "beginnings" were labeled. Markers from the *lys* to *xyl* region were not found when the "ends" were labeled with 5-bromouracil. The reason for this is unclear, but this could have been caused by the induction of a second origin by 5-bromouracil, as also suggested by Abe and Tomizawa's results.

A number of other investigators have obtained results which are similar to those of Abe and Tomizawa (1967) and Wolf et al. (1968). Espardellier-Joset et al. (1967) have used a procedure similar to the "beginnings" procedure of Wolf et al., and the results indicate that the origin is between 61 and 74 minutes, and that the direction of replication is clockwise in Hfr strain B11. The same origin was observed when thymine deprivation was used to induce premature initiation. A more extensive analysis has been conducted by Cerdá-Olmeda and Hanawalt (1968), who used a procedure that did not involve the use of 5-bromouracil. Synchronous chromosome replication was obtained by amino acid deprivation in strains that had a number of nutritional requirements. When chromosome replication was initiated by the addition of the required amino acids, the cells were mutagenized at intervals with short pulses of nitrosoguanidine. As demonstrated by Cerdá-Olmeda and Hanawalt, this mutagen acts primarily in the replication fork region of the chromosome. The times at which various mutations reverted to independence from nutritional requirements demonstrated that the *arg* G marker at 61 minutes was the first to be replicated, and that replication was clockwise. Similar results were obtained with 15T$^-$, K12 Hfr, and K12 F$^-$ strains, and with the 15T$^-$ strain when premature initiation was induced via thymine deprivation.

An objection against the above experiments that demonstrate that the origin is at 55 to 65 minutes on the *E. coli* linkage map and that replication occurs in a clockwise direction is that all of the experiments involved the use of amino acid deprivation or premature initiation. These procedures could conceivably induce properties in the replication system different from those found in untreated cells. That the replication origin in untreated cells is also in this part of the chromosome is demonstrated by recent experiments of Helmstetter (1968) and Pato and Glaser (1968). The rate of DNA synthesis and the induced rates of synthesis of several enzymes were determined in relation to the cell cycle in *E. coli* B/r. The cells were either growing in synchronous cultures or they were selected with respect to cell age. The times at which the rate of DNA synthesis increased indicated the times at which new rounds of replication were initiated (see p. 26). The times at which the induced rates of synthesis of the enzymes increased indicated the times at which the corresponding structural genes were replicated, since the rate of synthesis is related to the number of copies of the structural gene in the cell (Pittard and Ramakrishnan, 1964; Masters and Pardee, 1965). The results are consistent with the mode of replication discussed above.

To sum up, it now seems that the majority of *E. coli* strains have the replication origin between 55 and 65 minutes on the chromosome, and that replication is clockwise. This type of replication has been observed in 15T$^-$, B/r, and K12 Hfr and F^- strains. These results demonstrate that the integrated F factor does not usually influence the location of the replication origin. There are strains, however, that appear to have different origins and directions of replication. This is suggested by the results of Nagata (1963), Vielmetter et al. (1968), and Wolf et al. (1968). In addition, Berg and Caro (1967) mention that some of their strains appear to have different origins and/or directions of replication. The reason for these differences is unknown at present. They are most likely due to strain difference, but they may be due to differences in procedures and technique. The resolution of this problem will depend on the characterization of the origin and direction of replication in these atypical strains by a number of different procedures.

Bacillus subtilis

The sequential replication of the *B. subtilis* chromosome was demonstrated by Yoshikawa and Sueoka (1963a), who used an ingenious approach to this problem. They reasoned that if chromosome replication has a polarity, genes near the origin of replication would be twice as frequent in exponentially growing cultures as genes near the terminus of replication. Genes in between would have a predictable frequency, depending on their position on the chromosome. (For a detailed mathematical presentation of this argument, see Sueoka and Yoshikawa, 1965.) Transformation was used to determine the frequencies of various genetic markers in exponentially growing cells. To correct for the differences in the integration frequencies of the markers, the transformation frequencies were also determined using DNA from late stationary phase cells. It was assumed that the chromosomes in such cells would be in a completed form of replication, and that all genes would be present equally. When the values obtained from the exponentially growing cells were normalized with respect to those from stationary phase cells, it was found that the relative marker frequencies ranged from 2 to 1 in *B. subtilis* strain W23. The *ade* marker was twice as frequent as the *met* marker, indicating that the former was near the replication origin, and that the latter was near the terminus of replication. Since the *B. subtilis* chromosome is a long DNA molecule (Ganesan, 1963; Dennis and Wake, 1966, 1967), this nonrandom mode of replication strongly suggests that replication occurs sequentially along the chromosome.

When experiments of the above type were done with *B. subtilis* strain W168, the relative frequencies of the markers only varied from 1.3 to 1 (Sueoka and Yoshikawa, 1963). This suggested that replication in this strain was different from that in strain W23, possibly in a fashion analogous to the difference between *E. coli* K12 Hfr and F^- replication proposed by Nagata (1963). However, density-shift experiments similar to those of Meselson and Stahl (1958) indicated that replication in strain W168 was nonrandom and that a specific replication origin was utilized (Yoshikawa et al., 1964). Additional experiments demonstrated that stationary phase cells of strain W168 did not contain completed chromosomes, invalidating the normalization procedure using DNA from such cells. When spore DNA was used for the normalization procedure, the relative frequencies of the markers in exponentially growing cells was found to range from 2 to 1, and the sequence of replication was the same in strains W23 and W168.

The sequential replication of the *B. subtilis* chromosome has also been demonstrated by other procedures. If strain W23 cells are grown to stationary phase in heavy medium and then grown in light medium, the markers appear in hybrid DNA in the replication sequence predicted on the basis of the marker frequency analysis described above (Yoshikawa and Sueoka, 1963b). A sequential replication of the chromosome is also observed in synchronous cultures derived from germinating spores (Wake, 1963; Oishi et al., 1964) and from stationary phase cells (Masters and Pardee, 1965).

Other Bacteria

Nonrandom chromosome replication has been demonstrated in bacteria other than *E. coli* and *B. subtilis,* suggesting that sequential replication is a general property of bacterial chromosome replication. Stonehill and Hutchison (1966) used ultraviolet light induced mutation in synchronous cultures of *Streptococcus faecalis* to demonstrate a polarity in the replication of five markers. Chromosome replication in *Staphylococcus aureus* has been studied by means of nitrosoguanidine induced mutation (Altenbern, 1968). Synchronous chromosome replication was achieved by use of phenethyl alcohol, and a polarity in the replication of ten markers was observed. Finally, Tyeryar et al. (1968) used the marker frequency procedure of Yoshikawa and Sueoka to demonstrate nonrandom replication in *Bacillus licheniformis.* On the very reasonable assumption that the chromosomes of the above bacteria are long DNA molecules, these results strongly imply that replication occurs sequentially.

CHROMOSOME REPLICATION AND BACTERIAL CELL GROWTH

The Rate of Chromosome Replication

Maaløe and Kjeldgaard (1966) have proposed that the replication fork is always saturated with substrates, and as a result the rate of DNA synthesis at a replication fork is constant irrespective of the growth rate of the cell or of the stage in the cell or replication cycle. Support for this hypothesis comes from a number of observations made by these investigators. Following a shift of *S. typhimurium* from a "poor" to a "rich" medium, or vice versa, the rate of DNA synthesis remained constant for about 15 to 25 minutes (Kjeldgaard et al., 1958; Maaløe and Kjeldgaard, 1966). If substrates or other factors limited the replication rate under some conditions, some of the shifts should have led to a rapid change in the rate of DNA synthesis, such as is observed for RNA synthesis. A constant rate of synthesis at the replication fork leads to the prediction that the time required to replicate the chromosome should be the same at different growth rates. Direct evidence for this comes from autoradiographic experiments cited by Maaløe and Kjeldgaard (1966). When cells were grown with a 40-minute generation time, DNA synthesis was virtually continuous throughout the cell cycle. When the generation time was 100 minutes, DNA synthesis occurred over only 40 to 60 minutes of the cell cycle.

Cairns' autoradiographs demonstrate very convincingly that the rate of DNA synthesis at an individual replication fork is constant throughout the cell cycle. When cells were labeled with radioactive thymidine for three minutes, it was found that the length of DNA labeled was about 60 to 70 μ (Cairns, 1963a). The lengths of the labeled sections would have varied if the rate of replication changed during the cell cycle. Experiments conducted with synchronous cultures of *E. coli* B/r lead to the same conclusion. Clark and Maaløe (1967), Helmstetter (1967), and Helmstetter and Cooper (1968) have all observed that the rate of DNA synthesis changed at definite times during the cell cycle. As demonstrated by Clark and Maaløe, the increase in the rate of DNA synthesis was caused by the initiation of new rounds of replication and not by a change in the rate of replication at the replication fork. The rate of DNA synthesis was constant between these increases, and the change from one rate to another corresponds approximately to the integral change expected when the initiation and termination of rounds of replication change the number of replication forks per cell.

The changes in the rate of DNA synthesis that occur during the cell cycle have also been used to determine the time required to replicate the chromosome in *E. coli* B/r growing at different growth rates (Helmstetter, 1967; Helmstetter and Cooper, 1968). Using a variety of media to produce different growth rates, the times at which these changes occurred was determined. Assuming that the changes were caused by the initiation or completion of rounds of replication, the results indicate that the time required to replicate a chromosome is about 42 minutes when the generation time is between 22 and 60 minutes.

Since the replication of the chromosome requires approximately 42 minutes, a question arises as to how replication occurs when the generation time is less than 42 minutes. Unless the net rate of DNA synthesis increases in direct proportion to the growth rate, rapidly growing cells eventually would produce progeny cells which did not contain intact chromosomes. The manner in which replication occurs in rapidly growing cells was first demonstrated by Sueoka and co-workers (Yoshikawa et al., 1964; Oishi et al., 1964). When *B. subtilis* is grown in broth, a round of replication is initiated before the preexisting replication fork reaches the terminus. Consequently, the number of chromosomes can double in a time less than that required for a single replication fork to traverse a chromosome. This pattern of replication leads to chromosomes that contain more than one replication fork, and it is called dichotomous or multifork replication.

Helmstetter and Cooper (1968) propose that replication occurs in this fashion in rapidly growing *E. coli* B/r. Other evidence that dichotomous replication occurs in *E. coli* can be obtained from the premature initiation experiments of Pritchard and Lark (1964), and the autoradiographic experiments of Cairns (1963a). With respect to the autoradiographs, since only 60 to 70 μ of DNA were labeled in three minutes, only one-half of the chromosome could have been replicated in the 30-minute generation time with which the cells were growing. Thus, to double the chromosome content of the cells in one generation, replication must have been occurring at more than one replication fork per chromosome (Cairns, 1963a; Maaløe and Kjeldgaard, 1966). Bird and Lark (1968) have used a double-label technique to demonstrate that the chromosomes in rapidly growing cells emerging from amino acid deprivation undergo dichotomous replication (see p. 31). Under these conditions initiation occurred on the average at 30-minute intervals, and a single replication fork required 40 minutes to traverse the chromosome.

The time required for a single replication fork to traverse the *E. coli* chromosome is not always approximately 42 minutes, however. In slowly growing cells, DNA synthesis is not continuous throughout the cell cycle. If chromosome replication required 42 minutes in all media, the length of the period of no DNA synthesis should equal the generation time minus 42 minutes. Experiments conducted with *E. coli* 15T⁻ have demonstrated that this is not the case (C. Lark, 1966). When cells were grown on proline (180-minute generation) and acetate (260-minute generation) medium, DNA synthesis occurred over one-half of the cell cycle. Consequently, the times required for replication in these media were 90 and 130 minutes, respectively. Apparently the supply of substrates and/or essential enzymes is not saturating at the replication fork in slowly growing cells. Helmstetter (1967) has also reported that the period required for the replication of the B/r chromosome increased when the generation time was greater than 60 minutes.

The replication rate of the chromosome is influenced by the growth rate in *B. subtilis,* also. Eberle and Lark (1967) used autoradiography to study chromosome replication in cells with generation times between 27 and 580 minutes. DNA synthesis was continuous in cells with generation times of up to 80 minutes. As mentioned above, broth-grown cells (27-minute generation) replicate the chromosome in the dichotomous fashion. This indicates that there is a minimum time required for the replication of the *B. subtilis* chromosome. In succinate-grown (360-minute generation) and acetate-grown (580-minute generation) cells there was a period in the cell cycle when no DNA synthesis occurred. This period was approximately 0.3 to 0.35 of the cell cycle. Consequently, as the generation time increased, the time required for the replication of the chromosome increased.

The Time of Initiation of Chromosome Replication

Chromosome replication is initiated about halfway through the cell cycle in *E. coli* growing on glucose-minimal medium. This was first demonstrated by Forro and Wertheimer, who studied the time of initiation in *E. coli* 15T⁻ (Forro and Wertheimer, 1960; Forro, 1965). Labeled cells of different sizes (ages) were isolated by micromanipulation and grown into microcolonies of 50 to 100 cells. Autoradiography was then used to estimate the number of conserved units present in the initial cell. The results demonstrate that the number of conserved units increased about halfway through the cell cycle.

More recently the time of initiation of replication has been studied by Clark and Maaløe (1967), who determined the rate of DNA synthesis in synchronous cultures of *E. coli* B/r. These investigators observed that the rate of DNA synthesis increased twofold about halfway through the cell cycle when the cells were grown with a 45-minute generation time in glucose-minimal medium. The conclusion that this increase in the rate of DNA synthesis was caused by the initiation of new rounds of replication, was strengthened by experiments in which chloramphenicol was used to inhibit protein synthesis at different times in the cell cycle, and the subsequent increase in DNA was determined. On the assumption that 150 μg/ml of chloramphenicol inhibits the synthesis of the proteins required for the initiation of new rounds of replication, the amount of DNA synthesized should be maximum when the inhibitor is added at the beginning of a round of replication. The greatest amount of synthesis occurred when chloramphenicol was added 20 to 25 minutes before cell division.

The time of initiation is not always halfway through the cell cycle, however. Clark and Maaløe also determined the effect of chloramphenicol on subsequent DNA synthesis in glycerol-grown (65-minute generation) and succinate-grown (100-minute generation) cells. In these media, as in the glucose medium, the maximum amount of DNA was synthesized when chloramphenicol was added 20 to 25 minutes before the end of the cell cycle. This time is progressively later in the cell cycle as the generation time increases, and Clark and Maaløe propose that the time elapsing between initiation of replication and cell division is independent of the growth rate.

A different model for the relationship between the time of initiation and cell division has been proposed by Cooper and Helmstetter (Helmstetter and Cooper, 1968; Cooper and Helmstetter, 1968). These investigators conducted a detailed analysis of the time of initiation of a round of replication in cells growing with generation times between 22 and 60 minutes. They observed that the time of initiation changed with the growth rate in a definite fashion in rapidly growing cells. To explain their results, they propose that a round of replication requires approximately 40 minutes, and that cell division occurs 20 minutes after the completion of a round of replication. Thus, initiation always occurs 60 minutes before the cell division which segregates the daughter chromosomes (see Fig. 9). For example, in cells growing with a 40-minute generation time, initiation occurs halfway through the cell cycle before the cycle in which termination and the segregation of the daughter chromosomes occur. In cells with a 60- or a 20-minute generation time initiation occurs at the time of cell division. However, in cells with a 60-minute generation time, initiation occurs at the start of the cell cycle in which termination occurs, whereas in cells with a 20-minute generation time initiation occurs at the start of the cell cycle two cycles before the one in which termination and segregation occur.

The model of Cooper and Helmstetter, although consistent with the results obtained from cells with generation times between 22 and 60 minutes, is not consistent with results obtained from cells with longer generation times. According to this model, when cells are growing with a 100-minute generation time, initiation should occur when the cells are 40 minutes old, and termination should occur in the cell cycle in which initiation occurred. As mentioned above, Clark and Maaløe have observed that in succinate-grown cells initiation occurred 20 to 25 minutes before cell division, and termination thus occurred in the subsequent cycle. Similar results have been obtained by Helmstetter (1967). These results suggest that in slowly growing *E. coli* B/r initiation always occurs in the cell cycle preceding the one in which termination occurs, and that the model of Helmstetter and Cooper applies only to rapidly growing cells.

Under some conditions of growth, however, initiation and termination can occur in the same cell cycle. In her studies on the time required for chromosome replication

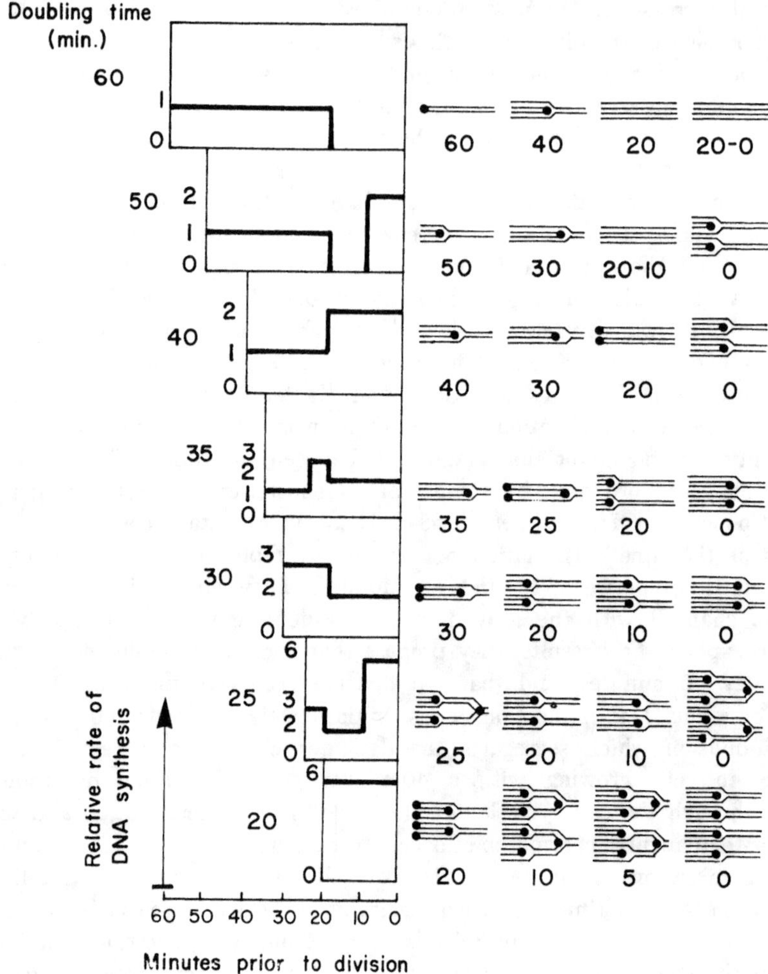

FIG. 9. Schematic illustration of the model of Cooper and Helmstetter (1968). The rate of DNA synthesis during the division cycle of *E. coli* B/r growing with various doubling times is shown on the left, assuming replication requires 40 minutes and cell division occurs 20 minutes after termination. The postulated chromosome configurations in these cells are shown at the right. The black dot indicates a replication point, and the numbers indicate the time in minutes prior to cell division at which the chromosome configuration is present in the cells. (From Cooper and Helmstetter. 1968. *J. Molec. Biol.*, 31:519.)

in slowly growing *E. coli* 15T⁻ cells, C. Lark observed that the average number of chromosomes per cell was a little more than one (C. Lark, 1966). This indicates that initiation and termination occurred in the latter half of the cell cycle. If initiation occurred just before cell division, replication would have been completed in the first half of the cell cycle. This would have produced an average of a little less than two chromosomes per cell.

Kubitschek et al. (1967) have also studied the time of initiation in slowly growing cells, and their results are in accord with those of C. Lark. *E. coli* 15 $T^-H^-U^-$ cells were grown in a chemostat on limiting glucose with doubling times of between 57 minutes and 12 hours. The cells were pulse-labeled with radioactive thymidine for 0.1 generation, separated in a sucrose gradient on the basis of size, and the radioactivity in the various fractions was determined. At the most rapid growth rate, 57 minutes, about 96 percent of the cells were labeled. The nonlabeled cells were among the smallest cells, suggesting that even in rapidly growing cells, replication is initiated soon after cell division. As the generation time increased, the onset of DNA synthesis was increasingly delayed, until a maximum delay of two-thirds of the cell cycle was observed for generation times of between 2 and 12 hours. In all of these experiments, the smallest cells were unlabeled. This indicates that a round of chromosome replication was complete before cell division; if replication were not complete before division, a large number of the youngest cells would have contained label.

The Number of Chromosomes Per Cell

In all the bacteria that have been investigated, the number of chromosome equivalents of DNA per cell varies with the growth rate. This was first observed by Schaechter et al. (1958), who grew *S. typhimurium* in a large number of different media and determined that the amount of DNA and the number of nuclei per cell were logarithmic functions of the growth rate. On the assumption that a visible nucleus contained one replicating chromosome, the results also indicate that the average number of chromosomes per cell varied continuously from 1.3 to 3 as the generation time changed from 100 to 21 minutes.

More recently, Lark and his colleages studied the chromosome content of *E. coli* 15T^- grown in different media (K. Lark and C. Lark, 1965; K. Lark and Bird, 1965b; C. Lark, 1966; K. Lark, 1966a; K. Lark et al., 1967). In these experiments the number of chromosome equivalents of DNA was determined by direct DNA determination (the amount of DNA in the chromosome was derived from Cairns' autoradiographs), and the number of chromosomes was estimated by autoradiographic determination of the number of conserved DNA units present per cell. The results indicate that cells grown with generation times of 120 minutes or more, 40 to 70 minutes, and 22 minutes contained enough DNA for one, two, and four replicating chromosomes, respectively.

The DNA content of *B. subtilis* also changes with the growth rate. Eberle and Lark (1967) observed that the DNA content of cells growing with generation times greater than 250 minutes was about one-half that found in cells growing with generation times of between 80 and 160 minutes. Cells with a 27-minute generation time (broth medium) contained 60 percent more DNA than glucose-grown cells (80-minute generation). As demonstrated by the work of Sueoka and co-workers (Yoshikawa et al. 1964; Oishi et al., 1964), this increase is at least partly caused by dichotomous replication. Assuming that the *B. subtilis* chromosome has a molecular weight of 3.9×10^9, Eberle and Lark propose that cells growing with 80- to 160-minute generation times contain two chromosomes, and that more slowly growing cells contain one.

The model proposed by Cooper and Helmstetter for the variation in the time of initiation with the growth rate provides a very plausible explanation for the variation in DNA content (Helmstetter and Cooper, 1968; Cooper and Helmstetter, 1968). According to this model chromosome replication requires approximately 40 minutes in *E. coli* B/r, and cell division occurs 20 minutes after the end of a round of replication. Consequently, cells growing with a 40-minute generation time will initiate replication halfway through the cell cycle (Fig. 9). Newly-born daughter cells will contain one replicating chromosome which is halfway through the replication cycle, and the average cell in an exponentially growing culture will contain 2.16 genome equivalents of DNA. As the generation time becomes less than 40 minutes, dichotomous replication occurs, and initiation also occurs earlier in the cell cycle. This results in a continous increase in the amount of DNA per cell as the generation time decreases, until newly-born cells growing with a 22-minute generation will contain a little less than three genome equivalents. The average cell at this growth rate will contain about four genome equivalents. It should be mentioned that, according to this model, rapidly growing cells which are about to divide can contain almost six genome equivalents, but this DNA will be contained in two multiforked chromosomes. Cooper and Helmstetter's model does not apply directly to cells with generation times greater than 60 minutes, since the time required for replication increases in such cells. As an explanation for the DNA content observed in slowly growing cells, Cooper and Helmstetter propose that initiation occurs just before cell division, and that replication occurs over a constant fraction of the cell cycle. If this fraction is half of the cell cycle, slowly growing cells at different growth rates would contain an average of 1.7 genome equivalents of DNA.

As a test of their model, Cooper and Helmstetter determined the average amount of DNA in cells growing at two different growth rates (Cooper and Helmstetter, 1968). The results fit the predictions of the model very well. Their data were also used to calculate the molecular weight of the *E. coli* chromosome and the value obtained is close to Cairns' estimate. In addition, Cooper and Helmstetter applied their model to the results of Schaechter et al. (1958) and found that the experimental and predicted values are similar. The variation in the amount of DNA observed in *B. subtilis* suggests that a similar relationship between the initiation time and the growth rate exists in this organism also. But *B. subtilis* has not been studied sufficiently to determine the nature of this relationship.

As a result of his studies on chromosome replication in *E. coli* 15T$^-$, Lark proposed a different model for the relationship between chromosome number and growth rate (K. Lark, 1966a; K. Lark et al., 1967). It was proposed that (1) initiation occurs at cell division in cells with generation times of 22 to 120 minutes, (2) the rate of DNA synthesis at a replication fork increases with the growth rate so that one generation is required for replication in cells with generation times between 22 and 120 minutes, (3) the increase in DNA with increased growth rate is caused by a greater number of chromosomes being replicated and not by a change in the initiation time or the number of replication forks per chromosome, and (4) cells growing with generation times greater than 120 minutes replicate the chromosome in the latter part of the cell cycle. The exception to these proposals is cells grown in succinate medium (70-minute generation). Newly-born succinate cells are proposed to contain two chromosomes which are replicated alternately. The initiation of replication of one of the chromosomes occurs at cell division and is complete about halfway through the cell cycle, at which time the initiation of replication of the other chromosome occurs. This pattern of replication can be contrasted with that for glucose cells (45-minute generation), which also are proposed to contain two chromosomes at the beginning of the cell cycle. Initiation occurs at cell division in glucose cells and both chromosomes are replicated simultaneously during the cell cycle.

Lark's observations on the amounts of DNA and the number of conserved units present per cell at different growth rates are consistent with the above model (K. Lark and C. Lark, 1965; K. Lark and Bird, 1965b; K. Lark, 1966; C. Lark, 1966; K. Lark et al., 1967). But as pointed out by Cooper and Helmstetter (1968), most of the data are also consistent with their model. A major difference between the models is the replication of chromosomes in succinate-grown cells. The first indication that replication in succinate-grown $15T^-$ might not be alternate, came from studies of Koch and Pachler (1967). Cells were labeled with tritiated thymine for about one-half generation and the sensitivity to tritium decay was determined. Koch and Pachler observed that the cells were inactivated in a first-order way and there was no tendency for the survival curves to level off. On the assumption that death due to tritium decay is recessive, this suggests that the cells did not replicate two chromosomes alternately.

More recently, Bird and Lark (1968) conducted extensive experiments which reconcile most of the differences that have been observed between $15T^-$ and B/r. The "beginnings" and the "ends" of chromosomes were specifically labeled, and the times at which these regions were replicated with respect to each other were determined. To do this, cells were deprived of required amino acids until DNA synthesis had ceased, and the "beginnings" were labeled with tritiated thymine when replication commenced after the readdition of the amino acids. After a brief labeling period the tritiated thymine and the amino acids were removed from the medium. When the round of replication was almost completed, ^{14}C-thymine was added to label the "ends" of the chromosomes whose "beginnings" were labeled with tritium. The required amino acids were then added to the medium, and replication commenced. Aliquots were removed from the culture at different times and briefly grown in the presence of 5-bromouracil, and the DNA from these cells was banded in CsCl gradients. The presence of tritium or ^{14}C in the hybrid DNA indicates that the "beginnings" or the "ends", respectively, were replicated at the time of the shift to 5-bromouracil.

Experiments of the above type done with succinate-grown cells demonstrate that 55 minutes elapsed between the replication of the "beginnings" and the "ends." This was followed by a 15-minute lag period, and then new rounds of replication were initiated. These data demonstrate that replication was not alternate in cells resuming DNA synthesis after amino acid deprivation. The labeled chromosomes were not replicated alternatively, since 70 to 80 percent of the labeled chromosomes participated in each replication cycle. Unlabeled chromosomes could not have been replicated in the 15-minute lag period, since this period is much too short for a replication cycle.

Other experiments of the above type demonstrate that the rate of DNA synthesis at the replication fork varied with the growth rate in a fashion similar to that observed in B/r. When the "beginnings" and "ends" were labeled in glucose medium and the cells were shifted to casamino-acid-supplemented medium (27-minute generation time) for subsequent growth, 40 minutes were still required for a replication fork to traverse the chromosome. Initiation occurred about every 27 to 30 minutes. Thus, there was a lower limit to the time required for the replication of the 15T⁻ chromosome, and cells growing with a generation time less than this lower limit replicated the chromosome dichotomously. Cells grown in aspartate medium (120-minute generation) required 80 minutes for chromosome replication. This increase in the replication time is similar to that observed in B/r grown with about this generation time (Helmstetter, 1967).

Although the major differences between 15T⁻ and B/r now seem to be resolved, there still are some points which require clarification. In particular, the relationship in 15T⁻ between the growth rate and the times of initiation and termination of replication are unknown. The period that elapses between the termination of a round of replication and cell division is probably longer in 15T⁻ than in B/r. An increase in this period would produce cells that contained more DNA, and glucose-grown 15T⁻ contains more DNA than B/r (Cooper and Helmstetter, 1968). K. Lark et al. (1967) also observed more DNA in broth-grown 15T⁻ than is predicted to be present in B/r grown at the same growth rate. Unfortunately, the times of initiation and termination of replication in 15T⁻ cannot be determined by the procedure used by Helmstetter and Cooper. 15T⁻ does not stick to the membrane filters used to obtain cells of specific ages, and a different procedure will have to be devised to study this.

CHROMOSOME SEGREGATION

In 1963, Jacob et al. proposed the replicon hypothesis, part of which dealt with the manner in which daughter bacterial chromosomes segregate from each other. Essentially, the model proposed that chromosomes are attached to the cell membrane, and that at the end of a round of replication the points of attachment of daughter chromosomes lie side by side (Fig. 10). Elements of the cell membrane

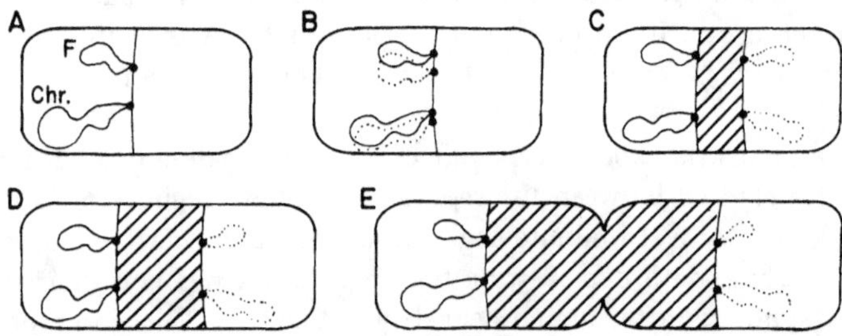

FIG. 10. A model for the equal distribution of DNA among daughter bacteria. The bacterium (F^+) carries two independent self-replication units: a chromosome and an independent sex factor. Replication occurs during A and B, segregation during C, D, and E. (From Jacob et al., 1963. *Cold Spring Harbor Symp. Quant. Biol.*, 28:329.)

then grow between the two attachment sites, moving the daughter chromosomes towards opposite poles of the cell. Other independently replicating genetic elements (replicons) present in the cell are attached to the cell surface in a similar fashion, and these elements are also segregated by the growth of the cell membrane. The replicon model was not specifically concerned with the control of cell division, but it indicated that the ultimate segregation of daughter chromosomes and episomes into daughter cells occurs as a result of septum formation in the region of growth between the attachment sites.

Since the formulation of the above model, a number of experiments have verified its basic tenets. As discussed above, it is now rather well established that the bacterial chromosome is associated in some fashion with the cell membrane. In addition, it has been demonstrated that cell growth and elongation can occur in a fashion which could readily lead to the segregation of daughter chromosomes that were attached to the cell surface. This was first shown by Cole and Hahn (1962), who used immunofluorescent methods to study cell wall synthesis in *Streptococcus pyogenes*. Their results demonstrate that the new cell wall was formed equatorially, and that the old material was conserved. If the daughter chromosomes were attached to the conserved units of the cell wall, segregation into daughter cells would have resulted. Cell membrane synthesis in *B. subtilis* occurs in an analogous fashion. Jacob et al. (1966) used tellurium crystals as a label to follow membrane synthesis. During the first generation of growth after labeling the surface, the crystals were driven to the poles of the cell, apparently by membrane growth in the middle regions of the cell. After the second generation, the cells were labeled at only one of the poles. Obviously, if some of the membrane growth which led to this distribution occurred between the attachment sites of daughter chromosomes and other replicons, segregation of the units would have resulted.

Other bacteria which have been studied do not synthesize new cell surface primarily in the middle regions of the cell. Only the poles of the cells are conserved in *E. coli* and elongation occurs by intercalation of new material (Beachey and Cole, 1966). Cell wall elongation also occurs by intercalation of new material in *S. typhimurium* (May, 1963). This type of synthesis is not as dramatic as that discussed above, but it could still lead to the segregation of chromosomes attached to the cell surface. Segregation would result if some of the newly synthesized material were laid down between the attachment sites.

It is not known whether the termination of a round of replication induces the synthesis of cell surface that occurs between daughter chromosomes, or whether this synthesis is only a part of the overall cell elongation. Either process could lead to the segregation of daughter chromosomes. Recent experiments demonstrate, however, that the termination of replication is involved in the control of septum formation and cell division. Clark (1968) and Helmstetter and Pierucci (1968) have both observed that if DNA synthesis was inhibited before the termination of a round of replication, cell division was inhibited even though cell elongation continued. If the inhibition of DNA synthesis occurred after the termination of a round of replication, cell division occurred. These results demonstrate that two completed chromosomes were required before the steps required for septum formation and cell division were initiated. This septum formation undoubtedly occurred between the two chromosomes, since both of the progeny from a cell which contains two chromosomes at cell division are viable.

The stability of the association between the chromosome and the cell surface is at present uncertain. Several different experiments indicate that it is stable, whereas other experiments indicate the opposite. Jacob et al. (1966) and Cuzin and Jacob (1967) have followed the distribution among progeny cells of an episome that could not replicate at a nonpermissive temperature. It was found that the nonreplicating episome segregated preferentially with chromosome strands that were present before the shift to the nonpermissive temperature. This result is consistent with the hypothesis that episome and chromosome strands are stably attached to conserved units on the cell surface, and if replicons once segregate together because they are attached to the same segregation unit, strands of the replicons will continue to segregate together in subsequent cell divisions.

Lark and co-workers have also obtained evidence that the association between the chromosome and the cell surface is stable. Eberle and Lark (1966) have examined the distribution of conserved chromosome units in chains of 16 to 32 cells derived from single, labeled cells. They observed that label was preferentially associated with the external poles of the chains. This suggests that when a chromosome strand becomes attached to a segregation unit at the pole of a cell, it can remain attached to the unit for as many as five generations. A similar stability of the attachment has been observed in *Lactobacillus acidophilus* by Chai and Lark (1967). In these experiments, the association between the chromosome and the cell surface was examined by a combination of autoradiography and fluorescent antibody techniques. In addition to suggesting a stable attachment between chromosome strands and conserved units of the cell surface, the results are consistent with the proposal that this attachment is formed when a strand is used as a template, and not before.

Other experiments have indicated that the association between the chromosome and cell surface is not stable, or that the attachment is to a segregation unit which assumes no particular position in the cell. Ryter and Jacob (1967) have conducted experiments similar to those of Eberle and Lark, and they have obtained contradictory results. Instead of finding a preferential association of the prelabeled strands with the external poles of a chain of cells, they observed that the labeled strands were randomly distributed. The chance of a strand associating with the external pole (as opposed to the internal pole) of a cell in a growing chain was 0.5 per division. Similar results have been obtained by Yoshikawa (1968), with the exception that there was an excess of labeled strands at the spore end of chains derived from germinating spores. At present there is no explanation for these conflicting results. These results raise some interesting questions, however, about the attachment of the chromosome to the cell surface. Since *B. subtilis* appears to have conserved membrane units (Jacob et al., 1966), these results suggest that if the chromosome is attached to these units, the attachment is unstable. Conversely, the attachment could be stable, but it would have to be to some part of the cell surface other than the conserved membrane units observed by Jacob et al., and this part of the cell surface would have to be distributed randomly at cell division.

If a cell contains two chromosomes at the time of cell division, the pattern of chromosome segregation is unambiguous. If viable daughter cells are to be formed, each cell must receive a chromosome. This simple pattern of segregation has been proposed for *E. coli* B/r by Cooper and Helmstetter (Helmstetter and Cooper, 1968; Cooper and Helmstetter, 1968). According to Cooper and Helmstetter's model, this segregation pattern also applies to rapidly growing cells which contain, at cell division, more than two nuclei and enough DNA for more than two chromosomes (Fig. 9). It is proposed that such cells still only contain two separate chromosome structures, but that these structures have as many as three replication forks and can contain as much as three genome equivalents of DNA. Also, the unfinished chromosomes are proposed to appear as nuclei before their replication is complete, so one such structure can appear as several nuclei. According to their model, the only conditions under which a cell might contain more than two such structures at division is when the generation time is 20 minutes or less. Since this is the lower limit of the *E. coli* generation time, it is possible that such a condition is never realized.

A more complex model for the pattern of chromosome segregation in *E. coli* 15T$^-$ has been proposed by Lark (K. Lark and Bird, 1965b; K. Lark, 1966a; K. Lark et al., 1967). The complexity of this model is derived from Lark's proposal that cells of this strain grown on glucose or succinate contain four completed chromosomes at cell division. Consequently, there are several possible segregation patterns. Segregation can be random, both progeny of a given parent chromosome can go to the same daughter cell, or the progeny of a given parent can segregate so that one goes to each daughter cell. It was proposed that segregation followed this last pattern (K. Lark and Bird, 1965b). In addition, this segregation was proposed to occur in a specific fashion (K. Lark, 1966a). Each daughter cell receives one chromosome which was synthesized on a template which had just been used for the first time and another chromosome which was synthesized on a template which had been used at least once previously. As mentioned above, each of these chromosomes is derived from a different parent chromosome. It is visualized that this difference between the template strands at segregation arises because a strand of the double helix becomes attached to the segregation structure when it is used as a template for the first time. Consequently, the strands that have been used as templates at least once previously have been segregated to the poles of the cell and do not end up in the same daughter cell.

The interpretation of the data that led to Lark's model was based directly on the assumption that chromosome replication occurs alternately in succinate-grown cells. As mentioned above, this assumption no longer seems valid. Lark's segregation data are consistent with the model of Cooper and Helmstetter, and segregation in 15T$^-$ and B/r is probably similar under most conditions. Strain 15T$^-$ might contain four chromosomes at cell division under some conditions, however, and segregation might then occur similarly to Lark's model. This might happen in glucose-grown cells, for example. The average DNA content of these cells (K. Lark and C. Lark, 1965; K. Lark, 1966a; K. Lark et al., 1967) increases from two to four genome equivalents during the cell cycle. Since dichotomous replication was not observed to occur in such cells (Bird and Lark, 1968), this suggests that the four genome equivalents present at the end of the cell cycle might be four completed chromosomes.

THE CONTROL OF CHROMOSOME REPLICATION

The observations discussed in the preceding sections allow certain deductions to be made concerning the control of chromosome replication. Given the information that protein synthesis is required for the initiation, but not the completion, of a round of replication; that initiation occurs at a specific locus on the chromosome; that the chromosome is attached to the cell surface; and that initiation is related to cell growth; a reasonable proposal for the control of replication can be made. Several models have been proposed on the basis of these and other observations, and these models, which provide a convenient framework for the available information, will be considered in this section. Unfortunately, the more explicit aspects of these models have not yet been tested experimentally, and new information subsequent to the formulation of the models is still lacking.

The first major model for the control of chromosome replication, the replicon model, was proposed by Jacob et al., in 1963. The model states that a genetic element that is replicated as a unit, such as a phage, episome, or bacterial chromosome, constitutes a replicon. The replication of a replicon is proposed to be under specific control, and the elements of the control system are contained in the replicon itself. One of the elements is a structural gene, which produces what is called an initiator protein. The exact nature of the initiator was left undefined, but the action of the initiator is specific and necessary for replication. The initiator is proposed to act in a positive fashion at the second specific site of the replicon, the replicator. The replicator is the operator of the replication system, and replication can only be initiated when the initiator is bound to it. Once replication is initiated, the DNA attached to the replicator is duplicated without further requirement for the initiator.

The replicon model also proposes that bacterial and episome chromosomes are attached to the cell surface, and that the attachment site functions in the coordination of cell growth and chromosome replication. A surface reaction is presumed to activate the replication system when the cell surface has grown to a certain point. Replication then occurs, and no further initiation is permitted until the bacterium has grown to the point that a new surface reaction occurs. As one example of how this activation might occur, Jacob et al. suggest that the initiator might be bound to some specific structure on the cell surface. This implies that the initiator is not sufficient by itself to cause replication, and the attachment of the chromosome occurs via the replicator region, at least at the start of a round of replication.

The replicon model is consistent with the basic facts that pertain to the control of chromosome replication. The model proposes that control is exerted only over the initiation, and not the continuation, of a round of replication. This is in accord with the observed relationship between protein synthesis and replication. The initiator protein could be one of the proteins which is required for the initiation of replication, and the specific structure on the cell surface could be the second required protein (C. Lark and K. Lark, 1964; K. Lark, 1966a). The attachment between the chromosome and the cell surface, although speculative when proposed by Jacob et al., has received subsequent substantiation. There is no evidence, however, that indicates whether this attachment functions in a control capacity in conjunction with surface changes. On the basis of theoretical arguments, Jacob et al. predicted that the replicator would be at the same position on the chromosome in all strains of *E. coli.* The replication origin, which would contain the replicator, is at 55 to 65 minutes in at least the majority of strains that have been investigated.

A more recent model for the control of chromosome replication has been proposed by Lark (K. Lark, 1966a; K. Lark et al., 1967). This model is similar to the replicon model in many respects, but it is more specific. According to this model, before the initiation of a round of replication the chromosome is a circular molecule that is attached by one strand to a protein called the replicator (Fig. 11).

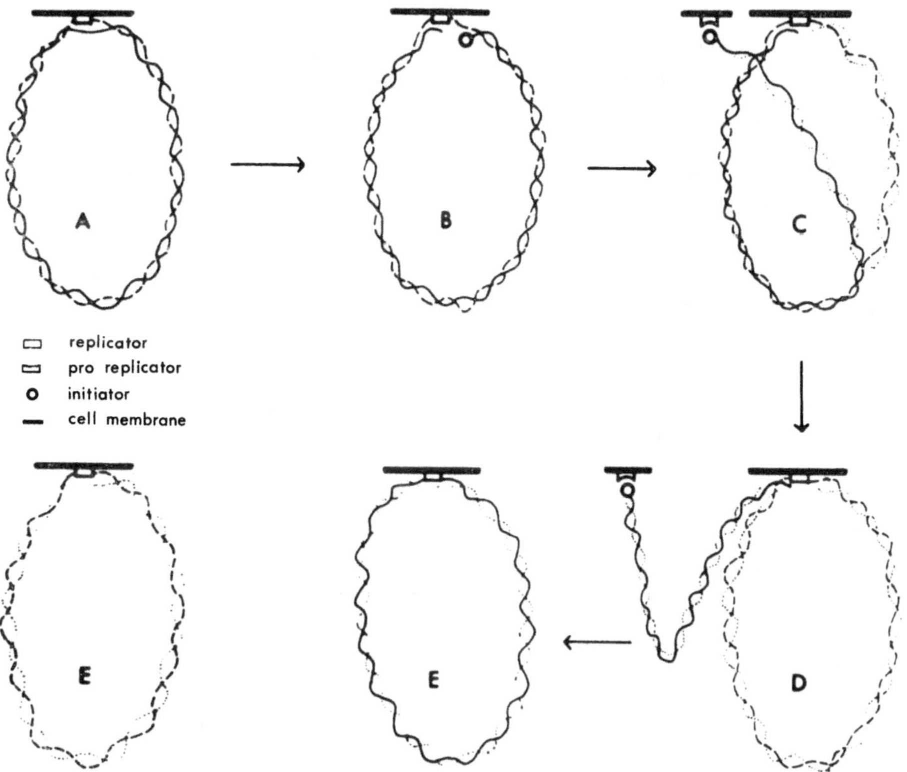

FIG. 11. Replication of a chromosome in *E. coli.* One strand of the chromosome is permanently attached to the cell membrane by a protein, the replicator. The other is detached by interaction with the initiator and is temporarily attached to the cell membrane via a protein, the proreplicator. Permanent attachment of this strand results when replication is completed, thus converting the proreplicator into a replicator. (From K. Lark, 1966. *Bact. Rev.*, 30:3.)

The replicator is attached to the cell wall-membrane complex. At the start of a round of replication, the other strand is broken, and one end of the broken strand becomes temporarily attached by means of the initiator protein to what is called a proreplicator. The proreplicator is attached to the cell surface. As a result of this attachment of the broken strand, a part of the double helix is opened, and replication is initiated. As replication proceeds, the proreplicator moves to its final position on the cell surface, which allows the daughter chromosomes to separate as replication is occurring. When the end of the chromosome is reached, the ends of the broken template strand are joined at the proreplicator, and the proreplicator is converted into a replicator. The newly synthesized strands may or may not have their ends joined, and the initiator is possibly released into the cell for future use. The chromosomes are then ready for a new round of replication.

With respect to the proteins required for initiation, Lark proposes that the initiator is the chloramphenicol sensitive protein and the proreplicator is the chloramphenicol insensitive protein. This is based on the following considerations. When cells that ceased DNA synthesis in the absence of required amino acids were incubated in a complete medium plus a low concentration of chloramphenicol, a reduced amount of initiation occurred (C. Lark and K. Lark, 1964). This is proposed to be caused by the synthesis of new proreplicator sites, and the utilization of preexisting initiators which were released after the completion of the preceding round of replication. This should lead to a linear increase in the amount of DNA. The observed increase lasted for only about one generation and the rate then rapidly decreased. It is possible that this was caused by the loss of unstable initiator. A different situation exists upon the incubation of cells in thymineless medium, which induces premature initiation. If a low concentration of chloramphenicol was present during the thymineless incubation, premature initiation was inhibited. Lark proposes that the chloramphenicol inhibited the synthesis of the initiator which normally was synthesized during thymine deprivation. When thymine was then added back to the treated cells, synthesis resumed only at the preexisting replication forks, since there was no excess of initiator. Proreplicator probably accumulated during the thymineless incubation in the presence of chloramphenicol, but this was insufficient to induce premature initiation.

As discussed previously, experiments using phenethyl alcohol also indicate that two proteins are required for the initiation of replication (K. Lark and C. Lark, 1966). In these experiments, the chloramphenicol sensitive protein that accumulated during phenethyl alcohol treatment would be the proposed initiator. The chloramphenicol resistant, phenethyl alcohol sensitive protein would be the proreplicator. The synthesis of the initiator was also sensitive to low concentrations of 5-fluorouracil (C. Lark and K. Lark, 1964) and to amino acid analogues (K. Lark et al., 1967). Proreplicator synthesis was not inhibited by these compounds. Although the evidence that two proteins are required for initiation is quite good, it may be that the proteins have functions entirely different from those of the initiator and proreplicator proposed by Lark.

Chai and Lark (1967) have tested the proposal that a chromosome strand becomes attached to the cell surface when it is used as a template for the first time. As mentioned above, their results are consistent with such an attachment in

L. acidophilus, but there is conflicting evidence as to whether a similar attachment occurs in other bacteria (Eberle and Lark, 1966; Ryter and Jacob, 1967; Yoshikawa, 1968). Also, Lark's model requires that the "beginnings" of the nascent daughter chromosomes be physically separated from each other. This is apparently contradicted by Cairns' autoradiographs, which demonstrate that the "beginnings" and the terminus of a replicating chromosome are joined together. But the degree of resolution in the autoradiographs (about 1 μ) is not sufficient to determine whether these ends are held together by cell wall or some other material. The electron micrographs of the replicating *M. hominis* (H. Bode and Morowitz, 1967) and mitochondrial (Kirschner et al., 1968) chromosomes demonstrate, however, that replication can occur in a circular chromosome without the separation of the ends of the growing chromosomes. If replication in *E. coli* follows a similar mechanism, the two proteins required for initiation would obviously have functions different from those proposed in Lark's model.

Lark's model for the control of chromosome replication also includes a proposal for how initiation is controlled with respect to time. It is proposed that the proreplicator sites are produced at a rate proportional to the growth rate, and that a completed proreplicator is needed for initiation. In rapidly growing cells, new proreplicator sites would be available as soon as a cycle of replication is completed, so initiation would occur immediately. In slowly growing cells, however, the synthesis of the proreplicator sites would not be completed when the end of a replication cycle is reached, since the rate at which DNA is synthesized would be greater than the rate of proreplicator synthesis. Thus, there would be a period of no DNA synthesis before initiation could occur. The model makes no specific proposal for the relationship between the rate of initiator synthesis and the growth rate, but it can be inferred that the rates are directly related.

A number of observations demonstrate that initiation is coordinated in some fashion with cell growth, and that cells must reach a certain stage before initiation can occur. In the experiments of Hanawalt et al. (1961) it was observed that if cells finished their rounds of replication in the absence of protein synthesis, not all the cells started replication immediately when the amino acids were added back to the culture. Instead, some cells started replication almost immediately, whereas other cells did not initiate replication for almost one generation. Once a cell initiated DNA synthesis, it was capable of synthesizing a complete round of replication. This heterogeneity in the population was most probably caused by the different stages of the cells when they were deprived of amino acids (Hanawalt et al., 1961; Maaløe and Kjeldgaard, 1966). Cells that had just started a round of replication finished that round, but the synthesis of other macromolecules ceased when the amino acids were removed. Cells that were about to finish a round of replication when protein synthesis was inhibited ended up with a finished chromosome and the rest of the cell was at almost the same stage of development. When protein synthesis resumed, the latter cells initiated replication almost immediately, whereas the former cells required almost one generation for initiation. This lag in initiation suggests that initiation proteins were produced at a rate corresponding to the growth rate, and that these proteins had to be produced in a sufficient amount before initiation could occur. On the basis of these and similar observations, Maaløe and Kjeldgaard (1966) have proposed that the initiator is produced in a constant, "derepressed" fashion, that the initiator has a high affinity for the DNA, and that initiation occurs when the DNA is saturated with the initiator. Of course, the data are also consistent with a model in which initiation is caused by a surface reaction that is coordinated with cell growth (Jacob et al., 1963), by the completion of a protein attachment site whose synthesis requires one cell cycle (K. Lark, 1966a), or by some other factor whose activity or synthesis is related to cell growth.

Experiments on premature initiation also indicate that initiation protein synthesis is coordinated with cell growth. When *E. coli* 15T$^-$ cells were starved for thymine for different periods of time, the amount of premature initiation that was induced was roughly proportional to the length of the starvation period (Pritchard and Lark, 1964). It required about one generation to achieve maximum premature initiation. This indicates that initiation proteins were synthesized during the period of thymine starvation, and that starvation for one generation was required so that cells which had just initiated a round of replication could synthesize enough initiation protein to initiate a new round. The experiments conducted with phenethyl alcohol lead to a similar conclusion (K. Lark and C. Lark, 1966). When cells were incubated in phenethyl alcohol, DNA synthesis was inhibited, but RNA and protein syntheses continued. The chloramphenicol sensitive protein accumulated under these conditions. The rate of synthesis of this protein is probably directly proportional to the growth rate, since the time required to achieve maximum premature initiation is longer in succinate medium than in glucose medium (K. Lark and C. Lark, 1965).

On the basis of the observation that thymine starvation results in premature initiation, it is tempting to propose that the thymidine phosphate pools are involved in controlling the initiation of replication. This does not seem very likely, however. As pointed out by Maaløe and Kjeldgaard (1966), these pools decreased with a half-life of about one to two minutes when cells were starved of thymine (Neuhard and Munch-Petersen, 1966), yet it requires about one generation to get maximum premature initiation. If the amounts of the thymine derivatives were important in inducing initiation, cells would have to maintain pools for much of the cell cycle that were too small to support DNA synthesis. It seems more likely that thymine starvation is simply another procedure which allows the accumulation of initiation proteins during cell growth in the absence of DNA synthesis. But this hypothesis does not fit all of the available data. Nalidixic acid, which inhibits DNA synthesis without inhibiting protein and RNA synthesis, affects replication in a manner analogous to thymine deprivation (Boyle et al., 1967). Cytosine arabinoside preferentially inhibits DNA synthesis, but it did not cause premature initiation (C. Lark and K. Lark, 1964). Novobiocin seems to act similarly, since the removal of the inhibitor did not lead to an increased rate of DNA synthesis (Smith and Davis, 1967). Novobiocin did not affect the deoxyribose triphosphate pools (Smith and Davis, 1967), and the effects of cytosine arabinoside and nalidixic acid on these pools would be of interest to determine if the pool levels affected initiator synthesis.

The most direct demonstration of the relationship between cell growth and initiation comes from an observation recently made by Donachie (1968). Using the data of Schaechter et al. (1968) for cell mass at different growth rates and the data of Cooper and Helmstetter (1968) for the times of initiation, he has estimated the cell mass at the time of initiation at different growth rates. Assuming that the relationship between growth rate and cell mass is the same in *S. typhimurium* and *E. coli,* the mass per replication origin at initiation is constant (Fig. 12). Cells with 1, 2, or 4 replication origins at the time of initiation have a relative mass of 1, 2, or 4, respectively. This relationship fits the proposal of Maaløe and Kjeldgaard (1966) very nicely, since this would be expected if the initiation proteins were produced constitutively, and they were bound to the DNA at the replication origin

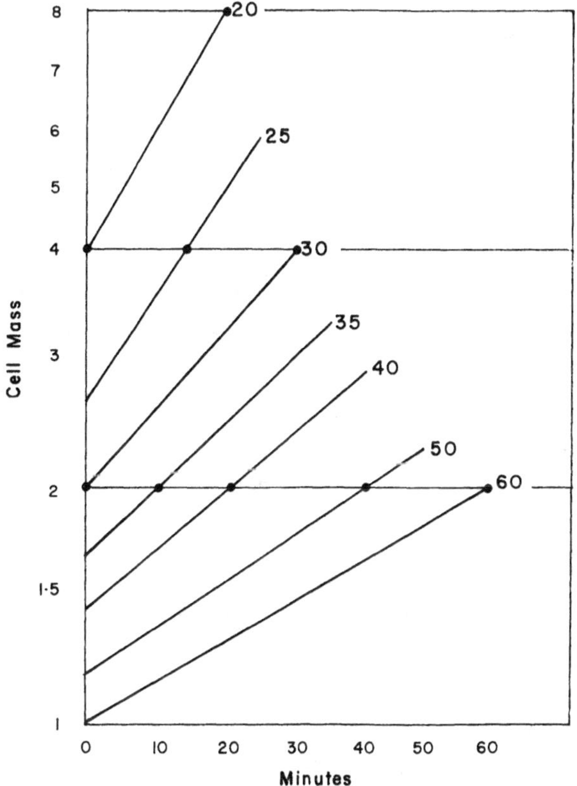

FIG. 12. Course of the increase in mass of individual cells with different rates of growth. The cells are assumed to grow exponentially over a single cell cycle which starts at the end of the previous division at 0 minutes. The next division, at the end of the cycle, takes place after the initial mass has doubled. The cell mass doubles in 20, 25, 30, 35, 40, 50, or 60 minutes. The initial mass at time 0 is taken to be proportional to the average mass of a population of cells growing at the same growth rate. Each line therefore shows the course of mass increase of individual cells from division at 0 minutes to the next division after the mass has doubled. Because there is a constant time (60 minutes for *E. coli* B/r) between the initiation of a round of DNA replication and cell division, it is possible to calculate the time when initiation occurs relative to cell division in cells growing at different rates. These times are marked as solid circles on the corresponding curves of mass increase. It can therefore be seen that the masses at which initiations take place are the same or multiples of the same cell mass for cells growing at all growth rates. (From Donachie, 1968. *Nature*, 219:1077.)

with a high affinity. But this relationship is also consistent with the proposals of Jacob et al. and of Lark, and it is not possible at present to discount any of these alternatives.

The hypothesis that initiation proteins take one generation to be formed implies that termination is not the event which controls initiation. This is shown directly by a variety of experiments. Hanawalt et al. (1961) observed that all of the cells in an amino acid deprived culture had completed chromosomes, yet they initiated replication at different times. If termination "triggered" the initiation of the next round of replication, all of the cells should have undergone initiation at the same time. The experiments of Pritchard and Lark (1964) lead to the same conclusion. If termination were necessary for initiation only a small percentage of a culture should undergo premature initiation following thymine deprivation. Finally, the dichotomous replication in *B. subtilis* and *E. coli* demonstrates that there is no direct cause and effect relationship between termination and initiation. In rapidly growing cells, initiation readily occurs at the origins of a partially replicated chromosome before the preexisting replication fork reaches the terminus.

The actual functions of the initiation proteins and the exact fashion in which they are synthesized during the cell cycle are still unknown. As mentioned above, the synthesis of these proteins is apparently related to the growth of the cell. But none of the available data actually demonstrates whether these proteins are synthesized continuously, or whether they are formed just before initiation, the synthesis being induced by some aspect of cell growth. As for the function of these proteins, about all that is known is that they are not the DNA polymerase or the enzymes required for the synthesis of the deoxyribonucleoside triphosphates (Billen, 1962). This state of ignorance should not last long. Procedures are now available for studying membrane-bound DNA (Smith and Hanawalt, 1967; Tremblay et al., 1968; Sueoka and Quinn, 1968), and the extent to which chromosome attachment is involved in initiation can now be studied. This should allow a direct test of Lark's model. In addition, a number of temperature sensitive mutants that are apparently blocked in the initiation of replication have been obtained (Fangman, 1966; Kohiyama et al., 1966; Mendelson and Gross, 1967; Kuempel, 1968; Copeland and Marmur, 1968). These mutants should prove of great benefit, since they can help elucidate the function of the initiator proteins and can also be used to determine when the initiation proteins are synthesized during the cell cycle.

THE MECHANISM OF CHROMOSOME REPLICATION

The mechanism of chromosome replication has been studied intensively since 1956, when Kornberg first reported at length on the *in vitro* enzymatic synthesis of DNA. After years of investigation, the mechanism of replication is still very poorly understood. A number of enzymes which act on DNA have been characterized, but the role played by these enzymes *in vivo* is still uncertain (Lehman, 1967; Howard-Flanders, 1968). Many of these enzymes undoubtedly function in processes other than replication, such as repair and recombination. But since only a few of these enzymes have been implicated in these other processes, it seems likely that some of the presently known enzymes function in replication. The demonstration of such a function has been difficult. In particular, the role played in the cell by the *E. coli* DNA polymerase is still uncertain.

The hypothesis that the *E. coli* DNA polymerase catalyzes the *in vivo* replication of the chromosome has been alternately accepted and rejected. Initial studies on this polymerase indicated that it has the properties expected of the "replicase" (see Kornberg, 1961, for a review). When the polymerase is incubated with the four deoxyribonucleoside triphosphates and template DNA, a net synthesis of DNA results. The product has the same base content as the template, even when templates with widely differing base compositions are used. Also, the product DNA has the same nearest neighbor frequency as the template. These results all indicate that the polymerase produces a faithful copy of the template, the copying probably occurring through base pairing as originally proposed by Watson and Crick.

But the polymerase also has properties that suggest it is not responsible for chromosome replication. One objection to the polymerase being the "replicase" is that the *in vitro* turnover number is much too low. A minimum estimate of the turnover number is 1,000 nucleotides per minute (Mitra and Kornberg, 1966). Assuming that two polymerase molecules are acting at the replication fork at a given instant, this turnover number predicts a rate of replication which is 100-fold less than that observed *in vivo*.

Since the *E. coli* cell contains about 400 polymerase molecules (Richardson et al., 1964b), it could be proposed that the polymerase functions *in vivo* at the observed *in vitro* turnover number, and that it still functions as the replicase. If half the molecules were simultaneously involved in DNA synthesis at the replication fork, replication would occur at the observed rate. The short segments produced by the polymerase molecules could be joined together subsequent to replication by an enzyme of the polynucleotide ligase type. This proposal is contradicted by the experiments of Okazaki et al. (1968), however. Whatever the enzyme that catalyzes replication, its turnover number is much greater than that so far observed for the DNA polymerase (see p. 46). The true turnover number of the polymerase has probably been severely underestimated, however. The *in vivo* template is undoubtedly different from that used so far *in vitro,* and this could lead to a large difference between the turnover numbers.

A second objection to the polymerase being the "replicase" has been that the *in vitro* product has no biological activity. When transforming DNA is used as the template to produce a net synthesis of DNA, there is no increase in transforming activity (Richardson et al., 1964a). Electron micrographs of product DNA demonstrate that it contains short branches or loops, and this could be the cause of the lack of biological activity. Apparently, as the polymerase copies one strand of the double-helical template DNA, the other strand is peeled off (Mitra and Kornberg, 1966). Before the first strand is completely copied the polymerase crosses to the strand which was being peeled off and starts copying it, producing a fork. The continuation of this produces a branched product. When the polymerase copies single-stranded DNA as the template, this branching does not occur until the initial template has been replicated once. Consequently, DNA with biological activity can be produced if the template is single stranded and replication is stopped at the appropriate time (Richardson et al., 1964a). By using suitable reaction conditions, Goulian et al. (1967) have synthesized ΦX174 DNA which is biologically active. These experiments demonstrate that the polymerase copies its template with great fidelity. Since the state of the template *in vivo* is probably different from that used so far *in vitro,* the lack of biological activity in DNA produced from a double-stranded template does not seem a strong objection to the polymerase being the "replicase."

A third objection to the polymerase being the replicase has been that the polymerase only synthesizes DNA in a 5′→3′ direction (Richardson et al., 1963). Cairns' autoradiographs demonstrate that chromosome replication occurs on both strands simultaneously, and it has seemed likely that one strand is synthesized in a 5′→3′ direction, and that the other strand is synthesized in a 3′→5′ direction. One possibility for how this "other" strand is synthesized is that the cell contains an as yet unidentified polymerase that uses 5′-triphosphates as precursors and causes chain elongation in a 3′→5′ direction. But searches in crude cell extracts have failed to provide evidence for such a mechanism (Mitra and Kornberg, 1966). These results do not rule out this possibility, however, since the noise level in the assay was relatively high. Another possibility for the replication of the "other" strand is that this strand is replicated by a polymerase which uses 3′-triphosphates as precursors. The synthesis of thymidine-3′-triphosphate has been detected in bacterial extracts (Canellakis et al., 1965). But Mitra and Kornberg (1966) report that Josse has detected no trace of polymerase activity using such substrates. This mechanism of replication is also contradicted by the experiments of Price et al. (1967), who have obtained evidence that the precursors employed in the synthesis of both strands *in vivo* are phosphorylated at the 5′ position.

The most likely possibility for the replication of the "other" strand is that it is catalyzed by the Kornberg DNA polymerase, or a polymerase similar to it. A model for chromosome replication based on this has been proposed by Mitra et al. (1967). As mentioned above, the *in vitro* replication of a double-stranded template peels off the strand that is not being copied. After an indefinite amount of this strand has been peeled off, the polymerase crosses to it and starts replicating it, producing a forked molecule. An analogous situation *in vivo* could lead to the replication of the chromosome by the polymerase (Fig. 13). Since the *in vivo* concentration of polymerase is much higher than that used *in vitro,* and since probably only the DNA at the replication fork is usually available as template, it seems likely that *in vivo* a free polymerase molecule would bind to this single-stranded segment as soon as a sufficient length had been exposed. Consequently, the forking would not occur. The polymerase molecule would replicate the short, exposed piece of template in a 5′→3′ direction, and the direction of replication would be the opposite of the direction of movement of the replication fork. Further synthesis by the polymerase molecule(s) moving in the same direction as the replication fork would expose more single-stranded template, and more polymerase molecules would bind to and replicate the short, exposed template. In this fashion the replication fork would move sequentially along the chromosome. One previous objection to this type of synthesis was that it would leave single-strand breaks in the growing daughter strands, at least in the newly synthesized 3′→5′ strand. But the discovery of the

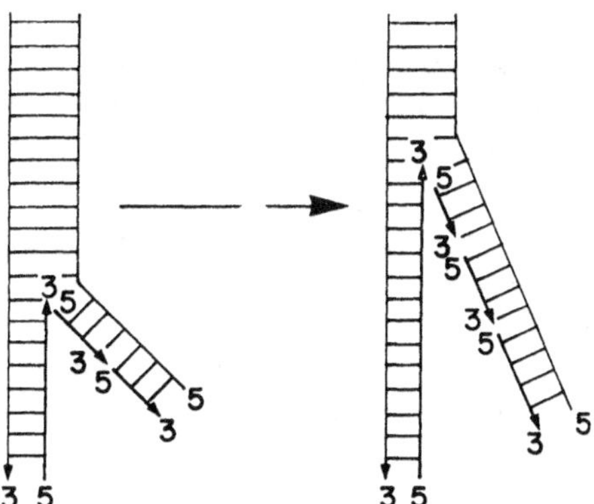

FIG. 13. Diagrammatic representation of scheme by which a number of polymerase molecules could replicate both strands simultaneously at the replication fork. Arrows show the direction in which the molecules are replicating. (From Mitra et al., 1967. *J. Molec. Biol.*, 24:429.)

bacterial polynucleotide ligase demonstrates a mechanism by which these breaks could easily be closed (Gellert, 1967; Olivera and Lehman, 1967; Zimmerman et al., 1967; Gefter et al., 1967).

Results obtained by Okazaki et al. (1968) are consistent with the above mode of replication. These investigators observed that if cells are labeled for a very brief period (5 seconds) with radioactive thymidine and the DNA then analyzed in alkaline sucrose gradients, the label is found in DNA with an average sedimentation rate of 11S. When the pulse time is 10 or 30 seconds, the radioactivity is found in the fast-sedimenting DNA as well as in the 11S fraction. With longer pulse periods, most of the radioactivity is in the fast-sedimenting DNA. The average sedimentation rate of this DNA increased gradually and was about 50S after a 10-minute pulse. These data suggest that newly synthesized DNA is not immediately covalently linked to the rest of the growing chains. Also, since most of the label (at least 70 percent) was found in the low molecular weight pieces following a 5- or 10-second pulse, it seems likely that breaks are present in both of the growing strands, not just the $3' \rightarrow 5'$ strand. Both growing strands in Figure 13 would therefore be composed of short pieces. This result is consistent with the hypothesis that at least several polymerase molecules act simultaneously in the synthesis of each strand.

An interesting property of the newly synthesized DNA is that it is largely found in the single-stranded form, even when a gentle extraction procedure is used which should not denature the DNA. This indicates that the DNA at the replication fork is present in some form that readily allows the separation of complementary strands that contain about 1,000 nucleotides. It is not known whether this separation results from some unusual secondary structure, or because the newly synthesized DNA is bound to protein or the cell membrane (Hanawalt and Ray, 1964; Ganesan and Lederberg, 1965; Smith and Hanawalt, 1967). This is a matter of great interest, since it is obviously related to the mechanism of replication.

Oishi (1968a) has also studied the properties of newly synthesized DNA and has independently obtained results similar to those of Okazaki et al. In addition, Oishi (1968b) has observed that this newly synthesized DNA is rapidly converted to another form which consists of double-helical DNA containing gaps in the newly synthesized strand. These gaps are not completely closed until after 10 to 15 percent of the cell cycle has elapsed.

One objection to the interpretation that the low molecular weight pieces of DNA are caused by replication in short segments has been that the pieces could arise during the extraction procedure from fragmentation of the DNA at the replication fork. Hanawalt and Ray (1964) have demonstrated that this DNA is very sensitive to shear. Evidence that the short pieces are related to replication and are not an artifact has been obtained by Sugimoto et al. (1968). When *E. coli* is infected with T4 phage which produces a temperature sensitive polynucleotide ligase, the short pieces accumulate at the nonpermissive temperature. The normal transition of the short pieces into high molecular weight DNA is observed when cells are infected at the permissive temperature or when cells are infected with wild-type phage at the elevated temperature. The phage ligase is evidently required for joining the segments together and adding them to the growing chains, and the bacterial ligase would be expected to function in a similar fashion.

Although the evidence in favor of the above model of replication is good, there is evidence that suggests this type of replication might be catalyzed by an enzyme other than the Kornberg DNA polymerase. Okazaki's data indicate that the turnover number of the enzyme synthesizing the 1,000 to 2,000 nucleotide long segments is much higher than the *in vitro* turnover number of the DNA polymerase. The experiments were done at 20°C, which reduces the rate of replication sixfold from the 37°C rate. The minimum estimate of the *in vitro* turnover number would be reduced to approximately 170 nucleotides per minute at 20°C. Within 30 seconds after the addition of radioactive thymidine to cells, at least one-half of the radioactivity was in fast sedimenting DNA. If the 1,000 to 2,000 nucleotide long segments are joined together by the ligase as soon as their synthesis is complete, the turnover number of the enzyme synthesizing the segments has to be 2,000 to 4,000 nucleotides per minute at 20°C (12,000 to 24,000 nucleotides per minute at 37°C) to fit the data. If the ligase reaction were the rate limiting step, the turnover number of the enzyme could be even higher. Since the true turnover number of the DNA polymerase probably has not yet been accurately determined, however, this is not a strong argument against the polymerase being the "replicase." This argument is of interest primarily since it sets an upper limit to the number of polymerase-like molecules that must be simultaneously engaged in replication.

Further studies on the DNA polymerase catalyzed replication of ΦX174 DNA have produced a more severe argument against the polymerase being the "replicase." The supernatant from a boiled cell extract was required for replication in this system (Goulian and Kornberg, 1967). Goulian (1968) has found that this requirement could be replaced by oligonucleotides and the oligonucleotides serve as a primer for the initiation of synthesis. In the model of replication discussed above, no primer would be available for the synthesis of the short segments. It is possible that the cell could supply oligonucleotides which served as this primer; but this creates other difficulties, such as the synthesis and specificity of the oligonucleotides. Since a primer would always be present in repair synthesis, one is tempted to relegate the *E. coli* polymerase to a repair function once again, and to seek some other enzyme which catalyzes replication.

As demonstrated by Ganesan and Lederberg (1965) and Smith and Hanawalt (1967), bacterial chromosome replication occurs at the cell surface. This suggests an explanation for the limited progress of previous studies on replication mechanisms and the enzymes involved. Since replication proceeds at such a high rate, and because of the requirements for almost absolute accuracy and continuity, it seems likely that replication is effected by a highly ordered complex of enzymes. This complex would contain the polymerases, ligases, and nucleases arranged in the configuration which optimized their various functions. The simple act of breaking open the cell could readily destroy this complex, with the result that some activities would be lost, and others might be altered. The previous approach to replication, although necessary, can probably be compared to breaking down the ribosome complex, and then wondering why protein synthesis does not proceed at the *in vivo* rate. Experiments on cells that have been treated so that the permeability barrier is partially lost, demonstrate how sensitive the replication machinery is. When *E. coli* is treated with Tris-EDTA or Tris-Mn^{++}, cell metabolism continues and exogenously added deoxyribonucleoside triphosphates are incorporated into DNA (Buttin and Kornberg, 1966). But this incorporation seems to result from a form of repair synthesis and not from the normal replication of the chromosomes, which is apparently inhibited by these treatments.

A different problem related to the mechanism of chromosome replication is the unwinding of the chromosome that occurs as the parental strands are copied and segregated into the daughter chromosomes. Since the strands of the double helix are plectonemically coiled and appear to be conserved in their entirety during replication, it is thought that the parental strands can only be separated from each other by the rotation of the parental double helix. With respect to circular chromosomes in particular, this rotation would cause difficulties. If the ends of the chromosome are not free to rotate relative to each other during replication, the chromosome would very quickly become a tangled mess incapable of replication.

As a solution to the rotation problem, Cairns (1963a, 1963b) proposed that the circular *E. coli* chromosome contains a swivel between the replication origin and terminus. More recently, Cairns and Davern (1966) have extended this model with the hypothesis that the swivel actively unwinds the chromosome. If the swivel rotated the parental double helix in a right-hand direction, the strands would unwind at the replication fork, where they could be exposed to and copied by the polymerase. As a test of this hypothesis, Cairns and Davern determined the effect of ^{32}P decays on DNA synthesis. The ^{32}P disintegrations would break the phosphodiester backbone of the chromosome, and this would result in a loss of transmission of the unwinding force. The observed rate of the inactivation of chromosome replication was one hit per 30 disintegrations. This result is consistent with the hypothesis but, as mentioned by these investigators, does not prove it. It simply demonstrates that a limited number of ^{32}P disintegrations lead to the cessation of DNA synthesis in a *rec*⁻ strain. Since cell-wide breakdown of DNA has been observed to occur following ^{32}P disintegrations in *rec*⁻ strains (Davern, 1968), it seems more likely that the incipient chromosome degradation caused the presence of a number of unrepairable breaks in the strands, and that replication ceased when it reached such a break.

With respect to the active unwinding model, it can also be argued that it would probably not be effective in a cell. Since newly synthesized strands contain gaps for some time after replication (Oishi, 1968b), and since the cell contains potent endonucleases, there would frequently be at least one break in the backbone between the replication fork and the terminus. These breaks would lead to a loss of transmission of the unwinding force.

The rotation of the chromosome would cause problems besides the twisting of circular chromosomes. The rotation would also interfere with RNA synthesis, and this would occur in both linear and circular chromosomes with a single swivel. Since the *E. coli* chromosome is replicated in 40 minutes, the chromosome would have to rotate at 10,000 rpm to allow unwinding. If the transcription of an operon occurred in the direction in which the replication fork was moving, transcription would have to occur at a rate greater than 1,670 nucleotides per second to avoid wrapping the nascent RNA around the rotating chromosome. The rate of ribosomal RNA synthesis is unknown, but an argument can be made that it approaches this rate (Maaløe and Kjeldgaard, 1966). The rate of messenger RNA synthesis is much lower than this, however. The rate of transcription of the tryptophan operon is estimated at approximately 20 nucleotides per second (Baker and Yanofsky, 1968). If the direction of transcription of an operon is the opposite of the direction of chromosome replication, the conditions for transcription seem equally severe. The RNA polymerase and nascent RNA would have to rotate around the axis of the chromosome at 10,000 rpm just to stay adjacent to the bases being copied.

The transcription of operons obviously poses no difficulties for the cell. Also, transcription can occur in both directions along the chromosome; operons can be inverted without affecting their function (Beckwith et al., 1966). This suggests that most of the chromosome does not rotate during replication. One solution to this problem is that the chromosome contains a number of single-strand breaks which act as swivels. These breaks would allow most of the chromosome to be relatively static during replication, since only the region between the replication fork and the nearest breaks would rotate as a result of replication. The swivels in the growing daughter strands could be the gaps observed by Okazaki et al. (1968) and Oishi (1968a, 1968b). The swivel in the parental double helix could be the result of random nuclease action, or the result of a nuclease which is part of the replication structure. Of course, such breaks would have to be closed before the region containing the break was replicated. This could be done by a ligase enzyme that was part of the replication apparatus.

If breaks occur in the chromosome, they do not occur at a very high frequency. Experiments conducted by Richards and Boyer (1966) demonstrate that the frequency of such breaks is not more than one break per 5 or 6 nucleotides. Unfortunately, however, their procedure did not allow the detection of breaks at a lower frequency. Hanawalt has conducted preliminary experiments on this problem, and he has obtained evidence that ^{32}P is incorporated into template strands during replication (Hanawalt, in discussion after Cairns and Davern, 1967). This incorporation could be caused by the closing of gaps that had acted as swivels. It was estimated that the incorporation occurred at a frequency of one phosphate per 1,000 nucleotides. To gain further information on this and other problems associated with the mechanism of replication, more studies on the properties of template DNA near the replication fork are obviously needed.

SUMMARY

Bacterial chromosomes consist of long double helices of DNA that are circular in at least *E. coli* and *M. hominis*. The *B. subtilis* chromosome is probably circular, but the evidence is still incomplete. No protein has been observed integrated into bacterial chromosomes; swivels or other mechanical elements apparently consist of DNA. The chromosome is associated with the cell membrane at the replication fork and possibly at the replication origin. Little is known concerning this association, and its study is of great importance since it probably functions in the control of initiation and in segregation as well as in replication.

The initiation of a cycle of chromosome replication requires the synthesis of specific proteins. Once initiation has occurred, replication can be completed in the absence of further protein synthesis. But the initiation of a new round of replication requires the synthesis of more initiation proteins. Initiation occurs at a specific locus on the chromosome, and replication proceeds in a definite direction. In the majority of *E. coli* strains the replication origin is between 55 and 65 minutes on the circular linkage map, and replication proceeds in a clockwise direction. The replication origin in *B. subtilis* is near the *ade* 16 locus.

The average amount of DNA per cell, the time during the cell cycle when initiation occurs, and the time required to replicate a chromosome vary with the growth rate. In *E. coli* B/r with generation times between 22 and 60 minutes, replication requires about 40 minutes, and cell division occurs 20 minutes after the termination of a round of replication. This simple relationship accounts for the variation in the initiation time and the average amount of DNA in such cells. When the generation time is greater than 60 minutes, initiation occurs approximately at cell division, and replication requires a constant fraction of the cell cycle. Other bacteria probably have analogous relationships between the replication time, the termination of replication, cell division, and the growth rate.

E. coli cells contain two chromosome structures at the end of the cell cycle and each daughter cell receives one such structure. In slowly growing cells these structures can be completed chromosomes. In rapidly growing cells they can be multiforked chromosomes containing up to three genome equivalents of DNA. The association between the chromosome and the cell membrane probably functions in this segregation. The evidence for this and the stability of the association is incomplete, however.

The initiation proteins appear to be synthesized at a rate proportional to the growth rate, and it requires one generation to synthesize enough to initiate a new round of replication. Such constitutive synthesis explains many of the relationships between replication and the growth rate. If the generation time equals the replication time, for example, enough initiation proteins are available at termination to immediately initiate a new round of replication. In rapidly growing cells sufficient proteins are available before termination and dichotomous replication results. In slowly growing cells there is a period following termination when initiation proteins accumulate to the point that initiation can occur again. The nature of the initiation proteins is unknown, and this is currently an active area of research.

The mechanism of chromosome replication is still unclear. Arguments can be made for and against the DNA polymerase being the "replicase." Both strands are probably replicated in a 5′→3′ direction, and at least several "replicase" molecules are simultaneously involved in the replication of each strand. Synthesis consequently occurs in short pieces that are probably joined together subsequent to replication by polynucleotide ligase. Newly synthesized DNA is usually obtained in a single-stranded form; this is undoubtedly related to the mechanism of replication. Much of the future elucidation of the mechanism of replication will depend upon the study of the replication system associated with the cell membrane.

REFERENCES

Abe, M., and J. Tomizawa. 1967. Replication of the *Escherichia coli* K12 chromosome. Proc. Nat. Acad. Sci. U.S.A., 58:1911–1918.

Altenbern, R. A. 1968. Chromosome mapping in *Staphylococcus aureus*. J. Bact., 95: 1642–1646.

Baker, R. F., and C. Yanofsky. 1968. The periodicity of RNA polymerase initiations: A new regulatory feature of transcription. Proc. Nat. Acad. Sci. U.S.A., 60:313–320.

Beachey, E. H., and R. M. Cole. 1966. Cell wall replication in *Escherichia coli*, studied by immunofluorescence and immunoelectron microscopy. J. Bact., 92:1245–1251.

Beckwith, J. R., E. R. Signer, and W. Epstein. 1966. Transposition of the *Lac* region in *E. coli*. Cold Spring Harbor Symp. Quant. Biol., 31:393–401.

Berg, C. M., and L. G. Caro. 1967. Chromosome replication in *Escherichia coli*. I. Lack of influence of the integrated F factor. J. Molec. Biol., 29:419–431.

Billen, D. 1962. Alteration in DNA synthesizing capacity in bacteria: an in vivo-in vitro study. Biochim. Biophys. Acta, 55:960–968.

Bird, R., and K. G. Lark. 1968. Initiation and termination of DNA replication after amino acid starvation of *E. coli* 15T⁻. Cold Spring Harbor Symp. Quant. Biol., 33:799–808.

Bode, H. R., and H. J. Morowitz. 1967. Size and structure of the *Mycoplasma hominis* H39 chromosome. J. Molec. Biol., 23:191–199.

Bode, V. C., and A. D. Kaiser. 1965. Change in the structure and activity in λ DNA in a superinfected immune bacterium. J. Molec. Biol., 14:399–417.

Bonhoeffer, F., and A. Gierer. 1963. On the growth mechanism of the bacterial chromosome. J. Molec. Biol., 7:534–540.

Boyle, J. V., W. A. Goss, and T. M. Cook. 1967. Induction of excessive DNA synthesis in *Escherichia coli* by nalidixic acid. J. Bact., 94:1664–1671.

Buttin, G., and A. Kornberg. 1966. Utilization of deoxyribonucleoside triphosphates by *Escherichia coli* cells. J. Biol. Chem., 241:5419–5427.

Cairns, J. 1962. A minimum estimate for the length of the DNA of *Escherichia coli* obtained by autoradiography. J. Molec. Biol., 4:407–409.

——— 1963a. The bacterial chromosome and its manner of replication as seen by autoradiography. J. Molec. Biol., 6:208–213.

——— 1963b. The chromosome of *Escherichia coli.* Cold Spring Harbor Symp. Quant. Biol., 28:43–46.

——— and C. Davern. 1966. Effect of ^{32}P decay upon DNA synthesis by a radiation-sensitive strain of *Escherichia coli.* J. Molec. Biol., 17:418–427.

——— and C. Davern. 1967. The mechanics of DNA replication in bacteria. J. Cell. Physiol., 70 (Suppl. 1):65–76.

Canellakis, E. S., H. O. Kammen, and D. R. Morales. 1965. Studies on the enzymatic synthesis of thymidine-3′-triphosphate. Proc. Nat. Acad. Sci. U.S.A., 53:184–187.

Cerdá-Olmeda, E., and P. C. Hanawalt. 1968. Mutagenesis of the replication point by nitrosoguanidine: Map and pattern of replication of the *Escherichia coli* chromosome. J. Molec. Biol., 33:705–719.

Chai, N., and K. G. Lark. 1967. Segregation of deoxyribonucleic acid in bacteria: Association of the segregating unit with the cell envelope. J. Bact., 94:415–421.

Clark, D. J. 1968. The regulation of DNA replication and cell division in *E. coli* B/r. Cold Spring Harbor Symp. Quant. Biol., 33:823–836.

——— and O. Maaløe. 1967. DNA replication and the division cycle in *Escherichia coli.* J. Molec. Biol., 23:99–112.

Cole, R. M., and J. J. Hahn. 1962. Cell wall replication in *Streptococcus pyogenes.* Science, 135:722–723.

Cooper, S., and C. E. Helmstetter. 1968. Chromosome replication and the division cycle of *Escherichia coli* B/r. J. Molec. Biol., 31:519–540.

Copeland, J. C., and J. Marmur. 1968. Temperature sensitive mutants in *Bacillus subtilis* affecting DNA synthesis. Bact. Proc., p.61.

Cutler, R. G., and J. E. Evans. 1967. Relative transcription activity of different segments of the genome throughout the cell division cycle of *Escherichia coli.* The mapping of ribosomal and transfer RNA and the determination of the direction of replication. J. Molec. Biol., 26:91–105.

Cuzin, F., and F. Jacob. 1967. Bacterial episomes as a model of replicon. *In* Regulation of Nucleic Acid and Protein Biosynthesis, Koningberger, V. V., and Bosch, L., eds., Amsterdam, Elsevier Publishing Company, pp. 39–50.

Davern, C. 1968. Effect of ^{32}P decay upon RNA transcription by a radiation-sensitive strain of *Escherichia coli.* J. Molec. Biol., 32:151–154.

Dennis, E. S., and R. G. Wake. 1966. Autoradiography of the *Bacillus subtilis* chromosome. J. Molec. Biol., 15:435–439.

——— and R. G. Wake. 1967. The *Bacillus subtilis* genome—Studies on its size and structure. *In* Replication and Recombination of Genetic Material, Peacock, W. J., and Brock, R. D., eds., Canberra, Australian Academy of Science, pp. 61–70.

Donachie, W. D. 1968. Relationship between cell size and time of initiation of DNA replication. Nature, 219:1077–1079.

Dubnau, D., C. Goldthwaite, I. Smith, and J. Marmur. 1967. Genetic mapping in *Bacillus subtilis.* J. Molec. Biol., 27:163–185.

Eberle, H., and K. G. Lark. 1966. Chromosome segregation in *Bacillus subtilis.* J. Molec. Biol., 22:183–186.

——— and K. G. Lark. 1967. Chromosome replication in *Bacillus subtilis* cultures growing at different rates. Proc. Nat. Acad. Sci. U.S.A., 57:95–101.

Espardellier-Joset, F., P. D. Harriman, J. Gots, and H. Marcovich. 1967. Localisation de l'origine de réplication végétative chez une souche Hfr d'*Escherichia coli* K12. C. R. Soc. Biol. (Paris), 264:1541–1544.

Fangman, W. L. 1966. Temperature-sensitive DNA synthesis in mutants of *Escherichia coli.* Bact. Proc., p. 81.

Fitz-James, P. C., and I. E. Young. 1959. Comparison of species and varieties of the genus *Bacillus.* Structure and nucleic acid content of spores. J. Bact., 78:743–754.

Forro, F. 1965. Autoradiographic studies of bacterial chromosome replication in amino acid deficient *Escherichia coli* 15T-. Biophys. J., 5:629–649.

——— and S. A. Wertheimer. 1960. The organization and replication of DNA in thymine deficient strains of *Escherichia coli.* Biochim. Biophys. Acta, 40:9–21.

Freese, E. 1958. The arrangement of DNA in the chromosome. Cold Spring Harbor Symp. Quant. Biol., 23:13–18.

Freifelder, D. 1968. Studies on *Escherichia coli* sex factors. IV. Molecular weights of the DNA of several F′ elements. J. Molec. Biol., 35:95–102.

Ganesan, A. T. 1963. Physical and biological studies of *Bacillus subtilis* deoxyribonucleic acid. Ph. D. Thesis, Stanford University, Stanford, California.

——— and J. Lederberg. 1965. A cell-membrane bound fraction of bacterial DNA. Biochem. Biophys. Res. Commun., 18:824–835.

Gefter, M. L., A. Becker, and J. Hurwitz. 1967. The enzymatic repair of DNA, I. Formation of circular DNA. Proc. Nat. Acad. Sci. U.S.A., 58:240–247.

Gellert, M. 1967. Formation of covalent circles of lambda DNA by *E. coli* extracts. Proc. Nat. Acad. Sci. U.S.A., 57:148–155.

Goldring, E. S., and R. G. Wake. 1968. A comparison of the segregation of chromosomes within microcolonies developing from single *Bacillus subtilis* and *Bacillus megaterium* spores. J. Molec. Biol., 35:647–650.

Goulian, M. 1968. Incorporation of oligodeoxynucleotides into DNA. Proc. Nat. Acad. Sci. U.S.A., 61:284–291.

——— and A. Kornberg. 1967. Enzymatic synthesis of DNA, XXIII. Synthesis of circular replicative form of phage ØX174 DNA. Proc. Nat. Acad. Sci. U.S.A., 58:1723–1730.

——— A. Kornberg, and R. Sinsheimer. 1967. Synthesis of infectious phage ØX174 DNA. Proc. Nat. Acad. Sci. U.S.A., 58:2321–2328.

Hanawalt, P. C., O. Maaløe, D. J. Cummings, and M. Schaechter. 1961. The normal DNA replication cycle, II, J. Molec. Biol., 3:156–165.

——— and D. S. Ray. 1964. Isolation of the growing point in the bacterial chromosome. Proc. Nat. Acad. Sci. U.S.A., 52:125–132.

Helmstetter, C. E. 1967. Rate of DNA synthesis during the division cycle of *Escherichia coli.* B/r. J. Molec. Biol., 24:417–427.

——— 1968. Origin and sequence of chromosome replication in *Escherichia coli.* B/r. J. Bact., 95:1634–1641.

——— and S. Cooper. 1968. DNA synthesis during the division cycle of rapidly growing *Escherichia coli* B/r. J. Molec. Biol., 31:507–518.

——— and O. Pierucci. 1968. Cell division during inhibition of DNA synthesis in *Escherichia coli.* J. Bact., 95:1627–1633.

Hickson, F. T., T. F. Roth, and D. R. Helinski. 1967. Circular DNA forms of a bacterial sex factor. Proc. Nat. Acad. Sci. U.S.A., 58:1731–1738.

Hopwood, D. A. 1967. Genetic analysis and genome structure in *Streptomyces coelicolor.* Bact. Rev., 31:373–403.

Howard-Flanders, P. 1968. DNA repair. Ann. Rev. Biochem., 37:175–200.

Jacob, F., S. Brenner, and F. Cuzin. 1963. On the regulation of DNA replication in bacteria. Cold Spring Harbor Symp. Quant. Biol., 28:329–347.

——— A. Ryter, and F. Cuzin. 1966. On the association between DNA and membrane in bacteria. Proc. Roy. Soc. (London), Ser. B, 164:267–278.

——— and E. Wollman. 1961. Sexuality and the Genetics of Bacteria. New York, Academic Press, Inc.

Kallenbach, N. R., and R. Ma. 1968. Initiation of deoxyribonucleic acid synthesis after thymine starvation of *Bacillus subtilis.* J. Bact., 95:304–309.

Kirschner, R. H., D. R. Wolstenholme, and N. J. Gross. 1968. Replicating molecules of circular mitochondrial DNA. Proc. Nat. Acad. Sci. U.S.A., 60:1466–1472.

Kjeldgaard, N. O., O. Maaløe, and M. Schaechter. 1958. The transition between different physiological states during balanced growth of *Salmonella typhimurium.* J. Gen. Microbiol., 19:607–616.

Kleinschmidt, A., A. Burton, and R. L. Sinsheimer. 1963. Electron microscopy of the replicative form of the DNA of the bacteriophage ØX174. Science, 142:961.

——— D. Lang, C. Plescher, W. Hellmann, J. Haas, R. Zahn, and A. Hagedorn. 1961. Über die intrazelluläre Formation von Bakterien-DNS. Z. Naturforsch. (B), 16b: 730–739.

Koch, A. L. 1966. On the difference between the lethal effects of ^{3}H and ^{32}P in bacteria. Radiat. Res., 29:18–32.

——— and P. Pachler. 1967. Evidence against the alternation of synthesis of identical chromosomes in *Escherichia coli* growing at low rates. J. Molec. Biol., 28:531–537.

Kohiyama, M., D. Cousin, A. Ryter, and F. Jacob. 1966. Mutants thermosensibles d'*Escherichia coli* K12. I. Isolement et characterisation rapide. Ann. Inst. Pasteur, 110:465–486.

Kornberg, A. 1956. Pathways of enzymatic synthesis of nucleotides and polynucleotides. *In* The Chemical Basis of Heredity, McElroy, W. D., and Glass, B., eds., Baltimore, Johns Hopkins Press, pp. 579–608.

——— 1961. Enzymatic Synthesis of DNA. New York, John Wiley and Sons, Inc.

Kubitschek, H. E., H. E. Bendigkeit, and M. R. Loken. 1967. Onset of DNA synthesis during the cell cycle in chemostat cultures. Proc. Nat. Acad. Sci. U.S.A., 57:1611–1617.

Kuempel, P. L. 1968. DNA synthesis in a temperature sensitive mutant of *Escherichia coli*. Bact. Proc., p. 60.

Lark, C. 1966. Regulation of DNA synthesis in *Escherichia coli*: Dependence on growth rates. Biochim. Biophys. Acta, 119:517–525.

——— and K. G. Lark. 1964. Evidence for two distinct aspects of the mechanism regulating chromosome replication in *Escherichia coli*. J. Molec. Biol., 10:120–136.

Lark, K. G. 1966a. Regulation of chromosome replication and segregation in bacteria. Bact. Rev., 30:3–32.

——— 1966b. Chromosome replication in *Escherichia coli*. *In* Cell Synchrony—Studies in Biosynthetic Regulation. Cameron, I. L., and Padilla, G. M., eds., Regulation, New York, Academic Press, Inc., pp. 54–80.

——— and R. Bird. 1965a. Premature chromosome replication induced by thymine starvation: Restriction of replication to one of the two partially completed replicas. J. Molec. Biol., 13:607–610.

——— and R. Bird. 1965b. Segregation of the conserved units of DNA in *Escherichia coli*. Proc. Nat. Acad. Sci. U.S.A., 54:1444–1450.

——— H. Eberle, R. A. Consigli, H. C. Minocha, N. Chai, and C. Lark. 1967. *In* Organizational Biosynthesis, Vogel, H. J., Lampen, J. O., and Bryson, V., eds., New York, Academic Press, Inc., pp. 63–89.

——— and C. Lark. 1965. Regulation of chromosome replication in *Escherichia coli*: Alternate replication of two chromosomes at slow growth rates. J. Molec. Biol., 13: 105–126.

——— and C. Lark. 1966. Regulation of chromosome replication in *Escherichia coli*: A comparison of the effects of phenethyl alcohol treatment with those of amino acid starvation. J. Molec. Biol., 20:9–19.

——— T. Repko, and E. J. Hoffman. 1963. The effect of amino acid deprivation on subsequent deoxyribonucleic acid synthesis. Biochim. Biophys. Acta, 76:9–24.

Lehman, I. R. 1967. Deoxyribonucleases: their relationship to deoxyribonucleic acid synthesis. Ann. Rev. Biochem., 36:645–668.

Levine, A. J., and R. L. Sinsheimer. 1968. The process of infection with bacteriophage ØX174. XIX. Isolation and characterization of a chloramphenicol-resistant protein from ØX-infected cells. J. Molec. Biol., 32:567–578.

Levinthal, C., and P. F. Davison. 1961. Degradation of deoxyribonucleic acid under hydrodynamic shearing forces. J. Molec. Biol., 3:674–683.

Maaløe, O. 1961. The control of normal DNA replication in bacteria. Cold Spring Harbor Symp. Quant. Biol., 26:45–52.

——— and P. C. Hanawalt. 1961. Thymine deficiency and the normal DNA replication cycle. I. J. Molec. Biol., 3:144–155.

——— and N. O. Kjeldgaard. 1966. Control of Macromolecular Synthesis, New York, W. A. Benjamin, Inc.

MacHattie, L. A., K. I. Berns, and C. A. Thomas, Jr. 1965. Electron microscopy of DNA from *Hemophilus influenzae*. J. Molec. Biol., 11:648–649.

Marvin, D. A., and H. Schaller. 1966. The topology of DNA from the small filamentous bacteriophage fd. J. Molec. Biol., 15:1–7.

Masters, M., and A. B. Pardee. 1965. Sequence of enzyme synthesis and gene replication during the cell cycle of *Bacillus subtilis*. Proc. Nat. Acad. Sci. U.S.A., 54:64–70.

May, J. W. 1963. The distribution of cell wall label during growth and division of *Salmonella typhimurium*. Exp. Cell Res., 31:217–220.

Mendelson, N., and J. D. Gross. 1967. Characterization of a temperature-sensitive mutant of *Bacillus subtilis* defective in DNA replication. J. Bact., 94:1603–1608.

Meselson, M., and F. W. Stahl. 1958. The replication of DNA in *Escherichia* coli. Proc. Nat. Acad. Sci. U.S.A., 44:671–682.

Mickel, S., C. L. Herschberger, and R. Rownd. 1968. Closed circular forms of R-factor DNA. Bact. Proc., p. 53.

Mitra, S., and A. Kornberg. 1966. Enzymatic mechanisms of DNA replication. *In* Macromolecular Metabolism, Boston, Little, Brown and Co., pp. 59–79.

——— P. Reichard, R. B. Inman, L. L. Bertsch, and A. Kornberg. 1967. Enzymatic synthesis of deoxyribonucleic acid. XXII. Replication of a circular single-stranded DNA template by DNA polymerase of *Escherichia coli*. J. Molec. Biol., 24:429–447.

Moran, P. A. P. 1963. *In* Sankhya, The Indian Journal of Statistics, Ser. A, Vol. 25, Part I, 65.

Nagata, T. 1963. The molecular synchrony and sequential replication of DNA in *Escherichia coli*. Proc. Nat. Acad. Sci. U.S.A., 49:551–559.

Neuhard, J., and A. Munch-Petersen. 1966. Changes in the amounts of deoxycytidine triphosphate and deoxyadenosine triphosphate in *Escherichia coli* 15T$^-$A$^-$U$^-$. Biochim. Biophys. Acta, 114:61–71.

Oishi, M. 1968a. Studies of DNA replication in vivo. I. Isolation of the first intermediate of DNA replication in bacteria as single-stranded NDA. Proc. Nat. Acad. Sci. U.S.A., 60:329–336.

——— 1968b. Studies of DNA replication in vivo. II. Evidence for the second intermediate. Proc. Nat. Acad. Sci. U.S.A., 60:691–698.

——— H. Yoshikawa, and N. Sueoka. 1964. Synchronous and dichotomous replications of the *Bacillus subtilis* chromosome during spore germination. Nature, 204:1069–1073.

Okazaki, R., T. Okazaki, K. Sakabe, K. Sugimoto, and A. Sugino. 1968. Mechanism of DNA chain growth. I. Possible discontinuity and unusual secondary structure of newly synthesized chains. Proc. Nat. Acad. Sci. U.S.A., 59:598–605.

Olivera, B., and I. R. Lehman. 1967. Linkage of polynucleotides through phosphodiester bonds by an enzyme from *Escherichia coli*. Proc. Nat. Acad. Sci. U.S.A., 57:1426–1433.

O'Sullivan, A., and N. Sueoka. 1967. Sequential replication of the *Bacillus subtilis* chromosome. IV. Genetic mapping by density transfer experiment. J. Molec. Biol., 27:349–368.

Pardee, A. B., and L. S. Prestidge. 1959. On the nature of the repressor of β-galactosidase synthesis in *Escherichia coli*. Biochim. Biophys. Acta, 36:545–547.

Pato, M. L., and D. A. Glaser. 1968. The origin and direction of replication of the chromosome of *Escherichia coli* B/r. Proc. Nat. Acad. Sci. U.S.A., 60:1268–1274.

Pittard, J., and T. Ramakrishnan. 1964. Gene transfer by F′ strains of *Escherichia coli*. IV. Effect of chromosomal deletion on chromosome transfer. J. Bact., 88:367–373.

Price, T. D., R. A. Darmstadt, H. A. Hinds, and S. Zamenhof. 1967. Mechanism of synthesis of deoxyribonucleic acid in vivo. J. Biol. Chem., 242:140–151.

Pritchard, R. H., and K. G. Lark. 1964. Induction of replication by thymine starvation at the chromosome origin in *Escherichia coli*. J. Molec. Biol., 9:288–307.

Richards, O. C., and P. D. Boyer. 1966. ^{18}O labeling of deoxyribonucleic acid during synthesis and stability of the label during replication. J. Molec. Biol., 19:109–119.

Richardson, C. C., R. B. Inman, and A. Kornberg. 1964a. The repair of partially single-stranded DNA templates by DNA polymerase. J. Molec. Biol., 9:46–49.

——— C. L. Schildkraut, H. Aposhian, and A. Kornberg. 1964b. Enzymatic synthesis of deoxyribonucleic acid. XIV. Further purification and properties of deoxyribonucleic acid polymerase of *Escherichia coli.* J. Biol. Chem., 239:222–231.

——— C. L. Schildkraut, and A. Kornberg. 1963. Studies on the replication of DNA by DNA polymerases. Cold Spring Harbor Symp. Quant. Biol., 28:9–18.

Roth, T. F., and D. R. Helinski. 1967. Evidence for circular DNA forms of a bacterial plasmid. Proc. Nat. Acad. Sci. U.S.A., 58:650–657.

Ryter, A. 1968. Association of the nucleus and the membrane of bacteria: A morphological study. Bact. Rev., 32:39–54.

——— and F. Jacob. 1966. Étude morphologique de la liaison du noyau à la membrane chez *E. coli* et chez les protoplastes de *B. subtilis.* Ann. Inst. Pasteur, 110:801–812.

——— and F. Jacob. 1967. Segregation des noyaux pendant la croissance et la germination de *Bacillus subtilis.* C. R. Soc. Biol. (Paris), 264:2254–2256.

Sanderson, K. E. 1967. Revised linkage map of *Salmonella typhimurium.* Bact. Rev., 31:354–372.

Schaechter, M., O. Maaløe, and N. O. Kjeldgaard. 1958. Dependency on medium and temperature of cell size and chemical composition during balanced growth of *Salmonella typhimurium.* J. Gen. Microbiol., 19:592–606.

Smith, D. H., and B. D. Davis. 1967. Mode of action of novobiocin in *Escherichia coli.* J. Bact., 93:71–79.

——— and P. C. Hanawalt. 1967. Properties of the growing point region of the bacterial chromosome. Biochim. Biophys. Acta, 149:519–531.

Soska, J., and K. G. Lark. 1966. Regulation of nucleic acid synthesis in *Lactobacillus acidophilus* R-26. Biochim. Biophys. Acta, 119:526–539.

Stahl, F. W. 1967. Circular genetic maps. J. Cell. Physiol., 70 (Suppl. 1):1–4.

Stonehill, E. H., and D. J. Hutchison. 1966. Chromosome mapping by means of mutational induction in *Staphylococcus faecalis.* J. Bact., 92:136–143.

Sueoka, N. 1966. Synchronous replication of the chromosome in *Bacillus subtilis. In* Cell Synchrony—Studies in Biosynthetic Regulation, Cameron, I. L., and Padilla, G. M., eds., New York, Academic Press, Inc., pp. 38–53.

——— 1967. Mechanisms of replication and repair of nucleic acid. *In* Molecular Genetics, Taylor, H. J., ed., New York, Academic Press, Inc., Part II, pp. 1–46.

——— and W. G. Quinn. 1968. Membrane attachment of the chromosome replication origin in *Bacillus subtilis.* Cold Spring Harbor Symp. Quant. Biol., 33:695–705.

——— and H. Yoshikawa. 1963. Regulation of chromosome replication in *Bacillus subtilis.* Cold Spring Harbor Symp. Quant. Biol., 28:47–54.

——— and H. Yoshikawa. 1965. The chromosome of *Bacillus subtilis.* I. Theory of marker frequency analysis. Genetics, 52:747–757.

Sugimoto, K., T. Okazaki, and R. Okazaki. 1968. Mechanism of DNA chain growth. II. Accumulation of newly synthesized short chains in *E. coli* infected with ligase-defective T4 phages. Proc. Nat. Acad. Sci. U.S.A., 60:1356–1362.

Taylor, A. L., and E. A. Adelberg. 1961. Evidence for a closed linkage group in Hfr males of *Escherichia coli* K12. Biochem. Biophys. Res. Commun., 5:400–404.

——— and C. D. Trotter. 1967. Revised linkage map of *Escherichia coli.* Bact., Rev., 31:332–353.

Tessman, E. S. 1966. Mutants of bacteriophage S13 blocked in infectious DNA synthesis. J. Molec. Biol., 17:218–236.

Treick, R. W., and W. A. Konetzka. 1964. Physiological state of *Escherichia coli* and the inhibition of deoxyribonucleic acid synthesis by phenethyl alcohol. J. Bact., 88:1580–1584.

Tremblay, G. Y., M. J. Daniels, and M. Schaechter. 1968. Isolation of a membrane-DNA-nascent RNA complex. Bact. Proc., p. 65.

Tyeryar, F. J., W. D. Lawton, and A. M. MacQuillan. 1968. Sequential replication of the chromosome of *Bacillus licheniformis*. J. Bact., 95:2062–2069.

Vielmetter, W., W. Messer, and A. Schütte. 1968. Growth direction and segregation of the *E. coli* chromosome. Cold Spring Harbor Symp. Quant. Biol., 33:585–598.

Wake, R. G. 1963. Sequential replication of DNA in synchronously germinating *Bacillus subtilis* spores. Biochem. Biophys. Res. Commun., 13:67–70.

Weber, M. J., and J. A. DeMoss. 1966. The inhibition by chloramphenicol of nascent protein formation in *E. coli*. Proc. Nat. Acad. Sci. U.S.A., 55:1224–1230.

Wolf, B., A. Newman, and D. Glaser. 1968. On the origin and direction of replication of the *Escherichia coli* K12 chromosome. J. Molec. Biol., 32:611–629.

Yoshikawa, H. 1965. DNA synthesis during germination of *Bacillus subtilis* spores. Proc. Nat. Acad. Sci. U.S.A., 53:1476–1483.

——— 1967. The initiation of DNA replication in *Bacillus subtilis*. Proc. Nat. Acad. Sci. U.S.A., 58:312–319.

——— 1968. Chromosomes in *Bacillus subtilis* spores and their segregation during germination. J. Bact., 95:2282–2292.

——— A. O'Sullivan, and N. Sueoka. 1964. Sequential replication of the *Bacillus subtilis* chromosome. III. Regulation of initiation. Proc. Nat. Acad. Sci. U.S.A., 52:973–980.

——— and N. Sueoka. 1963a. Sequential replication of *Bacillus subtilis* chromosome. I. Comparison of marker frequencies in exponential and stationary growth phases. Proc. Nat. Acad. Sci. U.S.A., 49:559–566.

——— and N. Sueoka. 1963b. Sequential replication of *Bacillus subtilis* chromosome. II. Isotopic transfer experiments. Proc. Nat. Acad. Sci. U.S.A., 49:806–813.

Zimmerman, S. B., J. W. Little, C. K. Oshinsky, and M. Gellert. 1967. Enzymatic joining of DNA strands: A novel reaction of diphosphopyrimidine nucleotide. Proc. Nat. Acad. Sci. U.S.A., 57:1841–1848.

David M. Prescott

Department of Molecular, Cellular, and Developmental Biology, University of Colorado, Boulder, Colorado

2

The Structure and Replication of Eukaryotic Chromosomes

INTRODUCTION

Progress in achieving a complete and detailed knowledge of the structure and replication of the eukaryotic chromosome has lagged well behind the advances made in the understanding of the basic structure and replication of viral and bacterial chromosomes. Certainly a major reason for the slower progress is the greater structural and functional complexity of the eukaryotic chromosome, although the degree of complexity has probably been sometimes overemphasized or overstated. For example, it is a common, uncritical practice to overcomplicate the problem by assuming that any molecules found with eukaryotic chromosomes must be considered to be an intrinsic part of the chromosome, and many studies of eukaryotic chromosome structure and reproduction have not recognized the need to discriminate between materials that may only be associated with chromosomes adventitiously, transiently, or temporarily and materials that are actually an integral and permanent part of chromosome structure. To consider all material present with the chromosome in the definition of the chromosome does not seem a useful approach, since the chromosome will then have many definitions, one for each of its many functional or configurational states. It is more sensible to search for one definition based on those molecular elements that are *always* a permanent part of its structure and which show the same restriction of turnover that characterizes deoxyribonucleic acid. Each functional and configurational state of the chromosome

Prepared during tenure of a research grant from the American Cancer Society.

could then ultimately be described in terms of the particular materials associated with the basic, invariant structure.

There has also sometimes been in the past a strong tendency to emphasize the differences between prokaryotic chromosomes (primarily those of bacteria and viruses) and eukaryotic chromosomes, rather than to exploit the concepts and knowledge gained for prokaryotic chromosomes as a basis for the design of experimental analyses of eukaryotic chromosomes. The disadvantage of such an attitude has become sharply evident with the very recent and profitable application of the concepts and information that have been established for prokaryotic chromosomes to problems of chromosome structure and replication in eukaryotic cells. Stressing basic similarities, the term chromosome is quite appropriately applied to the traditional chromosomes of the eukaryotic cell nucleus, to the genetic elements in prokaryotic cells, to the genetic elements in viruses and bacterial episomes, and probably to the DNA molecules of mitochondria and chloroplasts. Up to the present, the more that has been learned about basic properties of the chromosome, the more evident is a common set of principles of structure and replication in all of these situations.

Good and useful reasons warrant, nevertheless, the continued distinction between prokaryotic and eukaryotic types of chromosomes (Ris, 1967; Ris and Chandler, 1963). Chromosomes of bacteria, blue-green algae, and viruses lack associated histones, and I suspect the same probably will prove to be true for the DNA of mitochondria and chloroplasts. In all these situations the DNA is not separated from the cytoplasm by an enclosing envelope, and segregation at division does not occur by a mitotic type of arrangement. In bacteria and viruses the DNA behaves as a single replicating unit, i.e., as a replicon. Information about replicons is not yet available for blue-green algae or for mitochondrial and chloroplast DNA, although electron microscope study of mitochondrial DNA does suggest the presence of only a single replicon per chromosome (see Fig. 1) (Kirschner et al., 1968). All of these various properties, lack of histones, lack of a surrounding envelope, a nonmitotic mechanism for segregation, and the presence of a single replicon, can, at least for the present, be used to define the prokaryotic type of chromosome. By such a definition, therefore, the term legitimately includes not only the genetic elements of bacteria and blue-green algae but also the chromosomes of viruses, mitochondria, chloroplasts, and bacterial episomes.

It is reasonable to postulate that the larger, more complicated eukaryotic chromosomes have evolved from a simpler type, such as represented by contemporary prokaryotic-type chromosomes. The extent to which the principles of structure, replication, and segregation of the prokaryotic chromosome have been retained, modified, or supplemented with new principles in the evolution of the eukaryotic chromosome is yet to be fully determined, although there is now enough information to allow a reasonable beginning to speculation about certain aspects of the problem. For example, the mitotic mechanism of chromosome segregation in eukaryotes is manifestly different from segregation of prokaryotic chromosomes, but do the segregations in the two cases still operate with sufficiently similar principles to indicate an evolutionary connection? (See for example, recent papers from Lark's laboratory on the possibility of similar principles of DNA segregation in eukaryotes

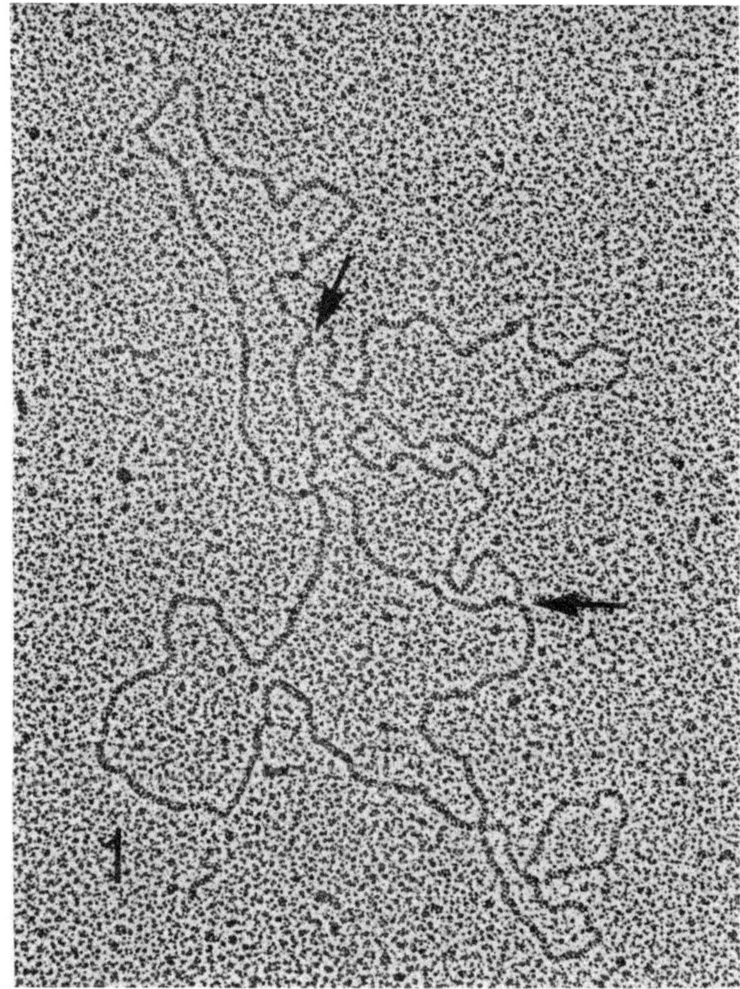

FIG. 1. A partially replicated, circular, prokaryotic chromosome of a mitochondrion from rat liver. Two forks, assumed to represent the initiation point and the replication point, are indicated by the arrows. The total length of the chromosome is 5.1 μ and probably contains only a single replicon. The method of preparation of DNA suggests that the end and beginning of the chromosome are joined together (to form a loop) by covalent bonding of DNA. A similar type of linkage is proposed as the means by which the multiple, linear replicons of a eukaryotic chromosome are joined in series. × 92,300. (From Kirschner et al., 1968. *Proc. Nat. Acad. Sci. U.S.A.*, 60:1466.)

and prokaryotes; Chai and Lark, 1967; Lark, 1967a; Lark et al., 1966.) In this regard, what is the significance of the fact that all eukaryotes have developed a multichromosomal state and a nuclear envelope, while chromosomes of the prokaryotic type occur only singly and without a nuclear envelope? Is the joining of the ends of certain prokaryotic chromosomes to form a loop related to the apparent end-to-end connection of replicons that occurs in multirepliconic eukaryotic chromosomes? Is the initiator protein hypothesis applicable to the regulation of replication of eukaryotic chromosomes?

In the discussion to follow, advantage is taken of the detailed information now available for prokaryotic-type chromosomes in interpreting data on eukaryotic chromosomes and in deriving hypotheses concerning the structure and replication of the eukaryotic chromosome. Of particular importance are the following points established for prokaryotic chromosomes (see Kuempel, Chapter 1; Lark, 1966; Lark, 1967b; Lark et al., 1967).

1. The DNA of the prokaryotic chromosome replicates semiconservatively.
2. Structural continuity of the prokaryotic chromosome is probably provided only by DNA, i.e., there is no evidence for linkers of RNA, protein, or other molecules within the chromosome.
3. The prokaryotic chromosome is composed of one DNA double helix in its lateral dimension, i.e., it is not polytenic, polynemic, or multistranded.
4. The DNA replicates as a single unit, i.e., as one replicon, normally beginning at a defined and fixed point and proceeding as a replication wave or fork to the end.
5. The initiation of DNA synthesis requires the synthesis of at least one new protein, and possibly more, which acts in an unknown manner to allow replication to be initiated. Protein synthesis is not needed for continuation and completion of replication of the replicon.

THE COMPONENTS OF EUKARYOTIC CHROMOSOMES

The word "chromosome" means colored body, a circumstance that is clearly remindful of the early emphasis placed on the eukaryotic chromosome as a cytological structure distinguishable during mitosis or meiosis. A long series of cytochemical studies, in particular on mitotic and meiotic chromosomes, dipteran polytene chromosomes, and amphibian lampbrush chromosomes, established the general, but vague, definition of the eukaryotic chromosome as a complex of DNA, RNA, and protein (see Kaufmann et al., 1960). At the present time a large amount of data has accumulated on the contents, the synthesis, the turnover, and the behavior of these various macromolecules found with chromosomes, but the manner in which these components are fitted together to form a chromosome is a most difficult problem that has only recently been partially solved. The major difficulty in accounting for the relatively enormous amounts of DNA, RNA, and proteins has been further complicated by the heterogeneity of these macromolecules, especially the proteins. Undoubtedly the complexity of the problem would be considerably reduced if it could be decided which of the macromolecules that are present with the chromosome are actually a part of its basic and permanent structure, and which of them are semipermanent and more or less loosely and transiently appended in connection with such functions as transcription, replication, and packaging into the meiotic, mitotic, and heterochromatic configurations.

The only reasonable approach, one which has been used consistently by constructors of chromosome models, is to build a concept of the chromosome that is reduced to the simplest possible molecular form consistent with available knowledge of structure, function, and replication. In model building this approach is easy; one need begin only with DNA and add the least number of other components (usually protein) to account for what is known about the arrangement and the behavior of DNA in various situations and under various experimental conditions. The actual analysis of real chromosomes is not so easy. Before the problem of chromosome structure can be stated in manageable terms, it will be necessary to remove from consideration all those molecules that are concerned with maintenance of particular functional states. The cytologists' view that the chromosome is composed of all those components that happen to be present in the polytene, mitotic, or meiotic chromosome, is no longer very useful. A clear distinction must be made between what is really part of the chromosome and what is serving only to maintain the chromosome in a particular functional state or configuration, or what is simply present adventitiously.

The primary general problems are then: to determine how the large amount of DNA is arranged in a eukaryotic chromosome, why there is so much more DNA than seems reasonable by traditional genetic considerations, how DNA replication and transcription are accommodated within the structural arrangement of the chromosome, and what roles the various proteins and RNAs might play in chromosome structure, replication, or function.

Experimental Materials for Chromosome Analysis

Cytological studies of eukaryotic chromosomes have concentrated on cells in three different situations in which chromosomes can be easily identified as complete and discrete entities: (1) cells in metaphase or anaphase of meiosis or mitosis, (2) dipteran cells with polytene chromosomes, and (3) amphibian oocytes with the chromosomes in early meiotic prophase (lampbrush chromosomes). These three situations present chromosomes in three different, specialized metabolic, configurational, and compositional states. Metaphase and anaphase chromosomes are tightly condensed, are not engaged in either DNA or RNA synthesis, and show no turnover of their associated proteins. Dipteran polytene chromosomes are in interphase, are active in both RNA and DNA synthesis (intermittently at least), and show turnover of associated proteins. Lampbrush chromosomes are partially condensed, are active in RNA synthesis, show protein labeling, but have no DNA synthesis. The ability to identify chromosomes cytologically in these materials has been a great advantage over the majority of chromosomes found in the interphase period.

Cytological techniques have demonstrated that substantial amounts of protein and RNA are associated with mitotic chromosomes. The roles of the proteins, particularly the histones, and RNA in the formation and maintenance of chromosome structure and function have been matters of considerable dispute.

Preparations of interphase nuclei and chromatin from such nuclei are frequently used to good advantage to attack problems of chromosome structure, replication, and function. Even though chromosomes cannot be identified as discrete structures, it seems reasonable to assume that any material somehow attached to the DNA is associated with the chromosome, by the same logic applied in the case of the non-DNA components of mitotic or meiotic chromosomes.

The recently developed techniques for isolating metaphase chromosomes (Chorazy et al., 1963; Somers et al., 1963; see references in Table 1) in relatively pure form have permitted a start on precise quantitative and qualitative studies of the composition of such structures, and have contributed to the clarification of a few issues about chromosome construction. Once preparations of isolated metaphase chromosomes are obtainable in highly purified form, it should be possible to apply a variety of analytical techniques to specific questions of structure and composition.

Quantitative analyses of DNA, RNA, and protein of mammalian metaphase chromosomes in several laboratories have given strikingly similar results even though the isolation media employed have varied considerably. These data are summarized in Table 1.

TABLE 1. Relative Contents of DNA, RNA, and Protein of Isolated Metaphase Chromosomes

	% DNA	% RNA	% PROTEIN
HeLa[a]	20	14	66
HeLa[b]	16	12	72
HeLa[c]	25	29	46
HeLa[d]	16	14	70
Ascites tumor[e]	14	14	72
L-cells[b]	17	13	70
Chinese hamster[b]	16	15	69
Syrian hamster[b]	17	15	68
Chinese hamster[f, g]			
Large size	28	5	67
Medium size	14	11	75
Small size	10	13	77

[a] *Salzman et al., 1966.*
[b] *Maio and Schildkraut, 1967.*
[c] *Franceschini and Giacomoni, 1967.*
[d] *Huberman and Attardi, 1966.*
[e] *Cantor and Hearst, 1966.*
[f] *Salzman and Mendelsohn, 1968.*
[g] *Mendelsohn et al., 1968.*

Chromosomal Proteins

The major component found in isolated mitotic chromosomes is clearly protein (Table 1). According to Huberman and Attardi (1966) about 40 percent of the protein in HeLa chromosomes is soluble in 0.2 N HCl (presumably this includes a major part of the histones) and 60 percent is acid insoluble. Maio and Schildkraut (1967) report that at least 50 percent of the chromosomal proteins of HeLa are acid soluble (0.2 N HCl). Salzman et al. (1966) found that only 26 percent of the proteins of HeLa cell chromosomes was soluble in 0.2 N H_2SO_4, although this treatment may not be sufficient to extract all histone molecules (Murray, 1965). Obviously much of chromosomal proteins must be other than histones.

Measurements of protein in chromatin from interphase nuclei and cytophotometric studies of heterochromatin in intact interphase nuclei have shown the presence of a substantial amount of nonhistone protein. Chromatin is reasonably assumed to represent interphase chromosomes. In pea seedling chromatin, 57 percent of the total material proved to be protein (Bonner and Huang, 1963) and in beef liver chromatin roughly 70 percent is protein (Izawa et al., 1963). In lampbrush chromosomes of newts, over 98 percent of the total chromosomal mass has been shown to be protein (Izawa et al., 1963). This very high proportion of protein has been ascribed to the unusually high amount of synthetic activity of this chromosome type. A variety of other reports have described the association of major amounts of acidic and residual proteins with DNA isolated from interphase cells.

The function of almost all of the large amount of proteins of mitotic and interphase chromosomes is not yet known, but, from a functional standpoint and based on current information, there are three possible roles for the proteins found with metaphase chromosomes: (1) maintenance of some degree of condensation, e.g., heterochromatin and mitotic chromosomes, (2) performance and control of such chromosomal functions as DNA and RNA syntheses, and (3) maintenance of chromosome integrity.

Condensation of Chromosomes. It seems virtually certain that the condensation of DNA that is observed in heterochromatin masses or interphase nuclei, in mitotic and meiotic chromosomes, and in the construction of sperm, must be due to complexing of the DNA with protein. Which proteins are involved in the three cases of DNA condensation is still a matter of conjecture. Histones are often postulated to be directly concerned with chromosome condensation both in connection with, and independent of, the thesis that histones are also involved with selective control of genetic transcription. Anderson (1956) has summarized most of the earlier evidence in support of a condensing function of histones. According to cytochemical studies (Swift, 1964, 1965), DNA is always found complexed with histones in eukaryotic cells, and most of the DNA in such cells has been reported to be condensed into genetically inactive heterochromatin (Bonner, 1965; Bonner and Huang, 1964; Frenster et al., 1963; Littau et al., 1964). However, histones are associated with dispersed or euchromatic chromosomal material as well as heterochromatin, according to cytochemical measurements on dipteran polytene chromosomes (Gorovsky and Woodard, 1967; Horn and Ward, 1957; Swift, 1962, 1964, 1965).

Swift (1964, 1965) has shown that the histone content apparently does not change when a band of a dipteran polytene chromosome undergoes puffing (a transition from a condensed to a dispersed state). In addition, cytophotometric measurements by Gorovsky and Woodard (1967) have shown that the ratio between histone and DNA does not change along the length of the chromosome, including interband regions. The presence of histones in the same amount in both dispersed and condensed chromatin may suggest that the amount of histones could not be the decisive factor in chromosome condensation, including mitotic condensation. The study of MacInnes and Uretz (1967) on the thermally induced loss of polarization of fluorescence in polytene chromosomes stained with acridine orange, is compatible with the idea that the activation of RNA synthesis may involve a loosening of the association of histone with DNA rather than an outright removal. Berlowitz (1965) has reported that the ratio of histone to DNA is higher in heterochromatin than in euchromatin in mealy bugs. Finally, Bonner and Huang (1963) suggest, on the basis of DNA melting profiles, that DNA active in support of RNA synthesis may not be complexed with any histones, or at least is complexed in a way that does not produce the elevation in melting temperature that is usual for DNA-histone complexes.

Allfrey et al. (1963) and Littau et al. (1965) have found that removal of lysine-rich histone from condensed chromatin of thymus nuclei causes dissociation into diffuse chromatin. Restoration of the histone produces chromatin condensation. Removal and addition of arginine-rich histones do not cause such changes. Since, according to these authors, both types of histones are combined with DNA through the phosphoric acid groups of DNA, they must combine in a different way. They conclude that the lysine-rich histones cause chromatin condensation by cross-linking of DNA while arginine-rich histones are combined along the DNA molecule. The conclusion and the experiments mentioned above are seemingly contradicted by the observation that the ratio between arginine- and lysine-rich histones is the same in both euchromatin and heterochromatin (Littau et al., 1965). Frenster (1965) has likewise found that, in calf thymus, the total amount of histone per amount of DNA is the same in both condensed and diffuse chromatin, with, in addition, no difference in the proportion of lysine-rich histones. There are, however, striking compositional differences between condensed and diffuse chromatin. According to Frenster (1965), diffuse chromatin contains twice as much nonhistone protein and five times as much RNA. This is similar to the observation of Swift (1964, 1965) that regions of polytene chromosomes active in RNA synthesis have the same amount of histone per amount of DNA, but an increased amount of nonhistone protein. Possibly some part of the nonhistone protein, acidic proteins, for example, may regulate interconversions between condensed and diffuse chromatin, as suggested by Anderson (1956) for condensation of chromosomes to the mitotic state.

Zubay (1964) has described a manner in which histones may bring about a very high degree of supercoiling of a very long DNA structure by cross-linking adjacent loops of coiled DNA with the histone attached in the major groove of the DNA double helix. Because of the wide looping of this model, the hypothetical DNA-histone fiber is an order of magnitude greater in diameter than has been observed with considerable consistency for DNA-histone fibers in many electron microscope studies. More recently, Pardon et al. (1967) have concluded from x-ray diffraction analyses that, in the DNA-histone complex, the DNA is arranged in a superhelix (coiled coil), with the axis of the DNA double helix inclined at an angle with reference to the axis of the DNA-histone fiber in such a way that a fiber (superhelix) of 120 Å diameter would be produced. The evidence leaves little doubt that at least some DNA of eukaryotic cells is arranged in a superhelix by complexing with histones, and the model presented by Pardon et al. (1967) has a diameter close to that sometimes observed for DNA histone fibers in the electron microscope. Pardon et al. (1967) cite particularly the electron microscope observations by Davies (1967, 1968) of hollow rods with an outside diameter of 150 Å in chromatin of chick erythrocytes.

MacInnes and Uretz (1966), however, have concluded, from measurements (with a method of polarized fluorescence microscopy) on interband regions of polytene chromosomes, that the axis of the DNA molecule runs exactly parallel to the axis of the chromosome, indicating the absence of a superhelical arrangement. An important question is whether histone is still present in the interband regions since the chromosomes were fixed with acetic acid. There have been conflicting reports (Busch et al., 1964; Swift, 1964) concerning the extraction of histone from nuclei by acetic acid. According to the more recent and detailed study on calf thymus tissue by Dick and Johns (1968), 45 percent acetic acid extracts three of five major histone fractions, including the lysine-rich histone; and ethanol-acetic acid extracts arginine-rich histones, but much less of the total. This agrees with the conclusion of Allfrey et al. (1968) concerning removal of arginine-rich histones by acetic acid. If histones are necessary for supercoiling of DNA, then the removal by acetic acid of some histone from interband regions of polytene chromosomes could account for the failure to detect the supercoiling observed by Pardon et al. MacInnes and Uretz (1968), however, find no evidence of supercoiling of DNA in the sperm of several insects using methods that would not be likely to dislodge histones.

Histones are found with mitotic and meiotic chromosomes, although these are by no means the only proteins present in these stages. The argument that histones are responsible for mitotic condensation, suffers the same weakness as it does in relation to condensed chromatin of interphase. Since most or perhaps all DNA in eukaryotes is apparently never without associated histones (Swift, 1964, 1965), it becomes difficult to attribute to histones the primary role in changing the degree of packing (condensation) of the DNA. Again it is necessary to consider the possible role of other proteins in condensation, for example, that of acting as cross-links between histone molecules already attached to DNA.

The idea that basic proteins function to condense DNA has also drawn support from the nuclear protein changes occurring during spermiogenesis. In a variety of organisms the lysine-rich histones are replaced first by arginine-rich histones (Alfert, 1956; Bloch and Hew, 1960) and then by protamine which contains up to 80 percent arginine (Felix et al., 1956). We can imagine that the replacement of histones with polypeptides extremely rich in arginine is both necessary for and the cause of the very dense packing of DNA in sperm (and at the same time the genetic inactivation of the DNA). During spermiogenesis in the mouse (Monesi, 1964) and *Drosophila* (Das et al., 1964), however, the replacement of histone apparently does not proceed to protamine, but only to the stage of arginine-rich histone. In several higher plants (Rasch and Woodard, 1959), in the frog (Vendrely, 1957; Bloch, 1962), and apparently in bovine spermiogenesis (Berry and Mayer, 1960), the usual somatic histones apparently do not undergo replacement to a more arginine-rich type during spermiogenesis but remain as more or less typical somatic histones. These latter observations show that the idea that DNA packing is a consequence of histone transition to much more basic polypeptides, must be an oversimplification. Although most of the protein in sperm is histone or protamine, a significant amount is nonbasic.

The role of the nonbasic proteins in spermiogenesis is not known but they may provide an essential function in the extreme condensation of DNA observed in sperm. Anderson (1956) has suggested that the control of chromosome condensation is mediated by the balancing of acidic and basic proteins associated with the DNA. Loss of acidic proteins would lead to condensation, and he suggests that the loss of protein from prophase nuclei (for which there is much evidence, see Prescott and Goldstein, 1968) may represent a shift in the balance toward condensation. Although chromosome condensation begins long before the mass exit of proteins from the nucleus in late prophase, the experimental evidence compels us to consider the possibility that proteins other than basic ones control the conversion of the chromosome from the diffuse state to various condensed states (mitotic and meiotic chromosomes, heterochromatin, sperm packaging). Histones may be necessary for condensation of DNA, but the evidence suggests that they are not sufficient. Both acetylation (Allfrey et al., 1968) and phosphorylation (Langan, 1968) of histones may be involved in activation of RNA synthesis, and therefore such modifications could be important in any role that histones might have in decondensation of chromatin.

Obviously, the processes of chromosome condensation and genetic repression are closely related. In the three situations in which condensation occurs (heterochromatin, mitotic and meiotic chromosomes, and sperm formation) genetic inactivation is a strikingly concomitant event. Clearly, heterochromatin formation is specifically controlled in a genetic sense just as repression is gene specific, although we do not have the explanation for selective heterochromatization, such as that of one of the two X chromosomes in mammalian female cells or of the entire paternal set of chromosomes in certain insects that occurs without any visible alteration of the female set of chromosomes (Nur, 1967).

There are many reports describing the binding of nonhistone proteins to DNA but there is no information and little or no speculation on the contribution of such proteins to the induction or maintenance of tertiary configurations of DNA. It is usually assumed that nonhistone proteins are involved in transcription and replication and the control of these two synthetic processes.

CHROMOSOMAL PROTEINS IN THE CONTROL OF DNA AND RNA SYNTHESIS. There is no doubt that transcription and replication of DNA require an intimate association of proteins with the DNA. Direct evidence about the details of binding of such proteins to DNA is relatively scarce. Wang (1967) has demonstrated the presence of DNA polymerase in the acid proteins bound to DNA in calf thymus. Holoubek and Crocker (1968) and Holoubek et al. (1966) have described DNA associated acidic proteins in Ehrlich ascites cells and in rat tissues. These authors suggest that such proteins are involved in the control of RNA synthesis because of a high rate of turnover of these proteins. Benjamin and Gellhorn (1968) have shown that the acidic proteins bound to DNA in mouse and rat liver are a heterogeneous mixture, and they too, suggest that such proteins are involved in control (repression) of the genome. Others have suggested that nonhistone protein associated with DNA serves as linker material, holding DNA molecules in series within the chromosome. Such a proposition is of doubtful validity in view of the extensive evidence against the existence of interruptions in the DNA molecule within a single chromosome (see p. 68).

Some of the nonhistone protein bound to DNA in eukaryotic cells must be

RNA polymerase and, although repressor proteins have not yet been identified in eukaryotic cells, probably a substantial part of the total nonhistone protein is concerned with the regulation of DNA transcription into RNA. In polytene chromosomes of *Drosophila,* the repressed (no RNA synthesis) regions of the chromosome contain the usual amount of histones but little nonhistone proteins in contrast to the high content of nonhistone protein in regions of the chromosome active in RNA synthesis (Swift, 1964, 1965). This is at least circumstantial evidence for a function of such proteins in RNA transcription and possibly transport of newly synthesized RNA away from the chromosome.

Because removal of histone from chromatin preparations leads to an increase in RNA synthesis and addition of histones leads to a suppression of synthesis, it has sometimes been assumed that DNA-histone complexes must necessarily be inactive in DNA and RNA synthesis. Although in cytochemical studies, no measurable amount of DNA in eukaryotic nuclei has been found to be free of histones (see Swift, 1964, 1965), up to 20 percent of DNA may be active in transcription. From the melting profiles of DNA, Bonner and Huang (1963) have concluded that 20 percent of the DNA in chromatin isolated from pea seedling may not be complexed with histone. But histones have been shown by cytochemical methods to be associated with genetically active euchromatin without blocking RNA synthesis. The melting profile experiments do not prove that histone is absent from some DNA, since histone could conceivably be complexed with this DNA in a way that does not affect the melting temperature. On the other hand, cytochemical methods would not detect highly restricted regions of histone-free DNA. Whether histone is complexed with DNA in the immediate region of transcription or replication are questions for which there are still no clear answers. It is probably necessary to assume that only certain kinds of histones are responsible for genetic repression, i.e., some fraction of those histones associated with heterochromatin, condensed chromosomes of mitosis, and sperm, but not present in euchromatin. For example, Allfrey et al. (1963) reported that arginine-rich histones are strong repressors of RNA synthesis and lysine-rich histones only weakly affect RNA synthesis. According to the more recent work from Allfrey's laboratory (Allfrey et al., 1968), the control of RNA synthesis may not involve the addition or removal of histones from chromatin but instead the reversible acetylation of histones. Increases in RNA synthesis are accompanied by such acetylation (see Allfrey et al., 1968). Langan (1968) has suggested that phosphorylation of histones may have a role in the control of RNA synthesis.

Finally, the number of different histone types so far identified is far short of the number needed to support the thesis that genetic repression can be mediated in a genetically specific manner by histones alone, but the number of histone types so far identified is adequate to cover a variety of different roles in developing different degrees of chromosome condensation or other configurations.

PROTEINS IN THE MAINTENANCE OF CHROMOSOME INTEGRITY. Discarding the older and clearly untenable proposition that proteins form the axis of the chromosome (see Callan, 1967; Swift, 1965), there are still two primary ways in which proteins can be postulated to have a role in the maintenance of chromosome structure, i.e., in lateral and serial linking of DNA molecules. In polyneme models of the eukaryotic chromosome, proteins have been envisaged to provide lateral cross-linking of DNA molecules that lie in parallel arrangement. If polynemy exists, the DNA helices must be held in lateral register by some mechanism, but perhaps this could be achieved by interaction of DNA molecules only. Certainly the DNA molecules must be held in parallel register in the polytenic chromosomes of dipterans and certain ciliated protozoa. Proteins could be involved in maintenance of the polytenic structure of these chromosomes, but speculation about such a role of structural proteins in polynemy of the chromosome seems quite pointless unless and until polynemy has been proven to exist. In fact, we have no information on the molecular architecture of the chromosome that permits any distinction between polynemy and polyteny. Presumably polynemy is meant to imply a different (more intimate?) type of lateral association of double helices than does polyteny. Justification for retention of both terms is a conceptual matter; polynemy is defined as a permanent and constant lateral redundancy of chromosomes, and polyteny is considered as an amplification of DNA restricted to certain special situations.

Swift (1962) has discussed the question of structural proteins in dipteran polytene chromosomes based primarily on the extensive observations made in his laboratory. The concentration of protein in the interband regions of polytene chromosomes is extremely low, and Swift suggests that it may represent only histones. Also, in bands inactive in RNA synthesis, the only protein present may be the histones. In regions active in RNA synthesis (puffs) the total protein content is increased by the addition of nonhistone proteins. Swift concludes that nonhistone proteins are present only in RNA puffs and that the presence of any structural nonhistone protein in the chromosome is therefore unlikely.

Concerning the question of protein linkers holding DNA molecules in a serial arrangement by tandem connections, there is definitive evidence against the hypothesis. Digestion of amphibian lampbrush chromosomes with either ribonuclease or proteinases does not interrupt the linear continuity of the chromosome, although deoxyribonuclease does (Macgregor and Callan, 1962). Lezzi (1965) has performed a similar study on isolated dipteran polytene chromosomes with similar results. DNase caused breakage along the chromosome but trypsin, chymotrypsin, or ribonuclease did not. A fuller consideration of the linker question is given in a later section (see pp. 75-80).

SYNTHESIS AND TURNOVER OF CHROMOSOMAL PROTEINS. Three isotope tracer studies on proteins present with mitotic chromosomes have shown that whatever the protein complement, it turns over rather rapidly in the course of the cell life cycle. Following labeling of Chinese hamster cells in culture with tritiated amino acids for long periods of interphase, the metaphase chromosomes contain radioactive proteins as detected by autoradiography (Prescott and Bender, 1963b). After withdrawal of the tritiated amino acids from the medium, radioactive proteins disappear from the metaphase chromosomes within two cell life cycles. The turnover of chromosomal proteins has been reported to be even more rapid in root meristems cells of *Vicia faba* following labeling with tritiated arginine (Prensky and Smith, 1964). In Steffenson's (1963) studies on *Lilium*, the $^{35}SO_4$ label present in pollen, presumably primarily in protein, was retained by endosperm nuclei for at least two cell cycles after fertilization. The retained proteins would most likely be nonhistone, since histone contains so little sulfur. The duration of these experiments was, however, too short to test for stable nuclear proteins.

Several studies (Bloch et al., 1967; Cave, 1966, 1967; Chernick and Davidson, 1968; Prensky and Smith, 1964; Prescott and Bender, 1963a; Shapiro and Levina, 1967) indicated that the proteins present with mitotic chromosomes are synthesized continuously throughout interphase and that the pattern of labeled proteins within the chromosomes is uniform and without any correspondence to the labeling pattern for DNA. Prescott and Stone (1965) have noted an apparent asynchrony in the labeling of proteins of the X chromosome in Chinese hamster cells that is similar to the asynchrony of ^{3}H thymidine labeling. Chernick (1968) has recently obtained a labeling pattern with ^{3}H arginine, supplied to female lymphocytes during the latter part of the S phase, that is quite similar to the pattern of late DNA synthesis, particularly with regard to the X chromosome. It certainly seems reasonable to expect that new histone will be added to a chromosome coordinately with DNA replication in a given chromosome in view of the parallel increases of total histone and DNA during the S phase. A particularly pertinent situation occurs in the macronucleus of the protozoan *Euplotes* in which the region of DNA synthesis at any instant during the S phase can be precisely identified; a major addition of newly synthesized histone occurs exactly at the point at which DNA replication is in progress (Gall, 1959; Prescott, 1966; Prescott and Kimball, 1961).

The segregation of labeled protein at mitosis is dispersive (Cave, 1966; Prescott and Bender, 1963a), in contrast to the semiconservative distribution of labeled DNA. Whether or not chromosomal proteins are synthesized directly on the chromosome or elsewhere has not been settled. Often it is assumed that short-term labeling of proteins will permit detection of the site of synthesis, and labeling of polytene chromosomes with tritiated amino acids has been interpreted as chromosomal synthesis of proteins. Because of the rate at which proteins could migrate from place to place within a cell, such observations cannot be accepted as evidence for chromosomal synthesis, and at this point it seems just as likely that all chromosomal proteins are synthesized in the cytoplasm (see Chapter 4 for discussion of nuclear protein synthesis). Kroeger et al. (1963) for example, have shown that implantation of protein-labeled chromosomes from salivary gland cells of *Chironomus* into unlabeled cells results in labeling of the host cell chromosomes. The interpretation of the finding that chromosomal proteins migrate between the nucleus and cytoplasm is reasonable. In *Amoeba proteus* the continuous and rapid migration of proteins between the nucleus and cytoplasm is well-documented (Goldstein and Prescott, 1967, 1968; Prescott and Goldstein, 1968), and it has been shown in experiments on several cell types that proteins synthesized in the cytoplasm are subsequently accumulated by the nucleus (Bloch and Brack, 1964; Robbins and Borun, 1967; Speer and Zimmerman, 1968; Zetterberg, 1966).

It is now clear that not only the nuclear accumulation of histones (Alfert, 1956; Bloch and Godman, 1955; Gall, 1959; McLeish, 1959; Woodard et al., 1961) but also the actual synthesis of histones in proliferating cells occurs tightly coupled to DNA replication (Bloch et al., 1967; Das and Alfert, 1968; Niehaus and Barnum, 1965; Prescott, 1966; Prescott and Kimball, 1961; Robbins and Borun, 1967;

Spalding et al., 1966; Yarbro, 1967). A conclusion from isotope incorporation data on regenerating liver that histone synthesis begins before the DNA synthesis period (Holbrook et al., 1962) has recently been contradicted. According to Takai et al. (1968), histone and DNA synthesis are tightly coupled in regenerating liver.

A number of isotope incorporation studies indicate histone turnover (e.g., Chalkley and Maurer, 1965; Daly et al., 1953; Goldstein and Prescott, 1968; Lee and Scherbaum, 1966; Prescott and Bender, 1963a) during non-DNA synthesis periods, but this does not conflict with the conclusion that the *net* synthesis of histones occurs only during DNA synthesis in proliferating cells. Yarbro's studies (1967) on inhibition of DNA synthesis with hydroxyurea in ascites tumor cells suggest that the continuation of histone synthesis is closely dependent upon continuation of DNA synthesis. This agrees with the experiments of Robbins and Borun (1967) on HeLa cells that show a rapid decline of histone synthesis (and disappearance of small polysomes apparently concerned with histone synthesis) immediately following inhibition of DNA synthesis. These observations appear to contradict the earlier suggestion in data of Flamm and Birnstiel (1964a, 1964b) that in tobacco plant cells histone synthesis and accumulation can continue even though DNA synthesis is blocked. Histone and DNA syntheses are probably tightly coupled in proliferating cells, but as Bloch et al. (1967) point out, during spermiogenesis the synthesis of arginine-rich histones proceeds without concomitant DNA synthesis.

Experiments of Gershon et al. (1965) suggest that the amount of host-cell DNA doubles in mammalian cells in culture without concomitant histone synthesis in the period shortly following the initiation of transformation with polyoma virus. Bloch et al. (1967) cite two other cases of eukaryotic cells in which DNA synthesis proceeds without concomitant histone synthesis, but no details are given.

There are also pronounced changes in the histone to DNA ratio connected with the change of cells from a nonproliferating to a proliferating condition. For example, in *Tetrahymena* (Stone, unpublished) the histone to DNA ratio during stationary phase of culture growth is close to 2:1. During the lag period following inoculation of cells into fresh medium, the histone to DNA ratio falls to a value close to 1:1, although it is not clear whether the fall is due to destruction of histone or to synthesis of DNA without concomitant histone synthesis. Similarly, in regenerating liver, the histone to DNA ratio falls from 3:1 to 1.35:1 prior to the initiation of DNA synthesis (Umana et al., 1964). These results must also mean that at some point, presumably in early stationary phase of culture growth, the rate of histone accumulation exceeds the rate of DNA synthesis. This agrees with the report of Flamm and Birnstiel (1964a, 1964b) of continued histone accumulation in plant cells after inhibition of DNA synthesis.

The turnover of histones of the chromosomes is possibly indicated in the studies of protein labeling and subsequent autoradiographic observation of the chromosomes in metaphase. According to two studies (Prensky and Smith, 1964; Prescott and Bender, 1963a), the labeled proteins disappear from the chromosome within a few cell cycles, at a rate that is too rapid to be accounted for by the reduction in the concentration of radioactive proteins through growth. In both cases the metaphase chromosomes were prepared with acetic acid-alcohol, and it is possible, but unlikely, that all histones were removed before the autoradiography (see Dick and Johns, 1968).

The turnover of arginine-rich histones has been demonstrated by electrophoretic fractionation in cells not engaged in DNA synthesis (Chalkley and Maurer, 1965). In this study the lysine-rich histone became labeled only during periods of DNA synthesis. Spalding et al. (1966), in confirming that histones show a net increase in HeLa cells only during DNA synthesis, showed that at least some of the histones undergo turnover during non-DNA synthesis periods. Byvoet (1966), however, has measured the rate of loss of radioactive label from DNA and histone in various rat tissues, and concluded that the two macromolecules "turnover at approximately the same rates." In many of the tissues studied by Byvoet, turnover rates were rapid (half-lives of DNA and histones of one to four days) because of rapid cell turnover; such circumstances are unfavorable for detection of a low rate of histone turnover that might occur in nonproliferating cells. In long term studies of some tissues (Byvoet, 1966), the rate of turnover of histone was an average of about 35 percent faster than that of DNA. Byvoet's isotope measurements on neoplasms in the rat showed more rapid turnover for DNA than for histones, a situation that is inconsistent with the results of other studies and suggests that comparative measurements of histone and DNA turnovers in these studies may contain unidentified variables.

The residual and acidic proteins associated with chromosomes also apparently undergo constant turnover, with the acidic proteins showing the higher rate (e.g., Holoubek and Crocker, 1968; Holoubek et al., 1966). The complete turnover of all nuclear proteins during a long period of cell growth without nuclear division has been demonstrated in *Amoeba proteus* (Goldstein and Prescott, 1968; Prescott and Bender, 1963a). Amebas can be kept in a continuous state of growth without cell division by amputation of cytoplasm at regular intervals; appropriately spaced amputations prevent cell division but do not inhibit growth. Starting with amebas heavily labeled with tritiated amino acids, it has been shown that more than 99.99 percent of the original amino acid label is lost during a period of cytoplasmic growth equivalent to 30 cell life cycles. About 5 percent of the total nuclear proteins in amebas are histones, and the protein turnover in these experiments must include turnover of the histones.

CONCLUSIONS CONCERNING CHROMOSOMAL PROTEINS. Proteins associated with the chromosome form a heterogeneous class of macromolecules. Aside from a few enzymes concerned with transcription or replication, the function of none of the proteins is known. No strictly structural proteins have been identified and the linear integrity of DNA throughout the length of the chromosome very clearly does not involve protein. A role of histones in configurational changes in DNA seems likely, but is unproven. The amount and type of protein associated with the chromosome change drastically as chromosomal activity changes, and *all* proteins associated with the chromosome undergo turnover, but at different rates. The turnover of chromosomal proteins and the high variability of the qualitative and quantitative features of such proteins sharply point up the futility of attempting to define "the" structure of a chromosome in which the proteins are considered a part of that structure. It is more useful to view the proteins as a variable component associated with the chromosome in accordance with its functional state and not as part of the chromosome itself.

RNA of Chromosomes

From cytochemical studies, it has long been known that RNA is associated with the chromosome. RNA is evident in the puffs of dipteran polytene chromosomes (e.g., Rudkin and Woods, 1959; Stevens and Swift, 1966) and is a major component found with the loops of lampbrush chromosomes (e.g., Gall and Callan, 1962). In both of these cases, the general conclusion is that this RNA is the product of transcription. RNA is also associated with mitotic and meiotic chromosomes (Boss, 1955; Jacobson and Webb, 1952; Kaufmann et al., 1948; LaCour, 1963; Love and Suskind, 1961) when all RNA synthesis has ceased; the type of RNA and/or its source have been long-standing questions. Many cytological studies have shown that nucleolar material tends to spread out along the chromosomes at the time of nucleolar dissolution (for example, more recently Brinkley, 1965; Heneen and Nichols, 1966; Hsu et al., 1965). This presumably contributes RNA to the mitotic chromosomes, but whether other sources of RNA are involved cannot be determined convincingly with cytochemical methods.

Autoradiographic studies of mammalian cells labeled with tritiated uridine and subsequently examined at metaphase also have shown the presence of labeled RNA with chromosomes, but with some indication that the labeled RNA was associated only with the surface of the chromosome, suggesting adventitious binding (Comings, 1966; Feinendegen and Bond, 1963; Prescott and Bender, 1963a). RNA labeled with tritiated adenosine (LaCour, 1963) or ^{14}C adenine or tritiated uridine (Woodard et al., 1961) has also been shown to be associated with metaphase chromosomes of plant root cells. Feinendegen and Bond (1963), in autoradiographic experiments, concluded that about 25 percent of nuclear RNA is retained with the chromosomes of HeLa cells through mitosis, but that this RNA is associated with the surface of the chromosome rather than with the chromosome proper. Freedman et al. (1967) have concluded from autoradiographic studies that no labeled RNA is retained with the metaphase chromosomes in amphibian embryonic cells. In autoradiographic studies of lymphocytes and fibroblasts of man, Comings (1966) found radioactive RNA associated with metaphase chromosomes in cells labeled during interphase, but concluded that such RNA "is a result of random, nonspecific adherence of labeled RNA to metaphase chromosomes."

A large number of reports (see King and Barnhisel, 1967) have confirmed that there is no RNA synthesis associated with the chromosomes from later prophase to telophase in a wide variety of cell types. Any RNA on metaphase and anaphase chromosomes must have stemmed from an earlier period of synthesis. Finally, it has been established that most of the RNA of the nucleus is released to the cytoplasm at late prophase (see, for example, Prescott and Bender, 1963a).

The isolation of large amounts of metaphase chromosomes in a relatively pure state has permitted a significant advance in the analysis of RNA associated with the chromosome. According to the data in Table 1, the amount of RNA is consistently very close to the amount of DNA. In analyses by Maio and Schildkraut (1967) more than 80 percent of the RNA of mitotic chromosomes of HeLa proved to be ribosomal (28S and 16S). Huberman and Attardi (1966) found that almost all of the RNA of isolated HeLa chromosomes was ribosomal, with a small amount of 45S RNA (presumably preribosomal) and 4S RNAs. Salzman et al. (1966) obtained very similar results and showed moreover than most, if not all, of the RNA probably stems from the adventitious binding of ribosomes from the cytoplasm during the chromosome isolation procedure. This conclusion is borne out by studies on the fractionation of isolated chromosomes of Chinese hamster cells. Salzman and Mendelsohn (1968) have separated the chromosome complement into three size classes by centrifugation. The largest chromosomes are the most effectively separated from cytoplasmic contaminants and these contain the least amount of RNA, i.e., RNA makes up only about five percent of the chromosomes.

The transient nature of the association of RNA with chromosomes is underscored by the complete absence of any RNA in the nucleus of late stages of spermiogenesis and in the finished sperm head (Bloch and Brack, 1964; Felix, 1959). All of these studies of RNA in metaphase chromosomes yield the general conclusion that some nucleolar RNA (45S) and ribosomes (yielding 28S and 16S RNA) are attached to the surface of the condensed mitotic chromosome as contaminants. There is no evidence that any RNA molecules form a part of the primary structure of the chromosome.

THE ORGANIZATION AND REPLICATION OF CHROMOSOMAL DNA

Semiconservative Replication of DNA

The semiconservative replication of DNA in bacteria and viruses is supported by enough evidence to be accepted as a proven generality, and demonstration of semiconservative replication has been extended to include the prokaryotic chromosomes of chloroplasts (Chiang and Sueoka, 1967) and mitochondria (Reich and Luck, 1966). In the evolution of the eukaryotic chromosome from a prokaryotic type it seems improbable that a principle as basic, and apparently successful, as the semiconservative mode of replication would have changed, although arguments might be made that the need to accommodate a vastly increased amount of DNA arranged as many replicons might lead to modification of the replication mechanism.

Direct evidence for semiconservative replication of eukaryotic DNA has been presented for chromosomes of mammalian cells (Chun and Littlefield, 1961; Djordjevic and Szybalski, 1960; Painter et al., 1966; Simon, 1961; Taylor, 1968) and cells of a higher plant (Filner, 1965). There seems no reason to suspect that the semiconservative mechanism does not obtain generally for chromosomes of the eukaryotic type.

The autoradiographic studies of the Taylor type (see Taylor, 1963a), in which the fate of labeled DNA is followed by observation of chromatid labeling at several successive metaphases, has sometimes been erroneously cited as evidence of semiconservative replication of DNA. The Taylor experiment proves that *segregation* of chromosomal DNA is semiconservative (see Fig. 2) and does not contribute any

FIG. 2. The diagram shows the sorting out of DNA double helices to form chromatids in a unineme (left side) and a bineme model (right side) of the chromosome. The sorting out must be restricted to the routes shown in order to be compatible with both semiconservative replication and segregation of DNA.

1. *Sorting out of radioactivity.* The first replication for each chromosome model results in labeling of one chain of every double helix (IIa and IIb) in both models. Radioactivity is represented by a dashed line. Both chromatids are labeled at the first metaphase after labeling (IIIa and IIIb). No mechanism is known to exist that would permit only one chromatid of a chromosome to be labeled when labeling has taken place in the DNA synthesis period immediately preceding the metaphase in question. In the case of unineme chromosome there is no alternative to the type of helix sorting shown in producing the daughter chromatids in IIIa. In a bineme model of the chromosome the four double helices of the replicated chromosome (in IIb) could segregate in various combinations of two to produce the two chromatids in IIIb. In order to retain semiconservative segregation of labeled DNA, however, the type of sorting of helices shown between IIIb and IVb must be followed, i.e., the two old chains with a 5′ phosphate end at the top must continue to be retained within the same chromatid. The "5′ chains" are shown attached to the kinetochore (indicated by a line across the middle of the chain).

In the G1 period following division, the unreplicated chromosomes are represented by the helices shown in IVa and IVb. All of the helices contain by arbitrary choice the chains ending with the 5′ phosphate at the top; these are indicated by a heavy continuous line. In IVb, the two double helices could have contained the newly synthesized chains (radioactive) ending in 5′, but the double helices in IVb could not be mixed, i.e., one containing an old 5′ chain and the other containing a newly synthesized 5′ end, because such mixing would not permit semiconservative segregation of DNA. The double helices shown in IVa and IVb undergo replication to produce the products in Vb and Vc. In the unineme model in Vb the two double helices produce the two chromatids at the second metaphase as shown in VIb. The DNA has segregated semiconservatively. In the bineme model the four double helices again sort out with the two grandparental 5′ chains segregating into one chromatid and the newly synthesized 5′ chains into the other chromatid. The result is the semiconservative segregation shown in the metaphase chromosome in VIc.The unlabeled chromatids in each model give rise to daughter chromosomes of the helix composition shown in VIIb and VIIc.

Va and Vd indicate how the sister chromatid exchanges (or isolabeling) shown in the chromosome in VIa and VId could have occurred. In a unineme model, there is no reasonable alternative to the exchanges between the double helices as drawn in Va. As explained in the text, if an exchange occurs within the terminal one micron of a chromosome arm, the result is an isoautoradiograph, which has been interpreted by some as isolabeling. Similarly, if two exchanges occur within 2 μ of one another (in the metaphase chromosome), an isoautoradiograph will be produced that could be interpreted as isolabeling. In Va, the two cases of apparent isolabeling are ascribed to sister chromatid exchanges. In the bineme chromosome in Vd, the isoautoradiographs are, for the sake of argument, ascribed to isolabeling. Two of several possible ways in which double helix sections could be exchanged to produce isolabeling are indicated. Retaining the rule that the grandparental 5′ chains segregate together would produce the labeling pattern in the chromatids in VId.

2. *Sorting out of a mutation.* A mutation in the form of a single base change in the chain ending in 5′ phosphate at the top is indicated by a small circle for the double helix in the unineme model (Ia) and one of the helices in the bineme model (Ib). The subsequent replication and segregation of the mutation is indicated throughout the diagram. The pattern is straightforward for the unineme model and the base change will be present in both chains of the one helix.

In the bineme model semiconservative segregation of DNA as shown in Ib through VIIc does not allow the construction of chromatids in which the mutation is present in both helices; a recessive mutation resulting from a single nucleotide change could never be expressed in a bineme chromosome unless a mechanism were invoked to introduce the mutation into the second helix. This could be accomplished by the type of exchange between *the two central helices* in Vd (see the helices in the daughter chromosome in VIId). This type of exchange would also produce isolabeling of chromatids as already discussed. Not all exchanges that lead to isolabeling can lead to the presence of a mutation in both helices, e.g., the exchange between the first and second helices in Vd. The types of exchanges that would or would not produce the presence of a mutation in both helices of a chromatid can be worked out by studying Vc for all of the possible exchanges.

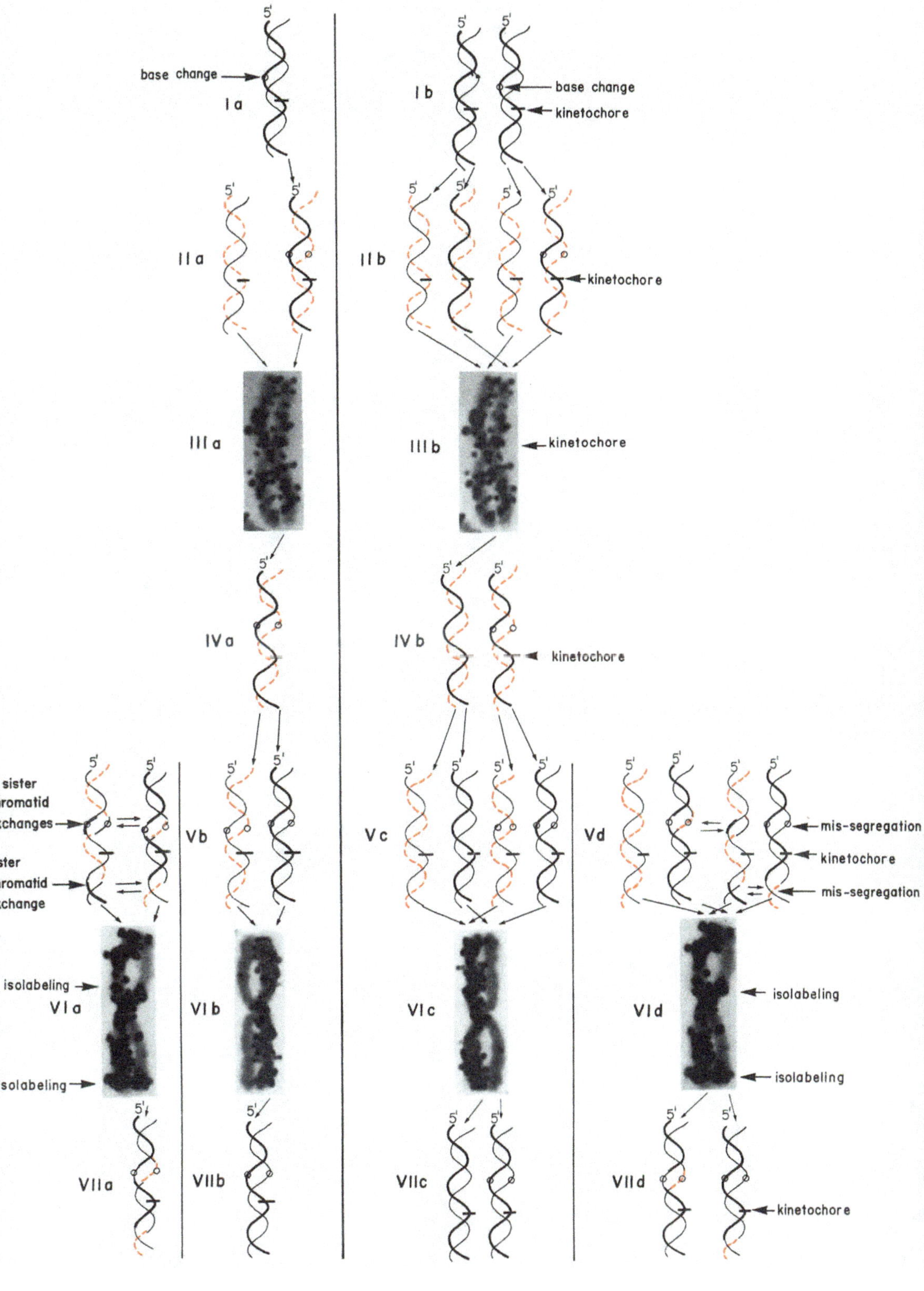

base change
Ia
Ib
base change
kinetochore
IIa
IIb
kinetochore
IIIa
IIIb
kinetochore
IVa
IVb
kinetochore
2 sister
hromatid
xchanges
ister
hromatid
xchange
Vb
Vc
Vd
mis-segregation
kinetochore
mis-segregation
isolabeling
VIa
isolabeling
VIb
VIc
VId
isolabeling
isolabeling
VIIa
VIIb
VIIc
VIId
kinetochore

direct evidence concerning the type of mechanism of DNA synthesis, primarily because the chromosome has not been proven beyond doubt to consist of only one DNA double helix in its lateral dimension (a unineme chromosome). If the chromosome proves to be unineme, then the Taylor experiment is direct proof of both semiconservative replication and semiconservative segregation of DNA (Fig. 2).

A major question still remaining concerns the accommodation of semiconservative replication within complex chromosomes containing relatively enormous amounts of DNA arranged in many replicating units (replicons) per chromosome.

Structural Continuity of the Eukaryotic Chromosome

The question is whether the accommodation of a very large amount of DNA in eukaryotic chromosomes involves some type of serial arrangement of DNA molecules by means of linker molecules or whether single DNA molecules extend from one end of the chromosome to the other. Evidence against the idea that proteins or RNA serve in such linking has been discussed in the section on proteins in the maintenance of chromsome integrity. A principal point is the failure of proteinases or ribonuclease to disrupt the linear integrity of amphibian lampbrush chromosomes (Macgregor and Callan, 1962) or dipteran polytene chromosomes (Lezzi, 1965). Linkers could exist that are affected by neither ribonuclease or proteinases, but there is virtually no evidence of their reality.

From time to time four different reasons for postulating linkers of DNA material to hold DNA molecules in tandem in eukaryotic chromosomes have been discussed: (1) The presumed instability *within the cell* of DNA molecules that exceed some critical length, (2) the presumption that replicating units of DNA within a single chromosome are not, or cannot, be directly connected to one another without interruption of the DNA double helix, (3) the need for a mechanism to relieve torsion on the DNA molecule developed in connection with semiconservative replication, and (4) as sites for genetic crossing over. These proposals are discussed in the sections immediately below.

1. The amount of DNA is sufficient to make a double helix several centimeters long in an average mammalian cell (Cairns, 1966) and up to one meter long for the very largest chromosomes of urodeles (Callan, 1963). The length of the DNA molecule would be, of course, shortened if polynemy exists, but each molecule would still be many centimeters long. In fact, in the chromosomes with the most DNA, i.e., urodeles, the evidence is strongest that the chromosome contains only one double helix in the lateral dimension. (Polynemy is discussed in a later section.) The idea of a DNA molecule centimeters, or even a meter long, apparently seems unreasonable to some researchers, but the rejection of the idea is evidently based on reasons of more interest to the psychologist than the biologist. Theoretical consideration for the maximum possible length of DNA that could survive physicochemical conditions in the cell cannot be taken very seriously because we know so little about the conditions immediately surrounding the DNA molecule. Obviously, a DNA molecule that is centimeters long must be folded to a great extent to fit into a nucleus 50 μ (in urodeles) or less in diameter, and such complex folding implies highly specialized con-

ditions around the chromosome; but this kind of circumstance applies equally to prokaryotic cells, for example, *E. coli,* in which a molecule 1,200 μ long must be highly folded to fit into a cell 2 μ long and 1 μ wide.

Direct measurements of the length of DNA molecules from eukaryotic cells is difficult, particularly because of the ease with which such enormously asymmetric molecules are sheared. Solari (1965) has found, with the electron microscope, pieces of DNA 93 μ long from sea urchin sperm. Cairns (1966) has made measurements of "DNA molecules" from HeLa cells by the indirect method of autoradiography and found lengths in excess of 500 μ, and Huberman and Riggs (1966) have reported autoradiographs of DNA from Chinese hamster cells up to 1.6 to 1.8 mm. The longest fibers found by the autoradiographic method are 2.2 cm (Sasaki and Norman, 1966), but the autoradiographs in this case indicated long bundles of fibers, and it cannot be said from this observation that single molecules extend for the full length of 2.2 cm.

2. The second reason for postulating linkers is also weak; it stems from the fact that the eukaryotic chromosome has multiple sites of initiation of DNA synthesis along its length, i.e., contains many replicons. It has been assumed that the presence of multiple replicons in series must mean implicitly that the individual units of replication along the chromosome must be separated from one another by non-DNA material (linkers) or that there must be many free ends of DNA. The assumption that the initiation of DNA synthesis in a replicon requires either free ends of the double helix or ends attached to non-DNA material is without evidence and seems quite unwarranted. Evidence on various kinds of prokaryotic chromosomes with the DNA arranged in a loop strongly indicates that initiation of DNA replication, although occurring at a restricted and specific site on the chromosome, does not require permanent free ends or attachment to non-DNA material.

It is not known whether nucleolar DNA within the lampbrush chromosomes of amphibians is composed of a closed loop. It seems unlikely but in any case the free-floating copies of nucleolar DNA produced by replication of this chromosomal DNA are in the form of loops. These loops are not broken by a variety of treatments, including digestion with proteinases and ribonuclease, but they are broken by deoxyribonuclease (Miller, 1966). The loops, moreover, are found in sizes that form a geometric progression, suggesting a variable number of component replicons. There is, in fact, no more compelling reason to postulate non-DNA linkers between replicating units of DNA (replicons) than between transcription units (operons).

3. The proposal that non-DNA linkers might serve to allow rotation of the DNA double helix during semiconservative replication is apparently no longer considered seriously, especially since it has been repeatedly demonstrated (e.g., Vinograd and Lebowitz, 1966) that a break in one chain of a DNA double helix allows the intact chain to act as a swivel and release torsion within the molecule.

Several years ago Bendich and Rosenkrantz (1963) reported that one serine molecule could be recovered from purified lymphocyte DNA for every 500 to 1,000 nucleotides and that some of the serine could be obtained as *O*-phosphoserine. They suggested that serine may be present in the DNA backbone, but no further published work on this matter is known to this writer. The report on phosphoserine led to speculation on the possibility of a swivel function for this molecule in DNA.

4. The need for non-DNA linkers to achieve genetic crossing over is not supported by any direct evidence; also, recombination in bacteria and viruses, as well as the insertion of viral or episomal DNA into the bacterial chromosome, occur without benefit of non-DNA linkers. It hardly seems reasonable to postulate linkers in connection with crossing over in eukaryotes.

Electron microscope observations of long segments of DNA prepared from sea urchin sperm by the Kleinschmidt technique do not give any evidence of changes in structure or normal discontinuities that might be interpreted as non-DNA linkers (Solari, 1965).

The suggestion of Hilgartner (1968) and Dounce and Hilgartner (1964), that the *in vitro* formation of gels by DNA-proteins isolated from rat liver nuclei reflects the presence of end-to-end linker proteins between DNA molecules, cannot be given much weight in view of other equally plausible explanations of the phenomenon; the formation of gels *in vitro* does not necessarily have anything whatsoever to do with the manner in which DNA is arranged in the intact chromosome. Their interpretation is, moreover, contrary to the evidence that digestion of proteins does not interrupt the linear integrity of the chromosome (Lezzi, 1965; Macgregor and Callan, 1962). Particularly pertinent are the studies of Kuehl (1962) on DNA solutions prepared by lysing of chicken erythrocytes in a way that minimizes the possibility of shearing of DNA molecules. Such DNA solutions are highly viscous and the DNA is sedimented by low centrifugal force. Treatments of such a solution with proteolytic enzymes or agents that disrupt disulfide bonds do not convert the DNA to a non-sedimentable form, although treatment with DNAase or subjection to shearing forces rapidly eliminate sedimentability.

The fact that undenatured DNA isolated from eukaryotic chromosomes tends to be in pieces that are one or more orders of magnitude too short in comparison with the length possible from consideration of the amount of DNA in the chromosome, is almost certainly due, at least in part, to shearing of DNA molecules during preparative procedures. As methods for isolation of DNA have become more and more gentle, the length of the molecules of undenatured DNA obtainable have become longer and longer, without evidence of the presence of any non-DNA linkers. Finally, in autoradiographic studies of DNA synthesis in Chinese hamster and HeLa cells, Huberman and Riggs (1968) have demonstrated the tandem or serial arrangement of many, relatively short replicons (see Fig. 3a, b, c, d). Prolonged treatment of DNA with pronase before autoradiography had no effect on the outcome of the experiments, thus clearly eliminating the possibility of protein linkers between replicating units of DNA.

The consistent but unjustified use of linker molecules in constructing models of the chromosome may also have had some influence in prolonging the life of the linker concept; models may offer explanations and produce insights but they do not provide evidence.

The concept of non-DNA linkers along the length of the chromosome is not supported by a single piece of convincing evidence, but there is clear evidence against their existence. The concept should be abandoned until, and unless, definite experimental evidence or other compelling reasons (conceptual or otherwise) demand its serious consideration. Linkers between replicons exist just as surely as linkers be-

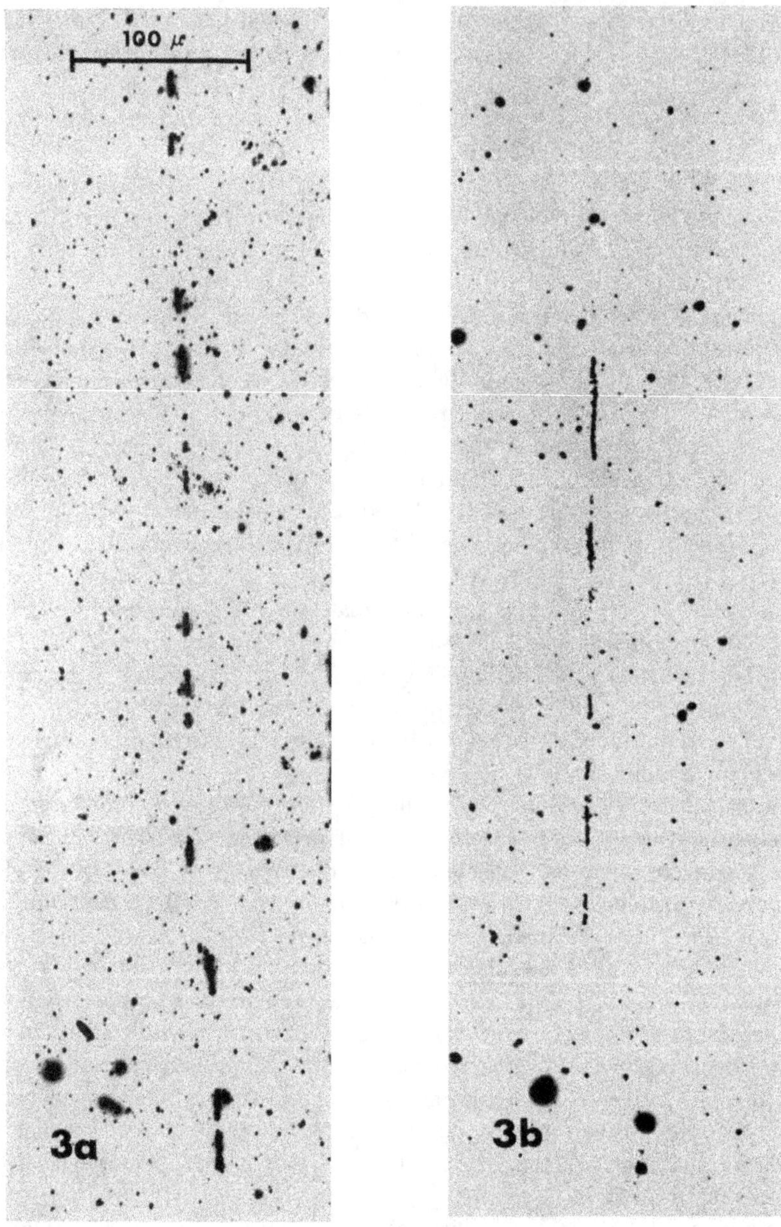

FIG. 3, a and b. Autoradiographs of DNA isolated from Chinese hamster cells after 30 minutes of labeling with tritiated thymidine at the beginning of the S period. The series of short autoradiographic lines represent the paired daughter double helices for several replicons joined together by unduplicated DNA or DNA duplicated prior to addition of tritiated thymidine.

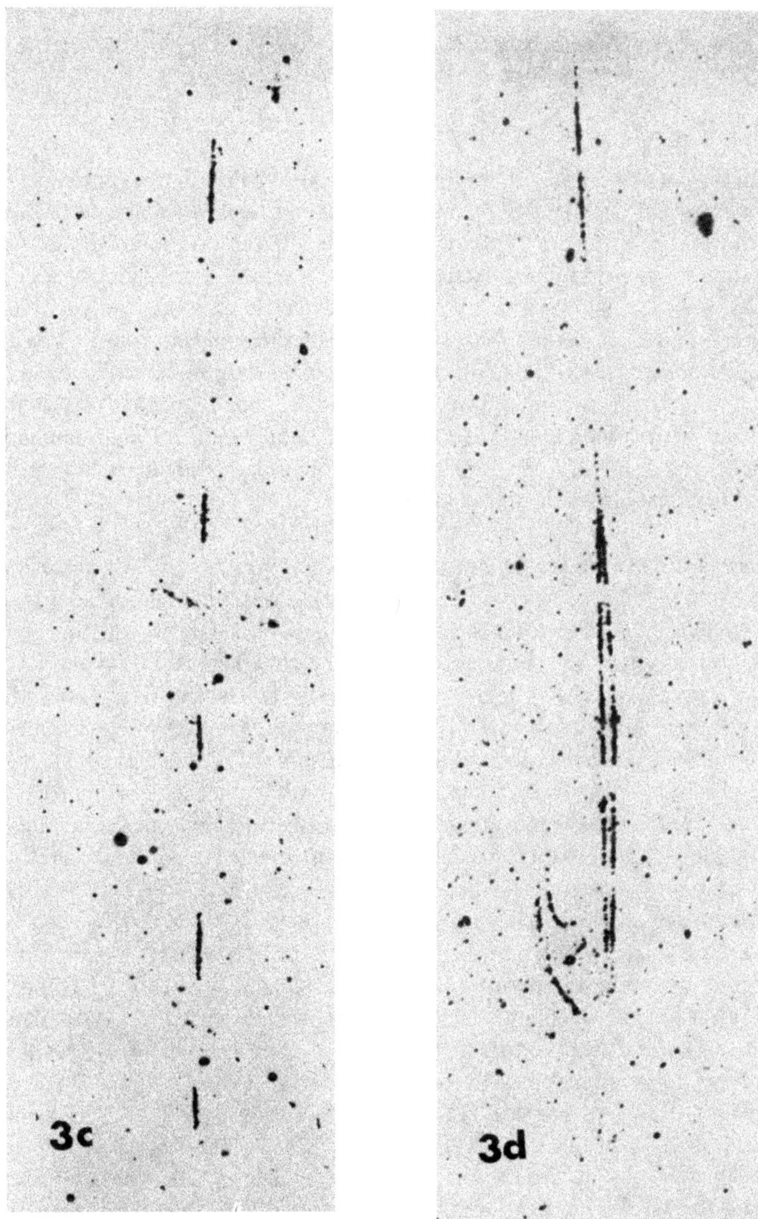

FIG. 3 (cont.), c and d. An autoradiograph of DNA isolated from Chinese hamster cells after a 30-minute incubation with tritiated thymidine (at the beginning of the S period) followed by a 45-minute chase with medium containing nonradioactive thymidine. Under these conditions the daughter double helices have separated sufficiently to produce two autoradiographic images lying side by side. The point at which two autoradiographic lines fuse may represent the replication fork. (From Huberman and Riggs, 1968. *J. Molec. Biol.*, 32:327. Courtesy of Academic Press, Inc.).

tween operons, but the linkers in both cases are very likely deoxynucleotide sequences that function as punctuation marks.

Multireplicons in Eukaryotic Chromosomes

A variety of experimental findings specifically point to the presence of many independently replicating units of DNA within the single eukaryotic chromosome. It is reasonable to speculate that the major step in the evolution of the eukaryotic type of chromosome from a prokaryotic type was the establishment of serial or tandem linking of single prokaryotic-type replicons to produce the relatively enormous eukaryotic chromosomes. The application of the replicon concept to eukaryotic chromosomes, in addition to its importance for understanding eukaryotic chromosome structure, has provided a way of looking at replication and its control that is consistent with virtually all of the experimental evidence and emphasizes the importance of discoveries on prokaryotic chromosomes for the understanding of the more complicated, eukaryotic counterpart.

IMPLICATIONS OF AUTORADIOGRAPHIC LABELING PATTERNS OF CHROMOSOMES FOR THE MULTIREPLICON CONCEPT OF CHROMOSOME STRUCTURE. The first evidence that the individual eukaryotic chromosome contains many replicating units, i.e., replicons, came from the autoradiographic observations by Taylor (1959) on metaphase chromosomes of Chinese hamster cells labeled with tritiated thymidine during the preceding interphase. With 10-minute labeling times some of the larger chromosomes showed several labeled sectors in a single arm at metaphase. Different chromsomes showed different patterns and in general supported the conclusion that "DNA may replicate simultaneously in various parts of the chromosome." Zakharov and Egolina (1968) have made refined studies of the X chromosome in Chinese hamster cells and concluded that even within heterochromatic regions there are multiple sites of DNA synthesis. Multiple initiation sites for DNA synthesis have been described by the autoradiographic method for human chromosomes by several authors (Bianchi and de Bianchi, 1965; Gilbert et al., 1962; Moorhead and Defendi, 1962; Morishima et al., 1962; Stubblefield and Mueller, 1962). Similarly, initiation of synthesis at a number of different sites has been reported for chromosomes of plant root cells (Evans, 1964; Wimber, 1961), germ cells of the grasshopper (Lima-de-Faria, 1961), and rat cells in culture (Bianchi, 1966). Hsu et al. (1964) and Hsu (1964) have used FUDR to block Chinese hamster cells at the beginning of the S phase. Using a five-minute pulse of tritiated thymidine for the initial relief of the FUDR block, they were able to observe, in the subsequent metaphase stages, the pattern of labeling that represented the first five minutes of the S phase. Synthesis of DNA started among many chromosomes at numerous sites on each chromosome. Similar results were obtained in the same type of experiment on human cells.

All of these results clearly show that the replication of DNA does not begin at one or the other, or both ends of the chromosome alone, or at the kinetochore alone, or at the kinetochore combined with initiation at chromosome ends, but that replication begins at at least several sites along the chromosome, and hence at least several replicons are present in each of the larger chromosomes of the eukaryotic type. Such evidence led Taylor (1963b) to propose a model of the chromosome in which many replicons are linked in tandem to form one chromosome. While the autoradiographic resolution of metaphase chromosomes is sufficient to establish that multiple sites of initiation of DNA synthesis exist (the multireplicon hypothesis), it is inadequate for determining the number of initiation sites by observations on the highly condensed metaphase chromosomes.

More detailed information about the multirepliconic structure of chromosomes has been obtained by autoradiographic analysis of the polytene chromosomes of *Diptera*. Several years ago Keyl and Pelling (1963) and Plaut (1963) reported that, following incubations with tritiated thymidine, the polytene chromsomes of dipteran salivary glands showed one of two distinct patterns: either the chromosomes were labeled generally throughout or only in the region of the kinetochores, which are heterochromatic. One of the figures of Keyl and Pelling shows a chromsome labeled throughout its entire length after a 30-minute incubation with tritiated thymidine. They conclude that replication of DNA begins generally throughout the chromosome, implying a high number of replicons, and ends with DNA synthesis restricted to the heterochromatic kinetochore regions. Gabrusewycz-Garcia (1964) has reported that the labeling pattern is more complicated in polytene chromosomes of *Sciara*. She suggests from short pulse-labeling with tritiated thymidine that chromosome replication begins with DNA synthesis in puff and prepuff sites, and perhaps other particular euchromatic regions; DNA synthesis then becomes more general throughout the chromosome (as in the work of Keyl and Pelling on *Chironomus*), and the replication cycle ends with synthesis in the kinetochore regions and other heterochromatic sites.

Howard and Plaut (1967; 1968) have also interpreted their results with *Drosophila* chromosomes as evidence for an initial discontinuous pattern of labeling, followed by a period of labeling throughout the chromosomes, and ending with a second period of discontinuous labeling. Nash and Bell (1968) and Rodman (1968), however, believe that replication in *Drosophila* begins with generalized DNA synthesis (multiple initiation sites) along the chromsome and ends with synthesis only at limited locations. Berendes (1966) and Plaut et al. (1966) had earlier arrived at a similar conclusion from their studies on *Drosophila* (but note the later reports by Howard and Plaut [1967; 1968] suggesting an initial period of discontinuous labeling). The initiation of synthesis at many sites at the beginning of the S period (continuous labeling pattern) would agree with the reported situation in mammalian chromosomes. Whether or not the beginning of chromosome replication is characterized by discontinuous synthesis (at a series of locations along the chromosome), all of the studies are in agreement in pointing to a high number of serially arranged replicons within the individual chromosome.

ESTIMATES OF THE NUMBER OF REPLICATING UNITS (REPLICONS) WITHIN EUKARYOTIC CHROMOSOMES. Plaut et al. (1966) have made a detailed analysis of the pattern of DNA replication in *Drosophila* and have provided a first approximation of the number of replicons in the haploid set of four chromosomes. Using incubations in tritiated thymidine of 10 to 15 minutes they have identified a minimum of 30 replicons in less than 15 percent of the cumulative length of the chromosomes. This suggests by extrapolation that the four chromosomes of *Drosophila* may be made up of more than 200 replicons. Plaut et al. (1966) also make the point that the replicons that are repeated laterally across the polytene chromosome replicate in synchrony. This latter discovery has important implication for the problem of the control of replicon replication discussed in a later section.

Cairns (1966) has obtained an estimate of the number of replicons in the average chromosome of the HeLa cell using a less direct method but one which is probably capable of much greater resolution. Cairns determined, by autoradiography of DNA prepared from HeLa cells that had been labeled with tritiated thymidine in a short pulse, that the rate of helix replication was 0.5 μ per minute. Knowing this rate and the fact that the average HeLa cell chromosome contains enough DNA to make a double helix 3 cm long, simple calculations show that the average chromosome must have more than 100 separate replication points in order to accommodate the total DNA synthesis within the normal S period of six to eight hours. The data imply an average replicon length of 300 μ or shorter, well under the length of 1,200 μ for the single replicon in the chromosome of *E. coli*. The experiment does not distinguish between a serial, tandem arrangement of all 100 replicons per chromosome and some degree of a parallel arrangement of replicons (polynemy). If the chromosome is polynemic, the number of DNA replicating units arranged in tandem would be correspondingly reduced. Huberman and Riggs (1968) have recently provided some evidence that replicons in mammalian chromosomes may replicate from both ends. This would have the effect of reducing by half the estimated number of replicons in Cairns' experiment on HeLa chromosomes. Nevertheless it is again directly clear in this experiment (Fig. 3) that some number of replicons must be arranged in tandem.

Painter et al. (1966) have approximated the number of replicons in a single nucleus of a HeLa cell in the following manner. HeLa cells were labeled with tritiated thymidine for 30 minutes followed by labeling with ^{14}C BUDR for 2 to 3 hours. The DNA was isolated, sheared, and the amount of DNA that was both labeled with tritiated thymidine *and* intermediate in density (equilibrium density-gradient centrifugation) between pure hybrid (one chain of the double helix labeled with ^{14}C BUDR) and normal density was measured. DNA of intermediate density that was labeled with tritiated thymidine can be assumed to contain the growing points of DNA molecules, that is, to contain a linkage of tritium labeled DNA with BUDR labeled DNA. With this information it could be calculated that a HeLa cell nucleus contains a minimum of 1,000 replicating sites at any one instant of time. Knowing the amount of DNA per nucleus this suggests that the average replicon must be no larger than 10^{10} daltons or about 5 mm of double helix. As Painter et al. (1966) point out, the number of replicons is probably higher than 1,000 and therefore the size of the average replicon is less than 5 mm in length because the number of DNA growing points determined at any one instant of time (1,000) is probably considerably fewer than the total number of replicating units in the nucleus.

Okada (1968) had measured the amount of DNA in mouse leukemic cells and the rate of replication of the DNA helix (by a double labeling method). This information combined with the knowledge of total DNA synthesis time for one cell cycle allowed the calculation that the number of replicons per cell is between 160,000 and 410,000 or between 4,000 and 10,000 per chromosome. According to these values the average replicon length must be between 13 and 34 μ. This estimate of replicon lengths fits remarkably well with the results of Huberman and Riggs (1968).

The rate of double helix replication in Chinese hamster cells has been calculated by Taylor (1968) to be 1 to 2 μ per minute, a somewhat faster rate than measured by Cairns for HeLa cell DNA. Since large segments of chromosomes in Chinese hamster cells replicate in 180 minutes or less, replicons in such regions must be between 180 and 360 μ in length. According to Taylor, several hundred replicons of such length could be contained in the longer arm of the X chromosome. Autoradiographic studies by Huberman and Riggs (1968) on Chinese hamster cells strongly indicate that the replicons replicate from both ends with the replication forks approaching each other. This would mean that in Taylor's experiments, even though the rate of double helix replication at the growing fork is 1 to 2 μ per minute, the rate of replicon replication is 2 to 4 μ per minute. This consideration has the net effect of doubling Taylor's estimated length of the average replicon and halving the estimated number of replicons per chromosome.

Huberman and Riggs' (1968) autoradiographic studies of patterns produced by DNA of Chinese hamster cells and HeLa cells pulse labeled with tritiated thymidine show in a striking manner that many individual replicons must be joined together in series (Fig. 3a, b, c, d) as proposed by Cairns (1966). Replicon lengths, measured directly from the autoradiographic patterns, were less than 30 μ in most cases. These experiments probably give the most accurate measure of replicon length even though the length may be slightly underestimated because of the arrangement of DNA in less than a perfectly straight line (small deviations from a straight line would not be detectable in light microscope autoradiography). Since the average mammalian chromosome contains enough DNA to make 3 cm of double helix, the average chromosome made up of replicons 30 μ or less in length would need to contain in excess of 1,000 such replicating units, or in excess of 50,000 in a single HeLa cell nucleus.

Finally, the phenomenon of gene amplification discovered in studies on the multiple, unattached nucleoli of amphibian oocytes reflects very clearly the multireplicon organization of the eukaryotic chromosome. Each nucleolus contains DNA (Miller, 1964; Peacock, 1965), which is apparently synthesized during the pachytene stage of meiosis long after the chromosomes have finished the last premeiotic DNA replication (Gall, 1968). Although not demonstrated directly, it is virtually certain from a variety of observations including annealing of ribosomal RNA with DNA (Gall, 1968) that the DNA of the hundreds of unattached nucleoli is derived by replication of the DNA of the nucleolar organizer region of one of the chromosomes. The need for such amplification by selective replication of a very limited region of a chromosome is reasonably explained by the requirement for a relatively enormous production of ribosomal RNA during oogenesis (Brown and Dawid, 1968; Gall, 1968; Perkowska et al., 1968). Gall (1968) has pointed out that this amplification phenomenon during oogenesis is widespread among animals, occurring not only in amphibians generally but also in Orthopteran oocytes (Kunz, 1967a, 1967b), in oocytes of the fly *Tipula* (Lima-de-Faria and Moses, 1966), and other insects (Bier et al., 1967).

Miller (1966) has made electron microscope measurements of the DNA isolated from amphibian oocyte nucleoli. The material has a diameter of approximately 30 Å and is therefore almost certainly only one DNA double helix. More pertinent for the present discussion is the fact that the DNA molecules are in the form of closed loops in which the molecular length of a population of such DNA loops varies according to a regular geometric progression. Miller describes loops with the DNA measuring 11.25, 22.5, 45, 90, and 180 μ in length. The results suggest that the DNA of the nucleolus can be replicated in single units of 11.25 μ (or smaller according to Miller) or that multiple rounds of replication can occur in such a manner that units of 11.25 μ are joined together end-to-end to form larger loops.

The configuration of nucleolar DNA in amphibian oocytes appears strikingly similar to the configuration described for mitochondrial DNA, in which a basic unit, presumably a replicon, is in the form of a continuous loop of DNA (see Fig. 1) (e.g., Dawid and Wolstenholme, 1967). These observations, taken together with the detailed study of the mechanism of loop formation in viral chromosomes (Rhoades et al., 1968; Thomas, 1967), provide an adequate basis for speculating on the means by which individual replicons may be held together in tandem within the eukaryotic chromosome. There is no simple way to accommodate replicons of DNA arranged in a tandem of closed loops within the chromosome; rather, it is probable that replicons are hooked together, one to another, end-to-end by the same type of linkage that allows the loop arrangement of DNA in amphibian nucleoli, in some mitochondria, in some viruses, and in prokaryotic chromosomes generally.

There has been one report of DNA in the form of closed loops obtained from eukaryotic chromosomes (pig sperm) (Hotta and Bassel, 1965). Until it can be demonstrated, however, that closed-loop DNA is a general or even a common attribute of eukaryotic chromosomes, it seems pointless to struggle with schemes to accommodate loop structures within models of the chromosomes.

Polynemy Versus Uninemy of Eukaryotic Chromosomes

Two terms exist to describe the presence of a parallel array of identical DNA molecules within a chromosome; these are polyteny and polynemy. Polyteny is the accepted and unambiguous term used to describe the structure of the giant interphase chromosomes in dipterans and in the developing macronucleus in certain ciliated protozoa (Fig. 4) (Ammermann, 1965; Perez-Silva and Alonso, 1966). It is generally accepted that such polytene chromosomes represent the close pairing of chromatids during and following multiple rounds of endoreduplication of the chromosome; pairing of homologous chromosomes is usually but not necessarily present. Polyteny arises occasionally in cells still capable of undergoing mitosis and is evident at metaphase in the form of endoreduplicated chromosomes that contain four chromatids. There seems to be no confusion about either the definition of polyteny or the fact of its existence.

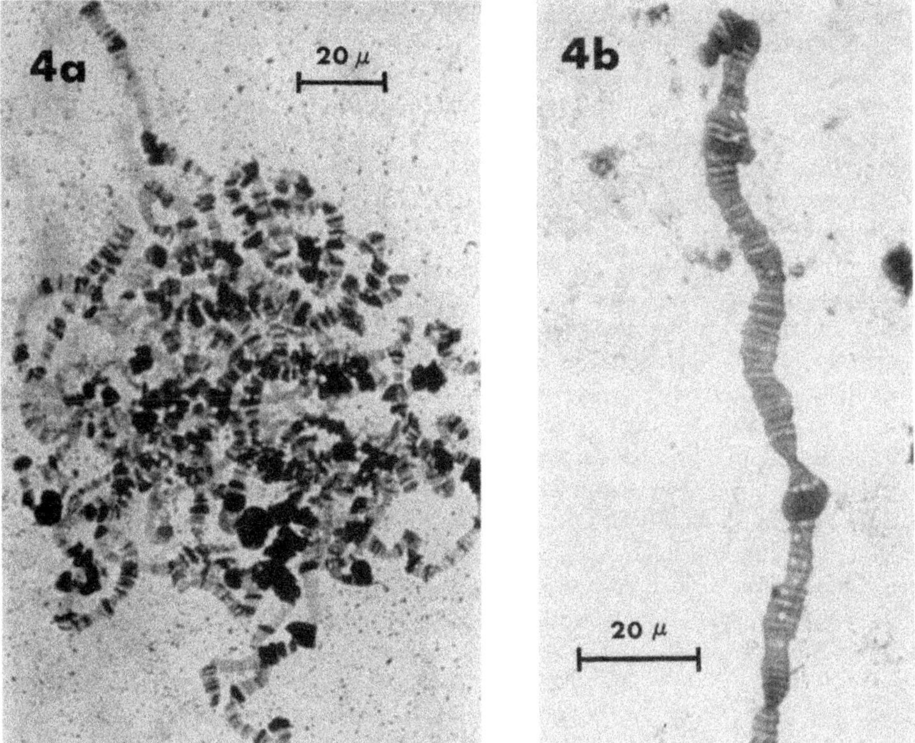

Fig. 4. Polytene chromosomes from the ciliated protozoan, *Stylonychia mytilus*. Polyteny is developed during reconstruction of the macronucleus after conjugation. The polytene chromosomes exist only transiently and are reduced to a simpler form by the loss of DNA. (Ammermann, 1965. The photographs were kindly provided by Dr. Ammermann.)

The term polynemy is meant to denote an unduplicated chromosome (or chromatid) in which two or more identical DNA double helices are arranged in parallel in such a tight association that they remain together in the formation of the chromatid at mitosis. The molecular difference between polynemy and polyteny presumably resides in the degree of intimacy with which the parallel double helices are bound together. Nothing is known about such binding, however, and it may be questioned whether the two terms really describe two different phenomena. The matter of polynemy (versus uninemy, or one DNA double helix per chromosome) has nevertheless been a principal source of debate with many premature claims that one or the other situation has now finally been proven beyond any doubt. Evidence for polynemy versus uninemy is presented and discussed in the following sections.

Evidence for Polynemy. Evidence that eukaryotic chromosomes may be polynemic falls into four categories: (1) microscopic observations of living and fixed mitotic or meiotic chromosomes at metaphase or anaphase that suggest the presence of two or four subunit strands in a single chromatid, (2) the formation of so-called subchromatid reunions in meiotic and mitotic chromosomes following ionizing radiation treatment, (3) the presence of wide differences in the total amount of chromosomal DNA in closely related species in which the chromosome numbers are the same or at least very similar, and (4) the autoradiographic detection of isochromatid labeling of chromosomes at the second metaphase following labeling with tritiated thymidine.

Claims of proof that the chromosome is polynemic have been made from electron microscope observations, but Ris (1967), who has made extensive electron microscope studies of chromosomes, has concluded that "It must be admitted that electron microscope studies up to now have told us very little about the arrangement of fibrils in the chromosome and have been of no help in deciding whether chromosomes are composed of a single strand or a bundle of fibrils." Observations on sections of interphase nuclei or mitotic chromosomes in the electron microscope have been virtually useless for deciphering the arrangement of DNA molecules or DNA-protein fibrils in the eukaryotic chromosome. Even the chromosome of *E. coli,* which has been proven to consist of one DNA double helix, shows, in sections, a mass of fibrils in the electron microscope that often appear in bundles and might easily be interpreted to represent a high degree of polynemy if such were the observer's inclination. Whole mounts of entire mitotic or meiotic chromosomes or interphase chromosomes prepared by the Kleinschmidt technique are similarly uninformative in distinguishing between polynemy and a complex pattern of folding of DNA-protein fibrils of a uninemic chromosome (see DuPraw, 1966; Ris, 1967; Wolfe and Hewitt, 1966; Wolfe and Martin, 1968). Obviously, a high degree of folding of DNA-protein fibrils is present in interphase nuclei (80 percent is in the form of heterochromatin), in sperm, and in mitotic or meiotic chromosomes.

Doubleness of Mitotic and Meiotic Chromatids. Several electron microscope studies of chromosomes indicate that the fiber of DNA-protein is about 250 Å in diameter (DuPraw, 1965; DuPraw and Rae, 1966; Gall, 1966; Rae, 1966; Ris, 1967; Wolfe, 1965, 1968; Wolfe and Hewitt, 1966). Whether such 250 Å fibers consist of one, two, or more DNA-protein fibrils of smaller diameter has been a matter of some dispute. Ris (1967) has proposed, on the basis of his own work and the observations of others, that the 250 Å fiber consists of two fibrils of 100 Å diameter each; the 100 Å fibrils are derived from different regions of the same fibril by a process of folding and are held together laterally under the influence of divalent cations. The 100 Å fibril is normally not uniform in thickness but in places is as thin as 25 Å (Ris, 1967). Digestion with pronase leaves a 25 Å fibril. Ris, therefore, concludes that the 100 Å fibril contains one DNA double helix combined with protein. This continuous, basic fibril forms very frequent loops that are collapsed to form the 250 Å unit, giving the impression that the 250 Å fiber is made up of two independent, 100 Å fibrils. The observation, as interpreted by Ris, shows how a chromosome consisting of one DNA double helix, is enormously shortened by the presence of a very large number of loops formed under the influence of proteins and divalent cations and may give a false impression of polynemy.

Reports of the longitudinal doubleness of anaphase chromosomes, visible with light microscopy, extend back to the last century (see discussion by Ris, 1967). The doubleness has been detected chiefly at the ends of chromosome arms, and it has never been demonstrated distinctly enough to convince many cytologists. Two recent papers, one on fixed mitotic chromosome of *Scilla non scripta* (Gimenez-Martin and Lopez-Saez, 1965) and the other on maize chromosomes using Nomarski interference contrast microscopy (Maguire, 1968), both provide suggestive evidence of bipartite anaphase chromosomes, but the evidence is not at all convincing. The significance of such observations has also been questioned on the grounds that the apparent doubleness may be an artifact caused by fixation. This criticism is avoided in Bajer's (1965) more recent observations on living cells of *Haemanthus*. Thus chromosomes of this plant are especially easy to see clearly in endosperm tissue and are unusually large. Bajer has described a bipartite structure for the anaphase chromosomes in living cells; his photomicrographs distinctly indicate doubleness not only at the tips of chromosome arms but, in favorable cases, along the length of the chromosome. While Bajer's photomicrographs are a clear improvement over earlier illustrations, general acceptance of this type of observation as proof of a bipartite structure for the unduplicated chromosome is not likely to occur until it has been shown that the units can be separated to the degree that chromatids of the metaphase chromosome are normally separable; only with such a degree of separation can arguments about optical illusions or artifacts of preparation be convincingly discounted.

In an effort to bring out more distinctly the possible subunit structure of the chromatid, Trosko and Wolff (1965) have digested metaphase chromosomes of the plant *Vicia faba* with trypsin. In Feulgen stained preparations each chromatid is clearly made up of two main axes of DNA, although the two axes touch each other at many points and there are numerous strands of DNA connecting the two axes. These studies show more clearly than any previously that there is a bipartite organization of metaphase chromatids, but they do not prove, by any means, that the chromatid is made up of two DNA double helices arranged in parallel. As in Bajer's studies of living chromosomes, the two apparent subunits are in close association with one another, and in no case has a complete and clear separation between the two axes been achieved. Wolfe and Martin (1968) have confirmed the presence of a double axis in metaphase chromatids of *V. faba* after trypsin treatment, but they were unable to detect clear doubleness in whole mount preparations viewed with the electron microscope.

If anaphase chromosomes are in fact bipartite and contain two double helices arranged in parallel subunits, these subunits *cannot* represent the incipient chromatids of the next metaphase. Semiconservative replication and semiconservative segregation of DNA would require that the chains of the two postulated double helices be reassorted as illustrated in Figure 2. One chain from each of the two double helices would necessarily have to be included in each chromatid at the next division. These complications do not disprove the doubleness of anaphase chromosomes, but the sorting out of subunits in a bineme or polyneme chromosome cannot be as simple and straightforward as has usually been assumed.

In summary, the microscopic observations on mitotic and meiotic chromosomes suggest, but fall far short, of proof of polynemy (binemy).

"Subchromatid reunions." The type of aberrations induced by exposure of interphase cells to ionizing radiation consistently fall into two categories (e.g., Bender, 1962; Wolff, 1961); these are chromatid-type and chromsome-type aberrations. Chromatid aberrations (only one chromatid of a metaphase chromosome is affected) result from irradiation delivered subsequent to the beginning of DNA synthesis. Irradiation prior to the initiation of DNA synthesis produces aberrations of the chromosome type, i.e., both chromatids of the metaphase chromosome have the particular aberration. In the latter case, it seems clear that an aberration induced in the unreplicated G1 phase chromosome is reproduced during the DNA synthesis period so that both chromatids (daughter chromosomes) have the same aberration at metaphase. These results mean that the chromosome behaves as a single linear unit before replication and as a double unit during or after replication; the singleness and the doubleness have been interpreted as evidence that the chromosome is one DNA double helix in width (unineme) before replication and two helices after replication. Two reports (Evans and Savage, 1963; Wolff and Luippold, 1964) suggest that the chromosome behaves in its response to radiation as if it were already a double structure (chromatid aberrations) late in the G1 phase just before DNA synthesis begins. This interpretation of the results is not secure, however, because of difficulties in precisely timing the synthesis of DNA at a particular site in the chromosome by autoradiography following a tritiated thymidine labeling.

Irradiation of cells in mitosis or meiosis produces aberrations, usually apparent as some form of anaphase bridge, that have been interpreted as evidence for the presence of subchromatids (a bineme or polyneme chromatid; LaCour and Rutishauser, 1954; Peacock, 1961; Wilson and Sparrow, 1960). The earlier observations are cited by Wilson et al. (1959). The primary point is that anaphase bridges appear to be made up of unions between chromosomes involving less than the full width of each chromosome or appear in configurations that suggest reunions between subunits within the daughter chromosomes. Because the fiber that forms the bridges between anaphase chromosomes is thinner than the chromosome proper, it can hardly be considered evidence of a subchromatid unit; such observations need not be evidence of anything more than the tension of the bridge has caused a stretching out of parts of the chromosome arms involved.

Ostergren and Wakonig (1954) have disputed the subunit interpretation because aberrations in the next metaphase are of the chromosome type rather than the chromatid type; chromatid aberrations would be expected if the original aberration had been of a subchromatid nature. These results of Ostergren and Wakonig are apparently contradicted by Peacock (1961), who found configurations at the second metaphase that did suggest that the initial aberration had been of a subchromatid nature. Peacock's results are in turn contradicted by Kihlman and Hartley (1967) who have induced aberrations with 8-ethoxycaffeine and concluded that their data provided "strong evidence in favor of the idea that the so-called subchromatid aberrations are in fact masked chromatid aberrations." Swanson and Young (1965) have obtained results that "are basically in accord with those of Ostergren and Wakonig" and therefore also in agreement with those of Kihlman and Hartley. In view of these latter results, it appears that none of the earlier observations of aberrant configurations of mitotic or meiotic chromosomes can be considered as very strong evidence of the existence of subchromatids.

Increase in the Amount of DNA to Chromosome Complements. Additional arguments for the existence of polynemy, still indirect, are based on the observation that the amount of DNA (and size of metaphase chromosomes) among members of a related group of organisms may vary over a wide range with little or no change in chromosome number or karyotype. Hughes-Schrader and Schrader (1957) have provided, for example, an extensive description of this type of situation in a large group of plants known as the *Nezara* complex, and they interpret the measurements of DNA amounts as evidence of evolution of polynemy. Rothfels et al. (1966) demonstrated an increase in the DNA content of a group of 22 related, diploid species of plants in the proportions of 1:8:12:16:20:24:40 without concomitant alteration of the common karyotype. In five species of the plant genus *Vicia,* Martin and Shanks (1966) have found DNA contents in the proportion of approximately 17:19:57:59:100. The chromosome number varies little or not at all from species to species and the karyotypes are somewhat similar, although a number of differences are clearly evident. The absence of the 1:2:4 relationship, that would be expected for straightforward development of polynemy, could be due to evolutionary superimposition of translocations and unequal crossovers; it is virtually certain that such changes must have occurred.

There are two possibilities to account for such large changes in DNA amount in the absence of little or no changes in chromosome number; either specific segments of DNA must be added interstitially at discrete points along the chromosome or all of the DNA must be uniformly involved through a polynemic increase in the number of DNA molecules in lateral (parallel) arrangement. Interstitial addition of DNA is strongly indicated in Keyl's studies (1965a, 1965b) with *Chironomus* (see below) and is the most reasonable mechanism for explaining evolution of one protein from another, for example, the apparent evolution of the different protein subunits of immunoglobulin from a single original protein.

Interstitial additions might result from localized mistakes in replicon duplication (Keyl, 1965a, 1965b) or could stem from unequal crossovers. On the other hand, at present no evidence compels the acceptance of polynemy to explain DNA increases. Rothfels et al. (1966), in their studies on plant chromosomes prefer the polynemic explanation, but they point out that the increases in DNA observed by them do not form a geometric series, requiring the somewhat awkward postulate that only some of the strands of the chromosome participated in the polynemic increase in DNA during evolution. The same postulate is required by the less than twofold differences in DNA between two species with a haploid number of six in the plant genus *Luzula* (Halkka, 1965).

In three species of the amphibian genus *Bufo* (*B. bufo, B. viridis,* and *B. calamita)* the haploid chromosome number is 11, and all three species have similar karyotypes (Ullerich, 1966). "All chromosomes of *B. bufo* contain significantly more than, but in no case twice as much DNA as their homologues in the other two species. Eight chromosomes of *B. bufo* contain 30 to 40 percent, three about 50 percent more DNA than their homologues in *B. viridis*." Differences in DNA content between *B. viridis* and *B. calamita* were present only in the larger chromosomes.

Ullerich concludes that the differences are not the result of the development of polynemy but the result from local increases in DNA in all of the chromosomes of *B. bufo* and only in the larger chromosomes of *B. viridis*. In a subsequent study Ullerich (1967) found that the DNA content of three species of *Rana,* all with a similar karyotype, i.e., *R. temporaria, R. arvalis,* and *R. esculenta,* varied in the proportion of 1:1.28:1.54; Ullerich considers these data as additional evidence of local increases in DNA in the chromosomes rather than of polynemy. Finally, Ohno et al. (1968) have made extensive analyses of chromosome number and DNA content among many species of fish and concluded that in certain groups, increases in DNA have occurred without changes in chromosome number by means of unequal exchanges and "regional redundant duplication of DNA molecules." In connection with the hypothesis of regional duplication of DNA, they cite the evolution of isozyme patterns, which are presumed to require occurrence of gene redundancy as an initial event.

Particularly pertinent to this problem of DNA increase (or decrease) in evolution are the measurements by Keyl (1965a; 1965b) on the polytene chromosomes of two subspecies of *Chironomus thummi*. Measurements on pairs of homologous bands in unpaired regions of salivary gland chromosomes of the hybrid between *C. t. thummi* and *C. t. piger* showed that the DNA content of some bands of *C. t. thummi* was 2, 4, 8, or 16 times greater than *C. t. piger*. The total DNA content of *C. t. thummi* in both salivary gland nuclei and in spermatocytes was 27 percent higher than in *C. t. piger*. It seems clear that the difference is the result of localized increases of DNA, and Keyl suggests that the DNA increase occurs by a stepwise elongation of replication units. Keyl (1965; 1965b) presents a scheme by which interstitial addition of DNA might occur. Such interstitial addition of DNA has abundant precedent among the chromosomes of viruses, bacteria, and bacterial episomes. Gene duplication with serial, interstitial addition of the increased DNA has become a standard part of hypotheses concerned with the evolution of closely related proteins, e.g., isozymes, immunoglobulins, and monomers of the hemoglobin molecule. Perhaps the clearest example of a very high degree of serial repetition of cistrons is present in the nucleolar DNA responsible for the synthesis of ribosomal RNA. Whether this type of redundancy occurs more generally throughout the chromosomes of eukaryotes is not yet known, but Britten and Kohne (1968) have obtained DNA hybridization evidence that certain segments of the DNA of higher eukaryotes are repeated many thousands of times within a single genome.

It seems unlikely that serial redundancy would involve all of the DNA since multiple copies of all the genes in a chromosome would create a circumstance that would be difficult to reconcile with many of the results from traditional genetic studies of mutation and inheritance. Primarily on the basis of cytological data on lampbrush chromosomes, Callan and Lloyd (1960) and Callan (1967) have proposed, however, that serial redundancy is present for every unit of genetic function. The conflict with genetic considerations is reconciled by proposing that each unit of serially redundant genes consists of one master and many slave copies of the same gene. At regular intervals the slave copies are matched against the master copy, and any mutations in the master copy are introduced into the slave copies by a corrective mechanism. The scheme is ingenious, but none of it is yet supported by direct evidence, and it encounters rather complicated traffic problems. Redundancy of DNA may be restricted to a limited number of genetic units (including, for example, ribosomal cistrons that do not follow conventional segregation and by chance have never been the focus of genetic analysis). Complicated mechanisms that reconcile redundancy with traditional genetics would in such cases not be necessary.

In summary, there is strong direct evidence that DNA increases within a chromosome occur by localized interstitial additions by redundant replication or unequal crossover, and this might account for all of the observations cited above; the evidence for increases by a polynemic mechanism is inferential and less indicative, but the possibility of polynemy cannot yet be discounted.

Isochromatid Labeling. A more recently developed argument for the existence of polynemy is based on the occurrence of isochromatid labeling of chromosomes at the second division following labeling of DNA for one S period (Peacock, 1963; 1965). As shown originally by Taylor et al. (1957) with autoradiography, the segregation of DNA in a wide variety of eukaryotic cells is semiconservative. In the first metaphase following labeling of DNA with tritiated thymidine, both chromatids are labeled, and within the limits of quantitative accuracy of the autoradiographic method the two chromatids are labeled equally (see Peacock, 1965). At the second metaphase after labeling, only one of the chromatids is labeled. The consequences of sister chromatid exchanges (Figs. 2, 5, and 6) are often apparent at the second metaphase but they do not disturb the conclusion that segregation is semiconservative.

Peacock (1963, 1965) has pointed out that sometimes at the second metaphase after labeling, both chromatids may be labeled at a given position along the chromosome. He has called this phenomenon "isolabeling." Isolabeling mostly involves only short segments of the chromosome and in most cases it occurs at the ends of the chromosome, "but in other cases (isolabeling) appeared to extend along the full length" of the chromosome. The crucial issue is whether isochromatid labeling really represents an exception to the semiconservative segregation of DNA, or whether this type of labeling is a consequence of sister chromatid exchanges combined with the normal image spread that attends autoradiography of tritium. Peacock feels that the occurrence of sister chromatid exchanges are insufficiently frequent to account for isolabeling, but the argument is not at all convincing. Three exchanges in a single chromosome in two cell generations are not unusual in Chinese hamster cells, and four exchanges are found occasionally (Figs. 5 and 6). In addition, the number of sister chromatid exchanges needed to produce apparent isolabeling is really smaller than usually assumed, if the normal image spread of autoradiographs is taken into account. The resolution of light microscope autoradiography with tritium is under the very best conditions one micron. The autoradiographs of second metaphase chromosomes (following labeling with tritiated thymidine) of Chinese hamster cells in Figures 2, 5, and 6 show several examples of isolabeling that has resulted from one or more sister chromatid exchanges. The autoradiographs published by Peacock (1963, 1965) do not include any examples that cannot be quite reasonably explained as the result of one or more sister chromatid exchanges combined with autoradiographic image spread. The isolabeling reported by Walen

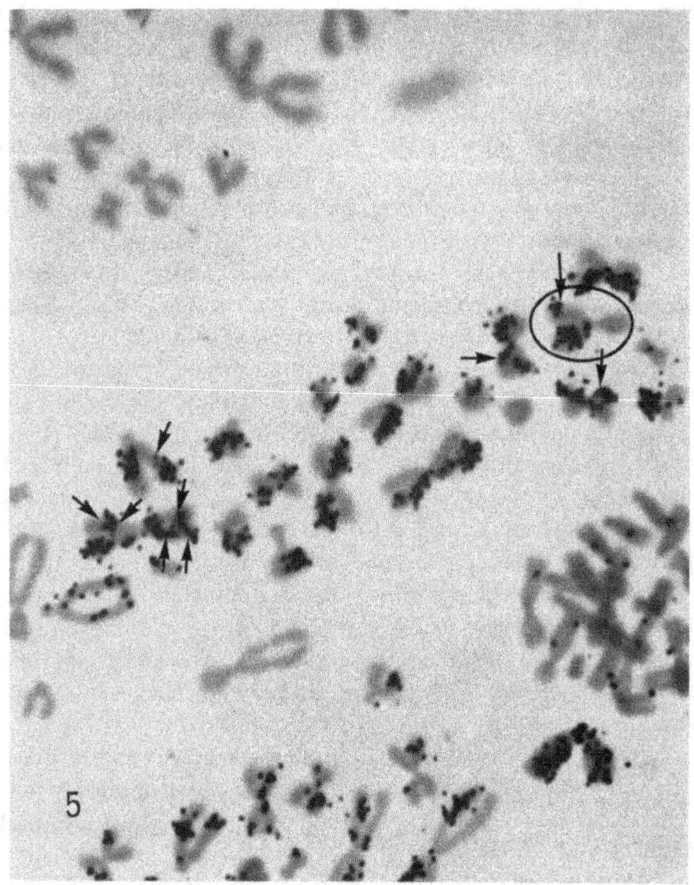

FIG. 5 An autoradiograph of Chinese hamster cell chromosomes in the second metaphase after labeling with tritiated thymidine. Several sister chromatid exchanges are present. One chromosome shows five exchanges. The X chromosome (circled) shows a short region of "isolabeling" at the end of the long arm that is probably the result of a single sister chromatid exchange. (From Prescott and Bender, 1963b. *Exp. Cell Res.*, 29:430.)

(1965) is very probably due to sister chromatid exchanges. Exchanges within at least the terminal micron of chromatid arms are particularly likely to give the impression of isolabeling. Image spread will produce autoradiographs over the terminal section of both sister chromatids (see Fig. 2); because of the terminal nature of the exchange, the usual criteria by which exchanges are identified cannot be applied. A slightly more proximal exchange would be identified as such by the absence of an autoradiographic image at the very tip of one of the chromatids (see Fig. 6).

If isolabeling represented an exception to semiconservative segregation of DNA, it would constitute very strong evidence that the unduplicated chromosome (chromatid) consisted of at least two DNA double helices in parallel (see Fig. 2). Until isolabeling has been proven to represent an exception to semiconservative

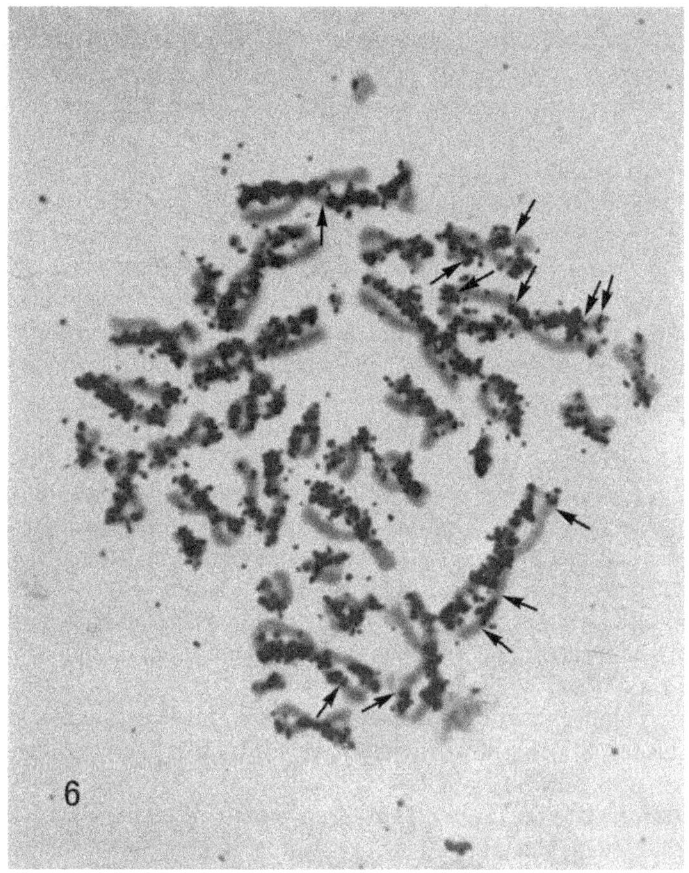

FIG. 6. An autoradiograph of a near tetraploid set of Chinese hamster chromosomes in the second metaphase after labeling with tritiated thymidine. A number of sister chromatid exchanges are present, and several chromosomes show autoradiographic patterns that could be interpreted as isochromatid labeling. Some of the exchanges are indicated by arrows. The circled chromosome contains an exchange that is just sufficiently subterminal to be identified as an exchange rather than isolabeling.

segregation, it need not be considered in the development of models or concepts of the eukaryotic chromosome.

EVIDENCE FOR UNINEMY. The evidence that the eukaryotic chromosome is uninemic (one DNA molecule in width) is derived from a variety of experimental evidence, some of which has already been briefly mentioned above. Additional points are discussed below.

Semiconservative Segregation of DNA. The semiconservative segregation of DNA that has been clearly demonstrated in a wide variety of eukaryotic cell types (Cave, 1966; Bender and Prescott, 1962; Prescott and Bender, 1963b; Schwarzacher and Schnedl, 1965; Taylor, 1958a, 1958b, 1963a; Taylor et al., 1957; Walen, 1965; Woods and Schairer, 1959) is most simply explained on the basis of a unineme chromosome as Taylor (1963a) has consistently maintained.

The observations therefore are consistent with, or at most, infer the unineme concept of the chromosome, but in themselves do not constitute direct evidence. The alternative is that semiconservative segregation and semiconservative replication of DNA in a bineme or polyneme chromosome require a more complex scheme for segregation of DNA double helices in the formation of the two chromatids at mitosis (see Fig. 2). To account for the autoradiographic data, of the two daughter double helices produced by semiconservative replication of each of the two double helices in the bineme chromosome, one must segregate into one chromatid and the other into the sister chromatid. We have as yet no experimental evidence of how such a segregation might be achieved and speculation seems hardly worthwhile until the chromosome has been proven to have a bineme or polyneme constitution.

A bineme or polyneme chromosome in which DNA synthesis and segregation are semiconservative raises a genetic problem. A recessive mutation in one of the double helices could never become homozygous within a chromatid since the rules of helix segregation demanded by semiconservative segregation of DNA within the chromosome would always produce a chromatid with one mutated and one unaffected helix (Fig. 2). For expression of such a recessive mutation it becomes necessary to invoke the further contrivance of either genetic suppression, normally of one of the two double helices in each chromosome, or the occurrence of errors in the semiconservative segregation of segments of DNA within a chromosome at mitosis. Such errors would be evident as isolabeling of chromatids as described by Peacock (1963, 1965).

These speculations are only demanded by a bineme or polyneme chromosome and until binemy or polynemy is proven, none of these contrived schemes should be considered with much seriousness.

Sister Chromatid Exchanges. The phenomenon of sister chromatid exchange, first discovered by Taylor et al. (1957), must be accounted for in any efforts to determine the arrangement of DNA within the chromosome. The chromosomes in the autoradiograph in Figures 5 and 6 were isolated from a cell in the second metaphase after labeling the DNA with tritiated thymidine; several sister chromatid exchanges are apparent. The exchanges are presumed to have occurred between DNA labeling and the first metaphase and during the next interphase. Using colchicine treatment at the first metaphase to produce a tetraploid cell, Taylor (1959) was able to examine the sister chromatid exchanges at next metaphase, at which time exchanges from both cell cycles would be accumulated. Exchanges that occurred prior to the first metaphase appeared as twin exchanges (two homologous chromosomes showing exchanges that were mirror images of one another) in the tetraploid cell, and exchanges that occurred between the first and second metaphase appeared as single exchanges (present in only one homologue). Taylor considered two possibilities for the development of exchanges: (1) a broken unit within a chromatid could rejoin with equal probability with any broken end within the same chromatid or with any broken end in its sister chromatid without regard to any possible differentiated nature of the broken subunits involved; or (2) the rejoining of a broken end could be restricted such that any broken end could only rejoin with the broken end of a *like* subunit. The theoretical ratio of twin to single exchanges in the first case is 1:10 and in the second case 1:2. The actual counts of twin versus single exchanges came very close to a ratio of 1:2 (Taylor, 1959), thus providing very strong evidence that if there are two or more subunits within a chromatid, these subunits are not identical with respect to rejoining of broken ends. The simplest interpretation, one to which Taylor had adhered (1963a), is that there are two subunits per chromatid and these are the two chains (opposite polarities) of a double helix of DNA and not two identical double helices as would be found in the bineme state.

Another scheme can also be postulated to account for the sister chromatid exchange data, but it requires that the two (or more) double helices in each chromatid be restricted with respect to exchanges between each other but each be able to exchange with only one particular helix of the pair in the sister chromatid. This more complicated explanation is less satisfactory because there is no evidence at all for differences between double helices comparable to that for the difference in the two DNA chains that would compel its consideration in preference to the explanation based on a unineme chromosome.

Finally, whether sister chromatid exchanges are normal, spontaneous events within the chromosome or whether they are caused by the very tritium radiation introduced into the chromosome in order to detect them, cannot be decided with finality on the basis of any experimental evidence yet produced. Marin and Prescott (1964) found no change in the frequency of sister chromatid exchanges in Chinese hamster cells over a 100-fold range of variation in the amount of tritiated thymidine incorporated. It might be argued that the induction of exchanges by the radiation might have reached a saturation level even with the lowest incorporation of tritium in these experiments, but this seems unlikely since visible aberrations were not produced even at the highest level of incorporation of tritiated thymidine. The data suggest that sister chromatid exchanges are spontaneous events, although Marin and Prescott did detect a slight increment in the frequency of exchanges following irradiation with 50 r of x-rays.

Sister chromatid exchanges do not occur randomly along the chromosome; in Chinese hamster chromosomes, one-fifth of all exchanges have been reported to occur at the kinetochore (Marin and Prescott, 1964); the frequency of exchanges in the Y chromosome is much higher at the middle of the long arm than other regions of either chromosome arm (Prescott, unpublished). The significance of these latter observations is not known.

Structure of Lampbrush Chromosomes. The most direct way to determine the number of double helices of DNA that are arranged in parallel in the chromosome would be by electron microscope observation. Unfortunately, it has been impossible, in every instance except one, to relate in any reasonable way the DNA-protein fibers observed in the electron microscope to the overall arrangement of DNA within the chromosome.

The exception occurs in the case of the lampbrush chromosomes of salamanders. These chromosomes are in prophase of the first meiotic division and two chromatids are present but not separated. The structure of these chromosomes is succinctly described by Callan and Lloyd (1960):

> "The axis of a lampbrush chromosome consists of a series of chromomeres connected to one another by short, thin, extensible and elastic filaments. Loops . . . are attached to the chromomeres, loops occurring in pairs (one loop from each chromatid). When lampbrush chromosomes are broken mechanically, each break occurs transversely across a chromomere, and separates the (successive) loop insertions in this chromomere in such a way that a pair of straightened-out lateral loops bridge the gap in the chromosome axis." p. 135.

From this observation Callan (1967) maintains that the loops are continuations of the main axis of the chromosome. Then these chromosomes are indeed composed of only one DNA molecule because Miller (1965) has demonstrated by electron microscopy that the loops, when divested of RNA and protein, consist of one fiber 20 to 30 Å in diameter. A fiber of such dimension cannot be composed of more than one DNA double helix. Miller also states that the connections between chromomeres in the main chromosome axis are between 30 and 50 Å, a diameter adequate to account for two double helices of DNA at most, one for each chromatid. These findings, added to the fact that lampbrush chromosomes are not broken by either proteinases or ribonuclease but are broken by deoxyribonuclease (Callan and Macgregor, 1958), constitute very strong evidence that these giant chromosomes are composed of a single, DNA double helix that must measure up to hundreds of centimeters in length.

Kinetics of Breakage of Lampbrush Chromosomes. Closely related to the mechanical breakage data and the electron microscope measurements just cited are the studies of Gall (1963) on the kinetics of deoxyribonuclease action of lampbrush chromosomes. From observations of the rate of appearance of microscopically visible breaks, Gall has calculated that there are 2.6 subunits in the loops and 4.8 subunits between chromomeres that must be attacked before a break becomes evident. The calculated numbers are probably slightly too high because of restriction of penetration of the enzyme through the matrix present on the chromosome to the DNA, and because the matrix may provide enough mechanical support to delay the appearance of breaks. The data clearly suggest that continuity of the loops is maintained by one double helix and continuity between chromomeres is maintained by two double helices (one from each chromatid). It might be argued that the data show only that the number of helices between chromomeres is twice the number in individual loops (4.8 to 2.6) without specifying the numbers of helices involved, but when Miller's electron microscope demonstration of only one double helix per loop is recalled, it seems certain that this interpretation is correct.

Conclusions Concerning Polynemy and Uninemy

In spite of all the evidence, it is not yet possible to come to a firm, unitary conclusion concerning the matter of polynemy, although the existence of polynemy seems unlikely. The evidence shows that some of the largest chromosomes known are uninemic, but even though the evidence is not very compelling, the possibility that some chromosomes may be polynemic under certain circumstances cannot be absolutely excluded. To save the binemic (polynemic) hypothesis in the face of the strong evidence for uninemy from studies on lampbrush chromosomes, it has been proposed that uninemy may be characteristic of meiotic chromosomes, but that binemy occurs in somatic (mitotic) chromosomes. If so, then the so-called sub-chromatid reunions induced by ionizing radiations of meiotic chromosomes must logically be admitted not to constitute evidence of binemy.

THE REGULATION OF CHROMOSOME REPLICATION

A primary reason for seeking a complete elucidation of the molecular arrangements within the eukaryotic chromosome is the reasonable expectation that such knowledge will surely be helpful, if not essential, for determining how chromosome replication occurs and how it is regulated. The regulation of chromosome replication is, in turn, a central issue in the study of cell reproduction, since it is now clear that, generally, regulation of cell reproduction is equivalent to the regulation of the exit of cells from the G1 phase and their entry into the phase of DNA synthesis in the cell life cycle.

At least three factors add complexity to the regulation of chromosome replication in eukaryotes in comparison to prokaryotes: (1) eukaryotic chromosomes apparently always consist of many replicons while prokaryotic chromosomes apparently always consist of one replicon; (2) eukaryotic cells always contain several to many different chromosomes while prokaryotes normally only contain one chromosome; (3) the regulation of cell reproduction—and therefore of chromosome replication—in the eukaryotic cells of multicellular organisms is controlled in a much more specific manner than in free-living cells (both prokaryotes and eukaryotes).

Autoradiographic Patterns of DNA Synthesis in Eukaryotic Cells

Autoradiography of metaphase chromosomes labeled with a brief pulse of ^{3}H thymidine provided the initial evidence that eukaryotic chromosomes are multi-repliconic (as already discussed above) and also that not all replicons of a single nucleus begin replication simultaneously. Although DNA synthesis begins at many points along most of the chromosomes in a variety of cell types, the autoradiographic patterns of chromosomes labeled at different parts of the S period suggest that there are parts of chromosomes in a complement that do not begin synthesis until later in the S phase. Most of these observations do not prove, however, that some of the replicons are out of synchrony with respect to initiation of replication. It could be supposed, for example, that all replicons begin synthesis at the same time, i.e., at the beginning of the S phase, but that different replicons finish replication at different times within the S phase, depending, for example, on the lengths of the various replicons. The labeling pattern of two chromosomes, however, is sufficiently unique to prove that at least some replicons of DNA begin synthesis at points later in the S phase. Taylor reported in 1960 that in male Chinese hamster cells the long arm of the X chromosome and the whole Y chromosome are duplicated in the last half of the interval of DNA synthesis. In female cells, one X chromosome "was all duplicated in the last half of the period of DNA synthesis." These findings have been variously confirmed in a variety of cell types in many laboratories during the years since Taylor's report (e.g., Bianchi, 1966; Bianchi and de Bianchi, 1966; Chang et al., 1965; Gartler and Burt, 1964; Gilbert et al., 1962; Hsu and Lockhart, 1964; Petersen, 1964; Priest et al., 1967; Sofuni and Sandberg, 1967; Takagi and Sandberg, 1968; Utakoji and Hsu, 1965). It is still not entirely clear, however, whether a delayed initiation of synthesis is characteristic only of heterochromatin or whether euchromatic segments of DNA also are late replicating. According to Taylor's estimate (1968), there may be as many as several hundred replicons in the late replicating (heterochromatic) arm of the X chromosome in Chinese hamster cells.

A very clear example (but still probably a special situation) of individual regulation of duplication of a replicons(s) is present in the amplification of DNA during oogenesis in which several hundred new groups of cistrons for ribosomal RNA are produced.

There are late replicating sections of DNA among the autosomes, but these generally appear to be only heterochromatic regions. In spermatogonia of the Chinese hamster the long arm of the X and Y chromosome are not late in replicating and are less heterochromatic and presumably now genetically active (Utakoji and Hsu, 1965). Also, in the grasshopper *Melanoplus,* the X chromosome is euchromatic in early spermatogonial cell generations and synchronous with the autosomes in its replication; in the final premeiotic interphase, the X chromosome is heterochromatic and now asynchronous in replication (Nicklas and Jaqua, 1965). These results support the view that heterochromatinization of DNA is responsible for a shift in replication to a later part of the S phase. Why the advent of heterochromatinization should cause a temporal alteration in the control of DNA synthesis is beyond our understanding at present. The change in control of DNA synthesis may not be a consequence of condensation of the chromosomes but rather related directly to the genetic inactivation of heterochromatin.

The simplest view for the present is that a majority of replicons within a chromosome complement all begin replication at the beginning of the S period and complete replication at various times up to late in the S period, depending upon the length of particular replicons. Whether all heterochromatic sections of chromosomes begin DNA synthesis at the same time in late S so as to constitute a second, single, major initiation point is not yet known, but there is some synchrony in the replication of various heterochromatic segments of chromosomes.

A large number of initiation sites along a single DNA molecule implies that there are many swivels, at least one for every replication point for the relief of torsion developed by unwinding of the double helix. Presumably, swivels are established, perhaps transiently, by scission of one chain of the double helix.

Taylor (1969) has found that a change occurs in the very long chains of double helices in mid to late G1 of the cycle in hamster cells that allows the isolation of a fairly homogeneous fraction of 100 $\mu\mu$ pieces of native DNA in weakly alkaline solution. The observation has been interpreted to mean that chain scissions may occur at 100 μ intervals in both chains but still separated by a sufficient number of nucleotide pairs to maintain a continuous molecule; possibly the scissions are introduced in the region of connection between two replicons. According to Taylor the 100 μ pieces are absent during mitosis, suggesting that the single chain scissions are closed at that time. At least these observations are consistent with the idea that precisely placed scissions are introduced into DNA in connection with replication, possibly to allow swivel action. Taylor also finds that the newly replicated DNA can be isolated as much smaller pieces of single chain DNA, similar to the pieces of newly replicated DNA obtained in *E. coli* (see Kuempel, Chapter 1). Evidently, DNA is synthesized in small pieces on a linearly intact template and these pieces are subsequently ligased together.

If the replicons were arranged such that the end of one was joined to the beginning of the next, then newly replicated chains would remain free at the initiation end until the adjacent replicon terminated replication and the new chains in each replicon became covalently linked. Free ends apparently exist during replication of prokaryotic chromosomes (Lark et al., 1967). The autoradiographic studies of Huberman and Riggs (1968), however, suggest that replicons are joined origin-to origin and terminus-to-terminus; in such a case, the occurrence of free ends of newly replicated chains would be restricted to cases of asynchronous initiation in adjacent replicons.

Observations of Stubblefield (1966) suggest that the timing of the replication of any particular segment of DNA is not influenced by its position on a particular chromosome but by some intrinsic property of the DNA itself. The translocation of an earlier replicating, euchromatic segment of chromosome One in a Chinese hamster cell to the late replicating Y chromosome did not change the timing of replication of the translocated piece.

Evidence for the sequential orderliness of DNA replication has also been obtained in the slime mold, *Physarum*, in labeling experiments with isolation and purification of DNA. The studies of Cummins and Rusch (1966) indicate that the S period consists of three to five overlapping subphases; most of the late replicating DNA (late subphase) cannot initiate synthesis until the earlier subphases of synthesis are finished. The experiments also contain evidence that the triggering of late DNA synthesis is dependent upon proteins that can only be synthesized in response to completion of early DNA synthesis. Braun and Wili (1968), extending the earlier observations of Braun et al. (1965), have proven by double labeling experiments in *Physarum* that "DNA molecules replicated in a small subfraction (about one-fifth) of the S period of one interphase were again replicated in the corresponding subfraction of the S period in the following interphase."

Preparations for DNA Synthesis

In most cell types, the period of DNA synthesis is separated from the previous cell division by a period of time of variable length called the G1 phase. No events have been identified that uniquely characterize the G1 phase, and our understanding of the basis of its existence is quite limited. The G1 phase is important for two reasons: (1) The temporary arrest of cell reproduction in free-living cells and the temporary or permanent arrest of cell reproduction in multicellular organisms is normally achieved by a coming-to-rest of the cells in the G1 phase of the cell life cycle. As a corollary, the slowing of the rate of cell reproduction is achieved primarily by a lengthening of the G1 phase (although in some instances the lengths of the

DNA synthesis phase and the G2 phase are increased as well). Therefore, the mechanisms by which cell reproduction is regulated are situated within the G1 phase. (2) The terminal part of G1, possibly the last few minutes or less, must contain events concerned with the initiation of DNA synthesis.

The arrest of cells in the G1 phase must be imposed through events that precede the immediate steps that initiate DNA synthesis, since the return of arrested cells to DNA synthesis following a stimulus or removal of inhibition requires many hours. The data of Killander and Zetterberg (1965) suggest in fact that the initiation of DNA synthesis is dependent upon the attainment of a particular cell mass. The implied hypothesis is an oversimplification because some cells in multicellular organisms remain arrested in the G1 phase in spite of a very large mass. Any explanation for the existence of G1 must also take into account the fact that some cells do not have a measurable G1 phase (e.g., Robbins and Scharff, 1967; see Baserga, 1965). During cleavage stages of embryogenesis, for example, the cell cycle has no G1 phase (Gamow, unpublished; Gaulden, 1956; Graham, 1966; Graham and Morgan, 1966), but G1 becomes a characteristic part of cell reproduction during early histogenesis of an embryo.

Some attempts have been made to attribute the synthesis of enzymes needed to support DNA synthesis as the basis for G1. It is true that in certain cell types, thymidine kinase activity rises near the beginning of DNA synthesis (e.g., Brent et al., 1965; Hotta and Stern, 1963; Littlefield, 1966; Sachsenmaier and Ives, 1965; Sachsenmaier et al., 1967; Stubblefield and Mueller, 1965; Stubblefield and Murphree, 1967; see also Adams et al., 1965). Similar rises in thymidylate kinases and deoxycytidine monophosphate deaminase have also been found (Johnson and Schmidt, 1966). A number of observations virtually eliminate a controlling role by such enzymes, by deoxyribonucleoside triphosphate pools, or by DNA polymerase in DNA synthesis. Thymidine kinase is not on the main pathway for the synthesis of thymidine triphosphate, and the enzyme is absent in some cell types capable of DNA synthesis. Thymidine triphosphate pools are carried over from one DNA synthesis period to the next in certain rapidly proliferating cells (Leach, 1968; Stone et al., 1965) and cannot normally be a limiting factor in DNA replication. DNA polymerase activity is high during periods of no DNA synthesis in L-cells (Adams et al., 1965; Gold and Helleiner, 1964; Littlefield et al., 1963) and can hardly be a limiting factor in initiation of synthesis. In regenerating liver, enzymes concerned with DNA synthesis are produced just prior to the beginning of synthesis, and it could be supposed that the appearance of one or more of these enzymes represents the final step in initiation; however, all of these enzymes continue to be present in high amounts in hepatocytes for some time after the DNA synthesis associated with regeneration has completely stopped. Clearly, the appearance of enzymes necessary to support DNA synthesis is coordinated with the initiation of synthesis, but it is unlikely that any of these enzymes is involved in regulation of the initiation of the S phase.

The Initiator Protein Concept

There are some experimental findings that suggest that the initiator protein hypothesis for the regulation of DNA synthesis in prokaryotes may be applicable in some form to the regulation of synthesis in eukaryotes. According to the hypothesis, the initiation of replication depends on the synthesis of a particular RNA message and the corresponding protein; presumably, initiator protein accumulates during cell growth and when present in sufficient amount it interacts with DNA at the replicator region of each replicon to make replication permissible. This is in agreement with Donachie's (1968) analysis of published data, on the relationship between bacterial growth and DNA synthesis, from which he concludes that the initiation of chromosome replication depends upon the attainment of a critical cell mass. Killander and Zetterberg (1965) have previously presented evidence in support of this hypothesis in their studies on the mass of mouse fibroblasts at the time of initiation of DNA synthesis. There are as yet no clues as to the means by which cell mass is sensed by the mechanism that initiates DNA synthesis.

The evidence for such an initiator protein mechanism in prokaryotes is relatively specific, detailed, and convincing (see Kuempel, Chapter 1) but for eukaryotes the evidence is of a more general nature. A requirement for protein synthesis to traverse the G1 phase and initiate DNA synthesis has been demonstrated with the use of inhibitors in several mammalian cell types *in vivo* and *in vitro* (Black et al., 1967; Gottlieb et al., 1964; Kishimoto and Lieberman, 1964; Mueller and Kajiwara, 1965; Powell, 1962; Terasima and Yasukawa, 1966; Terasima et al., 1968). Similar inhibition of DNA synthesis occurs if RNA synthesis is inhibited in G1 (Baserga et al., 1965a, 1966; Fujiwara, 1967; Rothstein et al., 1966). When protein synthesis is stopped after DNA synthesis is already in progress, DNA synthesis is rapidly depressed (Baserga et al., 1965b; Bennett et al., 1964; Lieberman et al., 1963; Littlefield and Jacobs, 1965; Terasima and Yasuka, 1966; Young, 1966). Whether the effect stems from the requirement for newly synthesized proteins to initiate DNA synthesis in segments of chromosomes that are late initiating, or reflects a need for continuous histone synthesis during DNA synthesis, cannot yet be decided. The idea of a requirement of histone synthesis in order to continue DNA synthesis is based primarily on the observation that histone synthesis is always an accompaniment of DNA synthesis in eukaryotic cells.

Synchrony and Asynchrony of Initiation of DNA Synthesis Between Homologous Chromosomes

If the regulation of DNA synthesis in eukaryotes does operate by an initiator protein mechanism, it would be expected that the response of the homologous replicons on homologous chromosomes would be synchronous. Several authors have commented on the pronounced tendency toward synchrony of replication of homologous chromosomes detected by autoradiography (German, 1964; Hsu et al., 1964; Moore and Uren, 1965; Morishima et al., 1962; Pflueger and Yunis, 1966). There are many reports, however, of some asynchrony between homologues (e.g., Atkins et al., 1966; Bianchi and de Bianchi, 1965; German, 1962; Lima-de-Faria et al., 1961, 1967; Stubblefield, 1966). Stubblefield (1966) suggests that the asynchrony between autosomal homologues may be a much less pronounced form of the same type of asynchrony that is present between the two X chromosomes, implying that autosomal asynchrony involves only very small regions of heterochromatin (present in one homologue but not the other). The synchrony of initiation of DNA synthesis between the paired homologues and among the multiple strands of dipteran polytene chromosome is, however, quite precise (Plaut, 1963). In spite of the established asynchrony among homologous autosomes (which may stem from heterochromatinization of one homologue), the similarity of labeling patterns between homologues argues that both chromosomes of a pair are responding in a nonrandom fashion in the initiation of DNA synthesis. On one hand, the observations tend to support a hypothesis of the initiator protein type; on the other hand, such a hypothesis seems contradicted by the differential behavior of euchromatin and heterochromatin.

Nuclear-Cytoplasmic Interaction in DNA Synthesis

A number of experiments and observations have shown that the regulation of DNA synthesis closely involves interactions between the nucleus and the cytoplasm. It is an old observation that nuclei in the same cytoplasm undergo mitosis in synchrony. More recently, synchrony of initiation of DNA synthesis of nuclei sharing a common cytoplasm has been demonstrated (e.g., Church, 1967; Harris and Watkins, 1965; Kimball and Prescott, 1962; Nygaard et al., 1960). There are, however, exceptions to this general situation; for example, the asynchrony in DNA synthesis between the micro and macronuclei in certain protozoa (McDonald, 1962; Prescott et al., 1962). The study of nuclear-cytoplasmic interaction in the regulation of DNA synthesis has been extended by nuclear transplantation between amebas in different stages of the cell life cycle (Prescott and Goldstein, 1967). Transplantation of a nucleus from an ameba in the G2 phase into an ameba still engaged in DNA synthesis results in the reinitiation of DNA synthesis in the transplanted G2 nucleus. The reciprocal type of transfer, an S phase nucleus into a G2 cell, results in a sharp reduction in DNA synthesis in the transplanted nucleus. The experiments can be interpreted to mean that the cytoplasm of an S phase cell either contains factors that actively support DNA synthesis or is deficient in an inhibitor of DNA synthesis; the reciprocal possibilities apply to the cytoplasm of a G2 phase cell.

De Terra (1967) has found a similar situation after cell grafting or nuclear transfer in *Stentor*. The nucleus of a cell in division of G1 initiates DNA synthesis when introduced into an S phase cell. Nuclei transferred to G1 cells cease DNA synthesis. She suggests "that DNA synthesis is regulated by the presence or absence of a cytoplasmic initiator," because fusion of an S phase cell with a G1 phase cell results in initiation of DNA synthesis in the G1 nucleus and does not cause cessation of DNA synthesis in the S phase cell. This interpretation based on an initiator is perhaps preferable in these experiments, but the results could still be explained on the basis of the presence or absence of an inhibitor of DNA synthesis. The importance of the cytoplasm for the active support of DNA synthesis has been decisively demonstrated by the injection of nuclei from liver, brain, and blood cells into unfertilized eggs of *Xenopus laevis* causing the subsequent reinitiation of DNA synthesis (Graham et al., 1966). Gurdon (1967) has shown that the cytoplasmic condition which supports DNA synthesis is developed at the time of rupture of the germinal vesicle and therefore the condition is probably derived directly or indirectly from the nucleus.

The cytoplasmic condition responsible for either permission or stimulation of DNA synthesis is not species, or even class, specific. In the experiments of Graham et al. (1966), mouse liver nuclei injected into *Xenopus* eggs underwent synthesis. Harris (1967), in addition, has fused a chicken erythrocyte with a HeLa cell with the result that the erythrocyte nucleus was reactivated to DNA synthesis. Finally, the importance of the cytoplasm for DNA synthesis has been demonstrated in a cell-free sysem (Thompson and McCarthy, 1968). When mouse liver nuclei are recombined with cytoplasmic preparations from cells active in DNA synthesis (ascites cells or L-cells), the synthesis of DNA is stimulated in the isolated liver nuclei. Again, the effect is not species specific (or order specific) since cytoplasmic preparations from HeLa cells stimulate DNA synthesis in mouse liver nuclei, and chicken erythrocyte nuclei are stimulated by cytoplasmic preparations from mouse TLT or L-cells. The cytoplasmic contribution is heat stable.

Friedman and Mueller (1968) have done similar experiments with HeLa cells but with somewhat different results. Cytoplasm from S phase or logarithmic phase cells appeared to contribute a *heat-labile* factor that was necessary for DNA synthesis in nuclei isolated from S phase cells, but such cytoplasmic preparations apparently did not initiate DNA synthesis in nuclei from cells not in the S phase. Obviously the cytoplasmic factor(s) in these experiments is of a different nature than the one(s) in the experiments of Thompson and McCarthy (1968).

The synchronous initiation of replication among many replicons at the beginning of the S period suggests that all of these replicating units have structurally common replicator regions at which synthesis is controlled. If asynchrony of initiation (late initiation) of synthesis is strictly a property of heterochromatin, then it is possible that all replicons of a nucleus possess the same replicator region. The lack of species specificity in the initiation of synthesis, in turn, suggests that several elements involved in the control of initiation of DNA synthesis have remained stable in evolution. This situation is in sharp contrast to the demonstrated difference in specificity of the initiator system for the F episome and the host chromosome in *E. coli.*

Any unitary hypothesis for the control of DNA synthesis in eukaryotes will have to take into account several variant situations. In some ciliated protozoa, for example, the micronucleus and macronucleus have completely separated S phases within the same interphase. Commonly, the micronucleus replicates during a very short interval beginning in late telophase, and the macronucleus does not begin DNA synthesis until well after the micronucleus has completed synthesis (McDonald, 1962; Prescott, 1960; Prescott et al., 1962). This asynchrony may be related to the fact that the DNA in the micronucleus is highly condensed and resembles heterochromatin and is inactive in RNA synthesis for most (or perhaps all, in some cases) of the cell cycle; it is conceivable that the asynchrony in this case is basically the same as exists for heterochromatin parts of nuclei generally.

The replication of DNA in mitochondria is not synchronous with replication of nuclear DNA. In the ciliate *Tetrahymena* (Cameron, 1966) and in the slime mold *Physarum* (Guttes et al., 1967), mitochondrial DNA synthesis is continuous throughout the cell cycle and the mitochondria are probably out of synchrony with each other. In a yeast, mitochondrial DNA has been shown to be made during a restricted period of the cycle but still out of phase with nuclear synthesis (Smith et al., 1968).

From these results it must be concluded that mitochondrial and nuclear DNA replications are controlled separately, either because of differences in specificity of the control mechanisms or because the mitochondrial membrane does not allow the necessary communication between cytoplasm and mitochondrial matrix. The evidence that mitochondria within a single cell replicate their DNA asynchronously (Cameron, 1966; Guttes et al., 1967) suggests that the mechanism for the regulation of DNA replication is entirely intramitochondrial.

In bacteria the replicator region appears to be attached to the plasma membrane (see Lark et al., 1966; Kuempel, Chapter 1). Recently, the initiation of DNA replication in human amnion cells has been demonstrated to be associated with the nuclear membrane (Comings and Kakefuda, 1968). In bacteria, the membrane attachment is believed to be important for chromosome segregation; although there is no such function for the nuclear membrane in eukaryotes, it may have some role in the initiation of replication in both bacteria and eukaryotic cells. One other well-documented case of initiation of DNA synthesis at the nuclear membrane occurs in the ciliated protozoan, *Euplotes* (Gall, 1959; Kimball and Prescott, 1962).

CONCLUSIONS

Chromosomal proteins are most likely concerned only with transcription, replication, and packaging of DNA and, in a genetic-biochemical sense, are not integral parts of the chromosome. The RNA associated with chromosomes is either an immediate product of transcription or, in the case of isolated mitotic chromosomes, a contaminant.

Several lines of evidence indicate that eukaryotic chromosomes, in general, consist of only one DNA double helix in the lateral dimension (are unineme). Meiotic urodele chromosomes, among the largest chromosomes known, have been shown by relatively direct analysis to be composed of only one DNA double helix. Polynemy may exist in some organisms, but the evidence is too indirect to be more than suggestive.

The linear integrity of the chromosome is maintained by a DNA molecule that extends from one end of the chromosome to the other. The DNA molecule is probably not interrupted by so-called linkers of non-DNA material. Linkers, if they exist, consist of DNA double helix.

Eukaryotic chromosomes are multirepliconic, i.e., consist of many units of replication joined in tandem. In mammalian chromosomes the average replicon is possibly as short as 30 μ in length, indicating that there are perhaps 1,000 replicons in an average chromosome.

A variety of observations suggest that the regulation of DNA replication in eukaryotes may be achieved in a manner similar to that outlined in the initiator protein hypothesis for replication control in prokaryotes. A major problem is encountered in attempts to explain the marked asynchrony of replication between dispersed and condensed DNA in eukaryotes.

Further work on the regulation and the molecular details of replication of prokaryotic chromosomes will certainly continue to have a profound impact on efforts to understand the organization and replication of eukaryotic chromosomes.

REFERENCES

Adams, R. L. P., R. Abrams, and I. Lieberman. 1965. Rise in deoxyribonucleic acid polymerase activity in the absence of deoxyribonucleic acid synthesis in cultured kidney cells. Nature, 206:512–513.

Alfert, M. 1956. Chemical differentiation of nuclear proteins during spermatogenesis in the salmon. J. Biophys. Biochem. Cytol., 2:109–114.

Allfrey, V. G., V. C. Littau, and A. E. Mirsky. 1963. On the role of histones in regulating ribonucleic acid synthesis in the cell nucleus. Proc. Nat. Acad. Sci. U.S.A., 49:414–421.

——— B. G. T. Pogo, V. C. Littau, E. L. Gershey, and A. E. Mirsky. 1968. Histone acetylation in insect chromosomes. Science, 159:314–316.

Ammermann, V. D. 1965. Cytologische und genetische Untersuchungen an dem Ciliaten *Stylonychia mytilus* Ehrenberg. Arch. Protistenk., 108:109–152.

Anderson, N. G. 1956. Cell division, I. The primeval mechanism, the initiation of cell division and chromosomal condensation. Quart. Rev. Biol., 31:169–199.

Atkins, L., J. A. Böök, and B. Santesson. 1966. Chromosome DNA synthesis in the cell of a human triploid/haploid mosaic. Hereditas, 55:55–67.

Bajer, A. 1965. Subchromatid structure of chromosomes in the living state. Chromosoma, 17:291–302.

Baserga, R. 1965. The relationship of the cell cycle to tumor growth and control of cell division: A review. Cancer Res., 25:581–595.

——— R. D. Estensen, and R. O. Petersen. 1965a. Inhibition of DNA synthesis in Ehrlich ascites cells by actinomycin D, II. The presynthetic block in the cell cycle. Proc. Nat. Acad. Sci. U.S.A., 54:1141–1148.

——— R. D. Estensen, and R. O. Petersen. 1966. Delayed inhibition of DNA synthesis in mouse jejunum by low doses of actinomycin D. J. Cell. Physiol., 68:177–184.

——— R. D. Estensen, R. O. Petersen, and J. P. Layde. 1965b. Inhibition of DNA synthesis in Ehrlich ascites cells by actinomycin D. I. Delayed inhibition by low doses. Proc. Nat. Acad. Sci. U.S.A., 54:745–751.

Bender, M. A. 1962. Chromosome breakage *in vitro*. *In* Mammalian Cytogenetics and Related Problems in Radiobiology. Pavan, C., Chagas, C., Frota-Pessoa, O., and Caldas, L. R. eds., New York, The MacMillan Company, pp. 87–107.

——— and D. M. Prescott. 1962. DNA synthesis and mitosis in cultures of human peripheral leukocytes. Expt. Cell Res., 27:221–229.

Bendich, A., and H. S. Rosenkranz. 1963. Some thoughts on the double-stranded model of deoxyribonucleic acid. *In* Progress in Nucleic Acid Research, Davidson, J. N., and Cohn, W. E., eds., New York, Academic Press, Inc., Vol. 1, pp. 219–230.

Benjamin, W., and A. Gellhorn. 1968. Acidic proteins of mammalian nuclei: Isolation and characterization. Proc. Nat. Acad. Sci. U.S.A., 59:262–268.

Bennett, L. L., Jr., D. Smithers, and C. T. Ward. 1964. Inhibition of DNA synthesis in mammalian cells by actidione. Biochim. Biophys. Acta, 87:60–69.

Berendes, H. D. 1966. Differential replication of male and female X chromosomes in *Drosophila hydei*. Chromosoma, 20:32–43.

Berlowitz, L. 1965. Analysis of histone *in situ* in developmentally inactivated chromatin. Proc. Nat. Acad. Sci. U.S.A., 54:476–480.

Berry, R. E., and D. T. Mayer. 1960. The histone-like basic protein of bovine spermatozoa. Exp. Cell Res., 20:116–126.

Bianchi, N. O. 1966. Chromosomes of the rat. II. DNA replication sequence of bone marrow chromosomes *in vivo*. Cytologia, 31:276–293.

——— and M. S. A. deBianchi. 1965. DNA replication sequence of human chromosomes in blood cultures. Chromosoma, 17:273–290.

——— and M. S. A. deBianchi. 1966. Shifting in the duplication time of sex chromosomes in the rat. Chromosoma, 19:286–299.

Bier, K., W. Kunz, and D. Ribbert. 1967. Struktur und Funktion der Oocytenchromosomen and Nukleolen sowie der Extra DNS während der Oogenese panoistischer und meroistischer Insekten. Chromosoma, 23:214–254.

Black, R. E., E. Baptist, and J. Piland. 1967. Puromycin and cycloheximide inhibition of thymidine incorporation into DNA of cleaving sea urchin eggs. Exp. Cell Res., 48: 431–439.

Bloch, D. P. 1962. Synthetic processes in the cell nucleus. I. Histone synthesis in non-replicating chromosomes. J. Histochem. Cytochem., 10:137–144.

——— and S. D. Brack. 1964. Evidence for the cytoplasmic synthesis of nuclear histone during spermiogenesis in the grasshopper *Chortophaga viridifasciata*. J. Cell Biol., 22: 327–340.

——— and G. C. Godman. 1955. A microphotometric study of the synthesis of DNA and nuclear histone. J. Biophys. Biochem. Cytol., 1:17–28.

——— and H. J. Hew. 1960. Schedule of spermatogenesis in the pulmonate snail *Helix aspersa,* with special reference to histone transition. J. Biophys. Biochem. Cytol., 7:515–531.

——— R. A. MacQuigg, S. D. Brack, and J.-R. Wu. 1967. The syntheses of deoxyribonucleic acid and histone in the onion root meristem. J. Cell Biol., 33:451–467.

Bonner, J. 1965. The template activity of chromatin. J. Cell. Physiol., 66 (Suppl. 1):77–90.

——— and R. C. Huang. 1963. Properties of chromosomal nucleohistones. J. Molec. Biol., 6:169–174.

——— and R. C. Huang. 1964. Role of histone in chromosomal RNA synthesis. *In* The Nucleohistones, Bonner, J., and Ts'o, P., eds., San Francisco, Holden-Day, Inc., pp. 251–261.

Boss, J. 1955. Mitosis in cultures of newt tissues. IV. The cell surface in late anaphase and the movements of ribonucleoprotein. Exp. Cell Res., 8:181–187.

Braun, R., C. Mittermayer, and H. P. Rusch. 1965. Sequential temporal replication of DNA in *Physarum polycephalum*. Proc. Nat. Acad. Sci. U.S.A., 53:924–931.

——— and H. Wili. 1968. Time sequence of DNA replication in *Physarum*. Biochim. Biophys. Acta, 174:246–252.

Brent, T. P., J. A. V. Butler, and A. R. Crathorn. 1965. Variations in phosphokinase activities during the cell cycle in synchronous populations of HeLa cells. Nature, 207:176–177.

Brinkley, B. R. 1965. The fine structure of the nucleolus in mitotic divisions of Chinese hamster cells *in vitro*. J. Cell Biol., 27:411–422.

Britten, R. J., and D. E. Kohne. 1968. Repeated sequences in DNA. Science, 161:529–161.

Brown, D. D., and I. B. Dawid. 1968. Specific gene amplification in oocytes. Science, 160:272–280.

Busch, H., W. C. Starbuck, E. J. Singh, and T. S. Ro. 1964. Chromosomal proteins. *In* The Role of Chromosomes in Development, Locke, J., ed., New York, Academic Press, Inc., pp. 51–71.

Byvoet, P. 1966. Metabolic integrity of deoxyribonucleohistones. J. Molec. Biol., 17:311–318.

Cairns, J. 1966. Autoradiography of HeLa cell DNA. J. Molec. Biol., 15:372–373.

Callan, H. G. 1963. The nature of lampbrush chromosomes. Int. Rev. Cytol., 15:1–34.

——— 1967. The organization of genetic units in chromosomes. J. Cell Sci., 2:1–7.

——— and L. Lloyd. 1960. Lampbrush chromosomes of crested newts *Triturus cristatus* Phil. Trans. Roy. Soc., B., 243:135–219.

——— and H. C. Macgregor. 1958. Action of DNase on lampbrush chromosomes. Nature, 181:1479–1480.

Cameron, I. L. 1966. A periodicity of tritiated-thymidine incorporation into cytoplasmic deoxyribonucleic acid during the cell cycle of *Tetrahymena pyriformis*. Nature, 209:630–631.

Cantor, K. P., and J. E. Hearst. 1966. Isolation and partial characterization of metaphase chromosomes of a mouse ascites tumor. Proc. Nat. Acad. Sci. U.S.A., 55:642–649.

Cave, M. D. 1966. Incorporation of tritium-labeled thymidine and lysine into chromosomes of cultured human leukocytes. J. Cell Biol., 29:209–222.

——— 1967. Chromosomal ^{3}H-lysine incorporation and patterns of deoxyribonucleic acid synthesis in human cells. Exp. Cell Res., 45:631–637.

Chai, N.-C., and K. G. Lark. 1967. Segregation of DNA in bacteria: Association of the segregating unit with the cell envelope. J. Bact., 94:415–421.

Chalkley, G. R., and H. R. Maurer. 1965. Turnover of template-bound histone. Proc. Nat. Acad. Sci. U.S.A., 54:498–505.

Chang, T.-H., V. Defendi, and P. S. Moorhead. 1965. DNA replication patterns in cultured female rat fibroblasts. Canad. J. Genet. Cytol., 7:571–582.

Chernick, B. 1968. Late synthesis of protein on the presumptive X chromosome of the human female lymphocyte. Nature, 220:195–196.

——— and L. Davidson. 1968. The incorporation of tritiated arginine by chromosomal proteins of the human lymphocyte. Exp. Cell Res., 50:257–264.

Chiang, K.-S., and N. Sueoka. 1967. Replication of chromosomal and cytoplasmic DNA during mitosis and meiosis in the eukaryote *Chlamydononas reinhardi*. J. Cell. Physiol., 70 (Suppl. 1):89–112.

Chorazy, J., A. Bendich, E. Borenfreund, and D. J. Hutchison. 1963. Studies on the isolation of metaphase chromosomes. J. Cell Biol., 19:59–69.

Chun, E. H. L., and J. W. Littlefield. 1961. The separation of the light and heavy strands of bromouracil-substituted mammalian DNA. J. Molec. Biol., 3:668–673.

Church, K. 1967. Pattern of DNA replication in binucleate cells occuring in mouse embryo cell cultures. Exp. Cell Res., 46:639–641.

Comings, D. E. 1966. ^{3}H-uridine autoradiography of human chromosomes. Cytogenetics, 5:247–260.

——— and T. Kakefuda. 1968. Initiation of deoxyribonucleic acid replication at the nuclear membrane in human cells. J. Molec. Biol., 33:225–229.

Cummins, J. E., and H. P. Rusch. 1966. Limited DNA synthesis in absence of protein synthesis in *Physarum polycephalum*. J. Cell Biol. 31:577–583.

Daly, M., V. G. Allfrey, and A. E. Mirsky. 1953. Uptake of glycine-^{15}N by components of cell nuclei. J. Genet. Physiol., 36:173–179.

Das, C. C., B. P. Kaufmann, and H. Gay. 1964. Histone-protein transition in *Drosophila melanogaster*. I. Changes during spermatogenesis. Exp. Cell Res., 35:507–514.

Das, N. K., and M. Alfert. 1968. Cytochemical studies on the concurrent synthesis of DNA and histone in primary spermatocytes of *Urechis caupo*. Exp. Cell Res., 49:51–58.

Davies, H. G. 1967. Fine structure of heterochromatin in certain cell nuclei. Nature, 214:208–210.

——— 1968. Electron microscope observations on the organization of heterochromatin in certain cells. J. Cell Sci., 3:129–150.

Dawid, I. B., and D. R. Wolstenholme. 1967. Ultracentrifuge and electron microscope studies on the structure of mitochondrial DNA. J. Molec. Biol., 28:233–245.

de Terra, N. 1967. Macronuclear DNA synthesis in *Stentor*: Regulation by a cytoplasmic initiator. Proc. Nat. Acad. Sci. U.S.A., 57:607–614.

Dick, C., and E. W. Johns. 1968. The effect of two acetic acids containing fixatives on the histone content of calf thymus deoxyribonucleoprotein and calf thymus tissue. Exp. Cell Res., 51:626–633.

Djordjevic, B., and W. Szybalski. 1960. Genetics of human cell lines. III. Incorporation of 5-bromo and 5-iododeoxyuridine into the deoxyribonucleic acid of human cells and its effect on radiation sensitivity. J. Exp. Med., 112:509–531.

Donachie, W. D. 1968. Relationship between cell size and time of initiation of DNA replication. Nature, 219:1077–1079.

Dounce, A. L., and C. A. Hilgartner. 1964. A study of DNA nucleoprotein gels and the residual protein of isolated cell nuclei: Relationship to chromosomal structure. Exp. Cell Res., 36:228–241.

DuPraw, E. J. 1965. Macromolecular organization of nuclei and chromosomes: A folded-fibre model based on whole-mount electron microscopy. Nature, 206:338–343.

——— 1966. Evidence for a 'folded-fibre' organization in human chromosomes. Nature, 209:577–581.

——— and P. M. M. Rae. 1966. Polytene chromosome structure in relation to the "folded-fibre" concept. Nature, 212:598–600.

Evans, H. J. 1964. Uptake of ^{3}H thymidine and patterns of DNA replication in nuclei and chromosomes of *Vicia faba*. Exp. Cell Res., 35:381–393.

——— and J. R. K. Savage. 1963. The relation between DNA synthesis and chromosome structure as resolved by X-ray damage. J. Cell Biol., 18:525–540.

Feinendegen, L. E., and V. P. Bond. 1963. Observations on nuclear RNA during mitosis in human cancer cells in culture (HeLa-S_3), studied with tritiated cytidine. Exp. Cell Res., 30:393–404.

Felix, K. 1959. The nucleoprotamines: their formation and their function. *In* Symposium on Molecular Biology, Zirkle, R. E., ed., Chicago, University of Chicago Press, pp. 163–177.

——— H. Fischer, and A. Krekels. 1956. Protamines and nucleoprotamines. Progr. Biophys., 6:1–23.

Filner, P. 1965. Semi-conservative replication of DNA in a higher plant cell. Exp. Cell Res., 39:33–39.

Flamm, W. G., and M. L. Birnstiel. 1964a. Inhibition of DNA replication and its effect on histone synthesis. Exp. Cell Res., 33:616–619.

——— and M. L. Birnstiel. 1964b. Studies on the metabolism of nuclear basic proteins. *In* The Nucleohistones, Bonner, J., and Ts'o, P., eds., San Francisco, Holden-Day, Inc., pp. 230–239.

Franceschini, P., and D. Giacomoni. 1967. Isolation and fractionation of metaphase chromosomes from HeLa cells. Atti Ass. Genet. Ital., 12:248–258.

Freedman, M. L., P. J. Stambrook, and R. A. Flickinger. 1967. The absence of labeled RNA on metaphase chromosomes of *Taricha* and *Rana* embryos. Exp. Cell Res., 47:640–643.

Frenster, J. H. 1965. Nuclear polyanions as de-repressors of synthesis of ribonucleic acid. Nature, 206:680–683.

——— V. G. Allfrey, and A. E. Mirsky. 1963. Repressed and active chromatin isolated from interphase lymphocytes. Proc. Nat. Acad. Sci. U.S.A., 50:1026–1032.

Friedman, D. L., and G. C. Mueller. 1968. A nuclear system for DNA replication from synchronized HeLa cells. Biochim. Biophys. Acta, 161:455–468.

Fujiwara, Y. 1967. Role of RNA synthesis in DNA replication of synchronized populations of cultured mammalian cells. J. Cell. Physiol., 70:291–300.

Gabrusewycz-Garcia, N. 1964. Cytological and autoradiographic studies in *Sciara coprophila* salivary gland chromosomes. Chromosoma, 15:312–344.

Gall, J. G. 1959. Macronuclear duplication in the ciliated protozoan *Euplotes*. J. Biophys. Biochem. Cytol., 5:295–308.

——— 1963. Kinetics of deoxyribonuclease action on chromosomes. Nature, 198:36–38.

——— 1966. Chromosome fibers studied by a spreading technique. Chromosoma, 20: 221–233.

——— 1968. Differential synthesis of the genes for ribosomal RNA during amphibian oogenesis. Proc. Nat. Acad. Sci. U.S.A., 60:553–560.

——— and H. G. Callan. 1962. ^{3}H uridine incorporation in lampbrush chromosomes. Proc. Nat. Acad. Sci. U.S.A., 48:562–570.

Gartler, S. M., and B. Burt. 1964. Replication patterns of bovine sex chromosomes in cell culture. Cytogenetics, 3:135–142.

Gaulden, M. E. 1956. DNA synthesis and x-ray effects at different mitotic stages in grasshopper neuroblasts. Genetics, 41:645.

German, J. L. 1962. DNA synthesis in human chromosomes. Trans. N. Y. Acad. Sci., 24:395–407.

——— 1964. The pattern of DNA synthesis in the chromosomes of human blood cells. J. Cell Biol., 20:37–55.

Gershon, D., P. Hausen, L. Sachs, and E. Winocour. 1965. On the mechanism of polyoma virus-induced synthesis of cellular DNA. Proc. Nat. Acad. Sci., 54:1584–1592.

Gilbert, C. W., S. Muldal, L. G. Lajtha, and J. Rowley. 1962. Time-sequence of human chromosome duplication. Nature, 195:869–873.

Gimenez-Martin, G., and J. F. Lopez-Saez. 1965. Chromosome structure in the course of mitosis. Cytologia, 30:14–22.

Gold, M., and C. W. Helleiner. 1964. Deoxyribonucleic acid polymerase in L cells. I. Properties of the enzyme and its activity in synchronized cell cultures. Biochim. Biophys. Acta, 80:193–203.

Goldstein, L., and D. M. Prescott. 1967. Proteins in nucleocytoplasmic interactions, I. The fundamental characteristics of the rapidly migrating proteins and the slow turnover proteins of the *Amoeba proteus* nucleus. J. Cell Biol., 33:637–644.

——— and D. M. Prescott. 1968. Proteins in nucleocytoplasmic interactions. II. Turnover and changes in nuclear protein distribution with time and growth. J. Cell Biol., 36:53–61.

Gorovsky, M. A., and J. Woodard. 1967. Histone content of chromosomal loci active and inactive in RNA synthesis. J. Cell Biol., 33:723–728.

Gottlieb, L. I., N. Fausto, and J. L. Van Lancker. 1964. Molecular mechanism of liver regeneration: The effect of puromycin on deoxyribonucleic acid synthesis. J. Biol. Chem., 239:555–559.

Graham, C. F. 1966. The regulation of DNA synthesis and mitosis in multinucleate frog eggs. J. Cell Sci., 1:363–374.

——— K. Arms, and J. B. Gurdon. 1966. The induction of DNA synthesis by frog egg cytoplasm. Develop. Biol., 14:349–381.

——— and R. W. Morgan. 1966. Changes in the cell cycle during early amphibian development. Develop. Biol., 14:439–460.

Gurdon, J. B. 1967. On the origin and persistence of a cytoplasmic state inducing nuclear DNA synthesis in frogs' eggs. Proc. Nat. Acad. Sci. U.S.A., 58:545–552.

Guttes, E. W., P. C. Hanawalt, and S. Guttes. 1967. Mitochondrial DNA synthesis and the mitotic cycle in *Physarum polycephalum*. Biochim. Biophys. Acta. 142:181–194.

Halkka, O. 1965. A photometric study of the *Luzula* problem. Hereditas, 52:81–88.

Harris, H. 1967. The reactivation of the red cell nucleus. J. Cell Sci., 2:23–32.

——— and J. F. Watkins. 1965. Hybrid cells derived from mouse and man: Artificial heterokaryons of mammalian cells from different species. Nature, 205:640–646.

Heneen, W. K., and W. W. Nichols. 1966. Persistence of nucleoli in short term and long term cell cultures and in direct bone marrow preparations in mammalian materials. J. Cell Biol., 31:543–561.

Hilgartner, C. A. 1968. The binding of DNA to residual protein in mammalian nuclei. Exp. Cell Res., 49:520–532.

Holbrook, D. J., Jr., J. H. Evans, and J. L. Irvin. 1962. Incorporation of labeled precursors into proteins and nucleic acids of nuclei of regenerating liver. Exp. Cell Res., 28:120–125.

Holoubek, V., and T. T. Crocker. 1968. DNA-associated acidic proteins. Biochim. Biophys. Acta, 157:352–361.

——— L. Fanshier, T. T. Crocker, and L. S. Hnilica. 1966. Simultaneous increase in labeling of nuclear RNA and of DNA associated acidic nuclear protein. Life Sci., 5:1691–1697.

Horn, E. C., and C. L. Ward. 1957. The localization of basic proteins in the nuclei of larval *Drosophila* salivary glands. Proc. Nat. Acad. Sci. U.S.A., 43:776–779.

Hotta, Y., and Bassel, A. 1965. Molecular size and circularity of DNA in cells of mammals and higher plants. Proc. Nat. Acad. Sci. U.S.A., 53:356–362.

——— and H. Stern. 1963. Molecular facets of mitotic regulation. I. Synthesis of thymidine kinase. Proc. Nat. Acad. Sci. U.S.A., 49:648–654.

Howard, E. F., and W. Plaut. 1967. Ordered chromosomal DNA synthesis in *Drosophila melanogaster*. J. Cell Biol., 35:59a.

——— and W. Plaut. 1968. Chromosomal DNA synthesis in *Drosophila melanogaster*. J. Cell Biol., 39:415–429.

Hsu, T. C. 1964. Mammalian chromosomes *in vitro*. XVIII. DNA replication sequence in the Chinese hamster. J. Cell Biol., 23:53–62.

——— F. E. Arrighi, R. R. Klevecz, and B. R. Brinkley. 1965. The nucleoli in mitotic divisions of mammalian cells *in vitro*. J. Cell Biol., 26:539–553.

——— and L. H. Lockhart. 1964. The beginning and the terminal stages of DNA synthesis of human cells with an XXXXY constitution. Hereditas, 52:320–324.

——— W. Schmid, and E. Stubblefield. 1964. DNA replication sequences in higher animals. *In* The Role of Chromosomes in Development, Locke, M., ed., New York, Academic Press, Inc., pp. 83–112.

Huberman, J. A., and G. Attardi. 1966. Isolation of metaphase chromosomes from HeLa cells. J. Cell Biol., 31:95–105.

——— and A. D. Riggs. 1966. Autoradiography of chromosomal DNA fibers from Chinese hamster cells. Proc. Nat. Acad. Sci. U.S.A., 55:599–606.

——— and A. D. Riggs. 1968. On the mechanism of DNA replication in mammalian chromosomes. J. Molec. Biol., 32:327–341.

Hughes-Schrader, S., and F. Schrader. 1957. The Nezara complex (Dentatomidae-Heteroptera) and its taxonomic and cytological status. J. Morph., 101:1–23.

Izawa, M., V. G. Allfrey, and A. E. Mirsky. 1963. Composition of the nucleus and chromosome in the lampbrush stage of the newt oocyte. Proc. Nat. Acad. Sci. U.S.A., 50:811–817.

Jacobson, W., and M. Webb. 1952. The two types of nucleoproteins during mitosis. Exp. Cell Res., 3:163–183.

Johnson, R. A., and R. R. Schmidt. 1966. Enzymic control of nucleic acid synthesis during synchronous growth of *Chlorella pyrenoidosa*. I. Deoxythymidine monophosphate kinase. Biochim, Biophys. Acta, 129:140–144.

Kaufmann, B. P., H. Gay, and M. R. McDonald. 1960. Organizational patterns within chromosomes. Int. Rev. Cytol., 9:77–127.

——— M. McDonald, and H. Gay. 1948. Enzymatic degradation of ribonucleo-proteins of chromosomes, nucleoli, and cytoplasm. Nature, 162:814.

Keyl, H.-G. 1965a. Demonstrable local and geometric increase in the chromosomal DNA of *Chironomus*. Experientia, 21:191–193.

——— 1965b. Duplikationen von Untereinheiten der Chromosomalen DNA wahrend der Evolution von *Chironomus thummi*. Chromosoma, 17:139–180.

——— and C. Pelling. 1963. Differentielle DNA Replikation in den Speicheldrusen Chromosomen von *Chironomus thummi*. Chromosoma, 14:347–359.

Kihlman, B. A., and B. Hartley. 1967. "Sub-chromatid" exchanges and the "folded fibre" model of chromosome structure. Hereditas, 57:289–294.

Killander, D., and A. Zetterberg. 1965. A quantitative cytochemical investigation of the relationship between cell mass and initiation of DNA synthesis in mouse fibroblasts *in vitro*. Exp. Cell Res., 40:12–20.

Kimball, R. F., and D. M. Prescott. 1962. DNA synthesis and distribution during growth and amitosis of the macronucleus of *Euplotes*. J. Protozool., 9:88–92.

King, D. W., and M. L. Barnhisel. 1967. Synthesis of RNA in mammalian cells during mitosis and interphase. J. Cell Biol., 33:265–272.

Kirschner, R. H., D. R. Wolstenholme, and N. J. Gross. 1968. Replicating molecules of a circular mitochondrial DNA. Proc. Nat. Acad. Sci. U.S.A., 60:1466–1472.

Kishimoto, S., and I. Lieberman. 1964. Synthesis of RNA and protein required for the mitosis of mammalian cells. Exp. Cell Res., 36:92–101.

Kroeger, H., J. Jacob, and J. L. Sirlin. 1963. The movement of nuclear protein from the cytoplasm to the nucleus of salivary cells. Exp. Cell Res., 31:416–423.

Kuehl, L. R. 1962. The state of deoxyribonucleic acid in the cell nucleus. Biochim. Biophys. Acta, 55:289–299.

Kunz, W. 1967a. Funktionsstrukturen im Oocytenkern von *Locusta migratoria*. Chromosoma, 20:332–370.

——— 1967b. Lampenbürstenchromosomen und multiple Nukleolen bei Orthopteren. Chromosoma, 21:446–462.

LaCour, L. F. 1963. Ribose nucleic acid and the metaphase chromosome. Exp. Cell Res., 29:112–118.

——— and A. Rutishauser. 1954. X-ray breakage experiments with endosperm. I. Subchromatid breakage. Chromosoma, 6:696–709.

Langan, T. A. 1968. Histone phosphorylation: Stimulation by adenosine 3′, 5′-monophosphate. Science, 162:579–580.

Lark, K. G. 1966. Regulation of chromosome replication and segregation in bacteria. Bact. Rev., 30:3–32.

——— 1967a. Nonrandom segregation of sister chromatids in *Vicia faba* and *Triticum bocoticum*. Proc. Nat. Acad. Sci. U.S.A., 58:352–359.

——— 1967b. The replication and segregation of DNA. Amer. Natur., 101:313–315.

——— R. A. Consigli, and H. C. Minocha. 1966. Segregation of sister chromatids in mammalian cells. Science, 154:1202–1205.

——— H. Eberle, R. A. Consigli, H. C. Minocha, N. Chai, and C. Lark. 1967. Chromosome segregation and the regulation of DNA replication. *In* Organization Biosynthesis, Vogel, H. J., Lampen, J. O., and Bryson, V., eds., New York, Academic Press, Inc., pp. 63–89.

Leach, W. M. 1968. The thymidine pool in grasshopper neuroblasts during mitosis. J. Cell Biol., 36:282–286.

Lee, Y. C., and O. H. Scherbaum. 1966. Nucleohistone composition in stationary and division synchronized *Tetrahymena* cultures. Biochemistry, 5:2067–2075.

Lezzi, M. 1965. Die Wirkung von DNase auf isolierte Polytän-Chromosomen. Exp. Cell Res., 39:289–292.

Lieberman, I., R. Abrams, N. Hunt, and P. Ove. 1963. Levels of enzyme activity and deoxyribonucleic acid synthesis in mammalian cells cultured from the animal. J. Biol. Chem., 238:3955–3962.

Lima-de-Faria, A. 1961. Initiation of DNA synthesis at specific segments in the meiotic chromosome of *Melanoplus*. Hereditas, 47:674–694.

——— N. O. Bianchi, and P. C. Nowell. 1967. Patterns of chromosome replication in a patient with chronic granulocytic leukemia. Hereditas, 58:31–62.

——— and M. J. Moses. 1966. Ultrastructure and cytochemistry of metabolic DNA in *Tipula*. J. Cell Biol., 30:177–192.

——— J. Reitalu, and S. Bergman. 1961. The pattern of DNA synthesis in the chromosomes of man. Hereditas, 47:695–704.

Littau, V. C., V. G. Allfrey, J. H. Frenster, and A. E. Mirsky. 1964. Active and inactive regions of nuclear chromatin as revealed by electron microscope autoradiography. Proc. Nat. Acad. Sci. U.S.A., 52:93–100.

——— C. J. Burdick, V. G. Allfrey, and A. E. Mirsky. 1965. The role of histones in the maintenance of chromatin structure. Proc. Nat. Acad. Sci. U.S.A., 54:1204–1212.

Littlefield, J. W. 1966. The periodic synthesis of thymidine kinase in mouse fibroblasts. Biochim. Biophys. Acta, 114:398–403.

——— and P. S. Jacobs. 1965. The relation between DNA and protein synthesis in mouse fibroblasts. Biochim. Biophys. Acta, 108:652–658.

——— A. P. McGovern, and K. B. Margeson. 1963. Changes in the distribution of polymerase activity during DNA synthesis in mouse fibroblasts. Proc. Nat. Acad. Sci. U.S.A., 49:102–107.

Love, R., and R. G. Suskind. 1961. Further observations on the ribonucleoproteins of mitotically dividing mammalian cells. Exp. Cell Res., 22:193–207.

Macgregor, H. C., and H. G. Callan. 1962. The actions of enzymes on lampbrush chromosomes. Quart. J. Micr. Sci., 103:173–203.

MacInnes, J. W., and R. B. Uretz. 1966. Organization of DNA in dipteran polytene chromosomes as indicated by polarized fluorescence microscopy. Science, 151:689–691.

——— and R. B. Uretz. 1967. Thermal depolarization of fluorescence from polytene chromosomes stained with acridine orange. J. Cell Biol., 33:597–604.

——— and R. B. Uretz. 1968. DNA organization in the mature sperm of several *Orthoptera* by the method of polarized fluorescence microscopy. J. Cell Biol., 38: 426–436.

Maguire, M. P. 1968. Nomarski interference contrast resolution of subchromatid structure. Proc. Nat. Acad. Sci. U.S.A., 60:533–536.

Maio, J. J., and C. L. Schildkraut. 1967. Isolated mammalian metaphase chromosomes. I. General characteristics of nucleic acids and proteins. J. Molec. Biol., 24:29–30.

Marin, G., and D. M. Prescott. 1964. The frequency of sister chromatid exchanges following exposure to varying doses of ^{3}H thymidine or X-rays. J. Cell Biol., 21:159–167.

Martin, P. G., and R. Shanks. 1966. Does *Vicia faba* have multistranded chromosomes? Nature, 211:650–651.

McDonald, B. B. 1962. Synthesis of deoxyribonucleic acid by micro- and macro-nuclei of *Tetrahymena pyriformis*. J. Cell Biol., 13:193–203.

McLeish, J. 1959. Comparative microphotometric studies of DNA and arginine in plant nuclei. Chromosoma, 10:686–710.

Mendelsohn, J., D. E. Moore, and N. P. Salzman. 1968. Separation of isolated Chinese hamster metaphase chromosomes into three size-groups. J. Molec. Biol., 32:101–112.

Miller, O. L., Jr. 1964. Extrachromosomal nucleolar DNA in amphibian oocytes. J. Cell Biol., 23:60a.

——— 1965. Fine structure of lampbrush chromosomes. Nat. Cancer Inst. Monogr., 18:79–99.

——— 1966. Structure and composition of peripheral nucleoli of salamander oocytes. Nat. Cancer Inst. Monogr., 23:53–66.

Monesi, V. 1964. Autoradiographic evidence of a nuclear histone synthesis during mouse spermiogenesis in the absence of detectable quantities of nuclear ribonucleic acid. Exp. Cell Res., 36:683-688.

Moore, R., and J. Uren. 1965. The pattern of DNA synthesis in short-term leukocyte cultures of *Protemnodon bicolor* males. Exp. Cell Res., 38:341–352.

Moorhead, P. S., and V. Defendi. 1962. Asynchrony of DNA synthesis in chromosomes of human diploid cells. J. Cell Biol., 16:202–209.

Morishima, A., M. M. Grumbach, and J. H. Taylor. 1962. Asynchronous duplication of human chromosomes and the origin of sex chromatin. Proc. Nat. Acad. Sci. U.S.A., 48:756–763.

Mueller, G. C., and K. Kajiwara. 1965. Regulatory steps in the replication of mammalian cell nuclei. *In* Developmental and Metabolic Control Mechanisms and Neoplasia. Symposium on Fundamental Cancer Research, Baltimore, The Williams and Wilkins Company, pp. 452–474.

Murray, K. 1965. The basic proteins of cell nuclei. Ann. Rev. Biochem., 34:209–246.

Nash, D., and J. Bell. 1968. Larval age and the pattern of DNA synthesis in polytene chromosomes. Canad. J. Genet. Cytol., 10:82–90.

Nicklas, R. B., and R. A. Jaqua. 1965. X chromosome DNA replication: Developmental shift from synchrony to asynchrony. Science, 147:1041–1043.

Niehaus, W. G., Jr., and C. P. Barnum. 1965. Incorporation of radioisotope *in vivo,* into ribonucleic acid and histone of a fraction of nuclei preparing for mitosis. Exp. Cell Res., 39:435–442.

Nur, U. 1967. Reversal of heterochromatinization and the activity of the paternal chromosome set in the male mealy bug. Genetics, 56:375–389.

Nygaard, O. F., S. Guttes, and H. P. Rusch. 1960. Nucleic acid metabolism in a slime mold with synchronous mitosis. Biochem. Biophys. Acta, 38:298–306.

Ohno, S., U. Wolf, and N. B. Atkin. 1968. Evolution from fish to mammals by gene duplication. Hereditas, 59:169–187.

Okada, S. 1968. Replicating units (replicons) of DNA in cultured mammalian cells. J. Biophys., 8:650–664.

Ostergren, G., and T. Wakonig. 1954. True or apparent sub-chromatid breakage and the induction of labile states in cytological chromosome loci. Botaniska Notiser, 4:357–375.

Painter, R. B., D. A. Jermany, and R. E. Rasmussen. 1966. A method to determine the number of DNA replicating units in cultured mammalian cells. J. Molec. Biol., 17: 47–55.

Pardon, J. F., M. H. F. Wilkins, and B. M. Richards. 1967. Superhelical model for nucleohistone. Nature, 215:508–509.

Peacock, W. J. 1961. Sub-chromatid structure and chromosome duplication in *Vicia faba.* Nature, 191:832–833.

——— 1963. Chromosome duplication and structure as determined by autoradiography. Proc. Nat. Acad. Sci. U.S.A., 49:793–801.

——— 1965. Chromosome replication. Nat. Cancer Inst. Monogr., 18:101–131.

Perez-Silva, J., and P. Alonso. 1966. Demonstration of polytene chromosomes in the macronuclear anlage of oxytrichous ciliates. Arch. Protist., 109:65–70.

Perkowska, E., H. C. Macgregor, and M. L. Birnstiel. 1968. Gene amplification in the oocyte nucleus of mutant and wild-type *Xenopus laevis.* Nature, 217:649–650.

Petersen, A. J. 1964. DNA synthesis and chromosomal asynchrony. J. Cell Biol., 23: 564–658.

Pflueger, O. H., Jr., and J. J. Yunis. 1966. Late replication patterns of chromosomal DNA in somatic tissues of the Chinese hamster. Exp. Cell Res., 44:413–420.

Plaut, W. 1963. On the replicative organization of DNA in the polytene chromosome of *Drosophila melanogaster.* J. Molec. Biol., 7:632–635.

——— D. Nash, and T. Fanning. 1966. Ordered replication of DNA in polytene chromosomes of *Drosophila melanogaster.* J. Molec. Biol., 16:85–93.

Powell, W. F. 1962. The effects of ultraviolet irradiation and inhibitors of protein synthesis on the initiation of DNA synthesis in mammalian cells in culture. Biochim. Biophys. Acta, 55:969–978.

Prensky, W., and H. H. Smith. 1964. Incorporation of ^{3}H arginine in chromosomes of *Vicia faba.* Exp. Cell Res., 34:525–532.

Prescott, D. M. 1960. Relation between cell growth and cell division, IV. The synthesis of DNA, RNA, and protein from division to division in *Tetrahymena.* Exp. Cell Res., 19:229–238.

——— 1966. The synthesis of total macronuclear protein, histone, and DNA during the cell cycle in *Euplotes eurystomus.* J. Cell Biol., 31:1–9.

——— and M. A. Bender. 1963a. Synthesis and behavior of nuclear proteins during the cell life cycle. J. Cell. Physiol., 62 (Suppl.): 175–194.

——— and M. A. Bender. 1963b. Autoradiographic study of chromatid distribution of labeled DNA in two types of mammalian cells *in vitro*. Exp. Cell Res., 29:430–442.

——— and L. Goldstein. 1967. Nuclear–cytoplasmic interaction in DNA synthesis. Science, 155:469–470.

——— and L. Goldstein. 1968. Proteins in nucleocytoplasmic interactions, III. Redistributions of nuclear proteins during and following mitosis in *Amoeba proteus*. J. Cell Biol., 39:404–414.

——— and R. Kimball. 1961. Relation between RNA, DNA, and protein syntheses in the replicating nucleus of *Euplotes*. Proc. Nat. Acad. Sci. U.S.A., 47:686–693.

——— R. Kimball, and R. F. Carrier. 1962. Comparison between the timing of micronuclear and macronuclear DNA synthesis in *Euplotes eurystomus*. J. Cell Biol., 13:175–176.

——— and G. E. Stone. 1965. Synthesis and turnover of nuclear proteins. *In* The Chromosome: Structural and Functional Aspects, In Vitro I, Tissue Culture Association, pp. 72–77.

Priest, J. H., J. E. Heady, and R. E. Priest. 1967. Delayed onset of replication of human X chromosomes. J. Cell Biol., 35:483–487.

Rae, P. M. M. 1966. Whole mount electron microscopy of *Drosophila* salivary chromosomes. Nature, 212:139–142.

Rasch, E., and J. W. Woodard. 1959. Basic proteins in plant nuclei during normal and pathological cell growth. J. Biophys. Biochem. Cytol., 6:263–281.

Reich, E., and D. J. L. Luck. 1966. Replication and inheritance of mitochondrial DNA. Proc. Nat. Acad. Sci. U.S.A., 55:1600–1608.

Rhoades, M., L. A. MacHattie, and C. A. Thomas, Jr. 1968. The P22 bacteriophage DNA molecule. I. The mature form. J. Molec. Biol., 37:21–40.

Ris, H. 1967. Ultrastructure of the animal chromosome. Biochim. Biophys. Acta, 10:11–21.

——— and B. L. Chandler. 1963. The ultrastructure of genetic systems in prokaryotes and eukaryotes. Cold Spring Harbor Symp., Vol. 28, 1–8.

Robbins, E., and T. W. Borun. 1967. The cytoplasmic synthesis of histones in HeLa cells and its temporal relationship to DNA replication. Proc. Nat. Acad. Sci. U.S.A., 57:409–416.

——— and M. D. Scharff. 1967. The absence of a detectable G_1 phase in a cultured strain of Chinese hamster lung cell. J. Cell Biol., 34:684–686.

Rodman, T. C. 1968. Relationship of developmental stage to initiation of replication in polytene nuclei. Chromosoma, 23:271–287.

Rothfels, K., E. Sexsmith, M. Heimburger, and M. O. Krause. 1966. Chromosome size and DNA content of species of *Anemone 1.* and related genera (*Ranunculaceae*). Chromosoma, 20:54–74.

Rothstein, H., J. Fortin, and D. Sonneborn. 1966. Inhibition of DNA synthesis and cell division by actinomycin D. Experientia, 22:294–295.

Rudkin, G. T., and P. S. Woods. 1959. Incorporation of ^{3}H cytidine and ^{3}H thymidine into giant chromosomes of *Drosophila* during puff formation. Proc. Nat. Acad. Sci. U.S.A., 45:997–1003.

Sachsenmaier, W., D. v. Fournier, and K. F. Gurtler. 1967. Periodic thymidine kinase production in synchronous plasmodia of *Physarum polycephalum*: Inhibition by actinomycin and actidione. Biochem. Biophys. Res. Commun., 27:655–660.

——— and D. H. Ives. 1965. Periodische Anderungen der Thymidin-kinase-Aktivität im Synchronen Mitosecylus von *Physarum polycephalum*. Biochem. Z., 343:399–406.

Salzman, N. P., and J. Mendelsohn. 1968. Isolation and fractionation of metaphase chromosomes. *In* Methods in Cell Physiology, Prescott, D. M., ed., New York, Academic Press, Inc., Vol. 3, pp. 277–292.

——— D. E. Moore, and J. Mendelsohn. 1966. Isolation and characterization of human metaphase chromosomes. Proc. Nat. Acad. Sci. U.S.A., 56:1449–1456.

Sasaki, M. S., and A. Norman. 1966. DNA fibres from human lymphocyte nuclei. Exp. Cell Res., 44:642–645.

Schwarzacher, H. G., and W. Schnedl. 1965. Endoreduplication in human fibroblast cultures. Cytogenetics, 4:1–18.

Shapiro, I. M., and L. Y. Levina. 1967. Autoradiographical study on the time of nuclear protein synthesis in human leucocyte blood culture. Exp. Cell Res., 47:75–85.

Simon, E. H. 1961. Transfer of DNA from parent to progeny in a tissue culture line of human carcinoma of the cervix. J. Molec. Biol., 3:101–109.

Smith, D., P. Tauro, E. Schweizer, and H. O. Halvorson. 1968. The replication of mitrochondrial DNA during the cell cycle in *Saccharomyces lactis*. Proc. Nat. Acad. Sci. U.S.A., 60:936–942.

Sofuni, T., and A. A. Sandberg. 1967. Chronology and pattern of human chromosome replication. VI. Further studies including autoradiographic behavior of normal and abnormal No. 1 autosomes. Cytogenetics, 6:357–370.

Solari, A. J. 1965. Structure of the chromatin in sea urchin sperm. Proc. Nat. Acad. Sci. U.S.A., 53:503–511.

Somers, C. E., A. Cole, and T. C. Hsu. 1963. Isolation of chromosomes. Exp. Cell Res., 9 (Suppl.): 220-235.

Spalding, J., K. Kajiwara, and Mueller, G. C. 1966. The metabolism of basic proteins in HeLa cell nuclei. Proc. Nat. Acad. Sci. U.S.A., 56:1535–1542.

Speer, H. L., and E. F. Zimmerman. 1968. The transfer of proteins from cytoplasm to nucleus in HeLa cells. Biochem. Biophys. Res. Commun., 32:60–65.

Steffenson, D. M. 1963. Conservation of the structural protein in *Lilium* chromosomes at fertilization and subsequent replication. Abstract, Second Annual Meeting of the American Society for Cell Biology.

Stevens, B. J., and H. Swift. 1966. RNA transport from nucleus to cytoplasm in *Chironomus* salivary glands. J. Cell Biol., 31:55–77.

Stone, G. E., O. L. Miller, and D. M. Prescott. 1965. ^{3}H thymidine derivative pools in relation to macronuclear DNA synthesis in *Tetrahymena pyriformis*. J. Cell Biol., 25:171–177.

Stubblefield, E. 1966. Mammalian chromosomes *in vitro* XIX. Chromosomes of Don-C, a Chinese hamster fibroblast strain with a part of autosome 1b translocated to the Y chromosome. J. Nat. Cancer Inst., 37:799–817.

——— and G. C. Mueller. 1962. Molecular events in the reproduction of animal cells. II. The focalized synthesis of DNA in the chromosomes of HeLa cells. Cancer Res., 22:1091–1099.

——— and G. C. Mueller. 1965. Thymidine kinase activity in synchronized HeLa cell cultures. Biochem. Biophys. Res. Commun., 20:535–538.

——— and S. Murphree. 1967. Synchronized mammalian cell cultures II. Thymidine kinase activity in colcemid synchronized fibroblasts. Exp. Cell Res., 48:652–656.

Swanson, C. P., and W. J. Young. 1965. Chromosome reproduction in mitotis and meiosis. *In* Reproduction: Molecular, Subcellular, and Cellular, Locke, M., ed., New York, Academic Press, Inc., pp. 107–127.

Swift, H. 1962. Nucleic acids and cell morphology in dipteran salivary glands. *In* The Molecular Control of Cellular Activity, Allen, J. M., ed., New York, McGraw-Hill Book Company, pp. 73–125.

——— 1964. The histones of polytene chromosomes. *In* The Nucleohistones, Bonner, J., and Ts'o, P., eds., San Fransisco, Holden-Day, Inc., pp. 169–181.

——— 1965. Molecular morphology of the chromosome. *In* The Chromosome: Structural and Functional Aspects, In Vitro I, Dawe, C. J., ed., Tissue Culture Association.

Takagi, N., and A. A. Sandberg. 1968. Chronology and pattern of human chromosome replication, VIII. Behavior of the X and Y in early S-phase. Cytogenetics, 7:135–143.

Takai, S., T. W. Borun, J. Muchmore, and I. Lieberman. 1968. Concurrent synthesis of histone and deoxyribonucleic acid in liver after partial hepatectomy. Nature, 219:-860–861.

Taylor, J. H. 1958a. Sister chromatid exchanges in tritium-labeled chromosomes. Genetics, 43:515–529.

——— 1958b. The mode of chromosome duplication in *Crepis capillaris*. Exp. Cell Res., 15:350–357.

——— 1959. The organization and duplication of genetic material. *In* Proc. 10th Intern. Congr. Genet., Toronto, University Toronto Press, Vol. 4, pp. 63-78.

——— 1960. Asynchronous duplication of chromosomes in cultured cells of Chinese hamster. J. Biophys. Biochem. Cytol., 7:455–464.

——— 1963a. The replication and organization of DNA in chromosomes. *In* Molecular Genetics, Taylor, J. H., ed., New York, Academic Press, Inc., Part I, pp. 65–111.

——— 1963b. DNA synthesis in relation to chromosome reproduction and the reunion of breaks. J. Cell. Physiol., 62 (Suppl. 1): 73–86.

——— 1968. Rates of chain growth and units of replication in DNA of mammalian chromosomes. J. Molec. Biol., 31:579–594.

——— 1969. Replication and organization of chromosomes. Proc. XII. Int. Congr. Genet., Tokyo, in press.

——— P. D. Woods, and W. L. Hughes. 1957. The organization and duplication of chromosomes as revealed by autoradiographic studies using tritium-labeled thymidine. Proc. Nat. Acad. Sci. U.S.A., 43:122–138.

Terasima, T., Y. Fujiwara, S. Tanaka, and M. Yasukawa. 1968. Synchronous culture of L cells and initiation of DNA synthesis. Cancer Cells in Culture, Proc. Int. Conf. Tissue Cult. in Cancer Res.

——— and M. Yasukawa. 1966. Synthesis of G_1 protein preceding DNA synthesis in cultured mammalian cells. Exp. Cell Res. 44:669–672.

Thomas, C. A., Jr. 1967. The rule of the ring. J. Cell. Physiol. 70 (Suppl. 1): 13–34.

Thompson, L. R., and B. J. McCarthy. 1968. Stimulation of nuclear DNA and RNA synthesis by cytoplasmic extracts *in vitro*. Biochem. Biophys. Res. Commun., 30: 166–172.

Trosko, J. E., and S. Wolff. 1965. Strandedness of *Vicia faba* chromosomes as revealed by enzyme digestion studies. J. Cell Biol., 26:125–135.

Ullerich, F.-H. 1966. Karyotyp und DNA-Gehalt von *Bufo, B. viridis, B. bufo* x *B. viridis* und *B. calamita* (Amphibia, Anura). Chromosoma, 18:316–342.

——— 1967. Weitere Untersuchungen über Chromosomenverhaltnisse und DNS-Gehalt bei Anuren (Amphibia). Chromosoma, 21:345–368.

Umana, R., S. Updike, J. Randall, and A. L. Dounce. 1964. Histone metabolism. *In* The Nucleohistones, Bonner, J., and Ts'o, P., eds., San Francisco, Holden-Day, Inc., pp. 200–229.

Utakoji, T., and T. C. Hsu. 1965. DNA replication patterns in somatic and germ-line cells of the male Chinese hamster. Cytogenetics, 4:295–315.

Vendrely, R. 1957. Données recentes sur la chimie de l'ADN et des desoxyribonucleo-proteines. Arch. Julius Klaus-Stiftung Verebungforschung Sozialanthropologie Rassen-hygiene, 32:538–553.

Vinograd, J., and J. Lebowitz. 1966. Physical and topological properties of circular DNA. J. Gen. Physiol., 49:103–125.

Walen, K. H. 1965. Spatial relationships in the replication of chromosomal DNA. Genetics, 52:915–929.

Wang, T. Y. 1967. Nonhistone chromatin proteins from calf thymus and their role in DNA biosynthesis. Arch. Biochem. and Biophys., 122:629–633.

Wilson, G. B., and A. H. Sparrow. 1960. Configurations resulting from isochromatid and iso-subchromatid unions after meiotic and mitotic prophase irradiation. Chromosoma, 11:229–244.

——— A. H. Sparrow, and V. Pond. 1959. Sub-chromatid rearrangement in trillium erectum, I. Origin and nature of configurations induced by ionizing radiation. Amer. J. Bot., 46:309–316.

Wimber, D. E. 1961. Asynchronous replication of deoxyribonucleic acid in root tip chromosomes of *Tradescantia paludosa*. Exp. Cell Res., 23:402–407.

Wolfe, S. L. 1965. The fine structure of isolated metaphase chromosomes. Exp. Cell Res., 37:45–53.

——— 1968. The effect of prefixation on the diameter of chromosome fibers isolated by the langmuir trough-critical point method. J. Cell Biol., 37:610–620.

——— and G. M. Hewitt. 1966. The strandedness of meiotic chromosomes from *Oncopeltus*. J. Cell Biol., 31:31–42.

——— and P. G. Martin. 1968. The ultrastructure and strandedness of chromosomes from two species of *Vicia*. Exp. Cell Res., 50:140–150.

Wolff, S. 1961. The doubleness of the chromosome before DNA synthesis as revealed by combined X-ray and tritiated thymidine treatments. Radiation Res., 14:517.

——— and H. E. Luippold. 1964. Chromosome splitting as revealed by combined X-ray and labeling experiments. Exp. Cell Res., 34:548–556.

Woodard, J., E. Rasch, and H. Swift. 1961. Nucleic acid and protein metabolism during the mitotic cycle in *Vicia faba*. J. Biophys. Biochem. Cytol., 9:445–462.

Woods, P. S., and M. U. Schairer. 1959. Distribution of newly synthesized deoxyribonucleic acid in dividing chromosomes. Nature, 183:303–305.

Yarbro, J. W. 1967. Relationship of histone synthesis to DNA synthesis in mouse ascites tumor $6C_3HED$. Biochim. Biophys. Acta, 145:531–535.

Young, C. W. 1966. Inhibitory effects of acetoxycycloheximide, puromycin, and pactamycin upon synthesis of protein and DNA in asynchronous populations of HeLa cells. Molec. Pharmocol., 2:50–55.

Zakharov, A. F., and N. A. Egolina. 1968. Asynchrony of DNA replication and mitotic spiralization along heterochromatic portions of Chinese hamster chromosomes. Chromosoma, 23:365–385.

Zetterberg, A. 1966. Protein migration between cytoplasm and cell nucleus during interphase in mouse fibroblasts *in vitro*. Exp. Cell Res., 43:526–545.

Zubay, G. 1964. Nucleohistone structure and function. *In* The Nucleohistones, Bonner, J., and Ts'o, P., eds., San Francisco, Holden-Day, Inc., pp. 95–107.

B. R. Brinkley and Elton Stubblefield

Department of Biology, Section of Cell Biology, The University of Texas M. D. Anderson Hospital and Tumor Institute at Houston, Houston, Texas

3

Ultrastructure and Interaction of the Kinetochore and Centriole in Mitosis and Meiosis

INTRODUCTION

The process of cell division in animals depends largely upon two components: the kinetochore (centromere), which is located on the chromosome, and the centriole, which is located at each pole of the spindle. Cytologists have long been aware that the kinetochore is fundamentally important in chromosome movement and that the centriole, among other functions, polarizes the direction of chromosome movement. Although both are structurally independent components separated by a distance of several microns, a dynamic interaction exists between the two which culminates in the formation and regulation of a division apparatus and, ultimately, in the equidistribution of genetic material into daughter cells.

The roles of these two components in mitosis have been established by various methods in a wide variety of organisms. The involvement of the kinetochore in chromosome movement was discovered at an early stage in cytological investigations. Schrader (1953), in his excellent review on mitosis, discusses most of the important early investigations. Lima-de-Faria (1958), in a more recent review, presents the important studies on the kinetochore before 1958 and the era of electron microscopy and molecular biology.

The centriole has been the subject of many cytological investigations and the literature on this organelle is voluminous. A comprehensive survey on the subject of centriole function and structure is beyond the scope of the present paper; for more detailed information, the reader is directed to the following review papers (Mazia, 1961; Went, 1966; Stubblefield and Brinkley, 1967). The present paper will be more concerned with recent studies on the structure of the kinetochore and the centriole, with particular emphasis on their interaction during the formation of the division apparatus in mitosis and meiosis.

On the basis of early electron microscope studies it was generally believed that, while the centriole had a distinct structure and organization, the kinetochore was a featureless region on the chromosome without definable ultrastructure. The use of improved methods of ultrastructural preservation, synchronization, and cell selection, developed in this and other laboratories has now produced clear indications that the kinetochore has a precise ultrastructure which can be studied throughout mitosis (Brinkley and Stubblefield, 1966; Jokelainen, 1967). An effort will be made in this paper to review studies of kinetochore ultrastructure and to relate them to current knowledge of kinetochore function obtained by other cytological techniques. A review of similar studies on the centriole in the light of current concepts of mitosis and meiosis will also be made. A concluding section will emphasize the interaction between kinetochores and centrioles and will speculate on their role in the formation of spindle microtubules.

THE KINETOCHORE IN MITOSIS

In discussing the role and structural differentiation of the kinetochore in mitosis, it is convenient to begin with the initial events of the division process. As Mazia (1961) has indicated, however, it is difficult to discuss the beginning of mitosis because the events which trigger the process are almost totally unknown. All available evidence suggests that mitosis per se is merely the physical culmination of a series of biosynthetic events which occurred in the preceding interphase. The appearance of a division apparatus which functions to regulate distribution of chromosomes to daughter cells is brought about largely by the assembly of protein subunits into spindle tubules. The kinetochore, however, plays an important role in directing the assembly of spindle subunits. We shall now consider its structure and behavior from the time, in prophase, when it first can be detected until it disappears, in telophase.

Prophase

The gross structural events of prophase have been described in many studies with both light and electron microscopes. Generally the nuclear aspects of prophase are characterized by the transition of chromatin from heterogeneous masses of condensed and diffused fibers to irregular condensed masses recognizable as individual chromosomes. Figure 1 shows an electron micrograph of a typical early prophase nucleus in a rat kangaroo (strain PtK_1) tissue culture cell. The nucleolus undergoes structural transitions at this time but remains active in RNA synthesis throughout early prophase (Hsu et al., 1965). In the kangaroo cells the nucleolus usually disintegrates bv metaphase but may persist throughout mitosis in other mammalian cells in tissue culture (Hsu et al., Brinkley, 1965; Haneen and Nichols, 1966).

The chromosomes appear as a loose array of 200 to 250 Å filaments which are packaged in irregular bundles less than 0.5 μ wide. As the prophase chromosomes become more condensed a small portion usually remains associated with the nuclear envelope. The prophase attachment site in mammals is usually confined to two or three stalks at the extreme end of the chromosome (Fig. 2). In plants several such attachment sites have been reported. In a careful study of thin serial sections, Sparvoli et al. (1964) have demonstrated that in *Tradescantia* the chromonemata of prophase chromosomes remain attached to the nuclear envelope at five sites.

Figure 2 shows the earliest stage at which we have been able to detect the kinetochore in mammalian cells. It first appears as a "patch" of condensed filaments located in a small constriction on each side of the chromosome. Each patch, presumably a kinetochore, is a spherical structure approximately 0.4 μ in diameter composed of densely packed filaments 50 to 80 Å in diameter.

Although we have not obtained sufficient serial sections to be entirely certain, preliminary micrographs suggest that the sister kinetochores are structurally separate from each other at this time. In addition to being separate, they appear to develop synchronously through prophase, prometaphase, and metaphase, and remain structurally symmetrical throughout mitosis. Our observations therefore have not supported the concept of asynchronous differentiation of sister kinetochores put forth by Jokelainen (1965). There is, however, evidence of functional asynchrony of sister kinetochores during prometaphase movement, which will be discussed in another section.

As prophase continues there is a transition in kinetochore structure from dense spherical masses to less dense bands or filaments which extend for a distance of 0.1 to 0.5 μ along each side of the chromosome (Figs. 3, 4, and 5). At this time the kinetochore takes on an appearance much like that reported by Harris (1965) in sea urchin eggs, Brinkley and Stubblefield (1966) in Chinese hamster cells, Robbins and Gonatas (1964) in HeLa cells, Krishan and Buck (1965) in mouse L cells, Barnicot and Huxley (1965) in cultured newt cells, Chang and Gibbley (1968) in rat hepatoma cells, Jokelainen (1967) in embryonic rat cells, and Murray et al. (1965) in rat lymphocytes. In longitudinal profile it contains two distinct components, a dense axial element 200 to 300 Å in diameter and a less-dense area 200 to 600 Å on each side of the axial elements. In some cases the less-dense area may extend out as much as 1,000 Å. A more detailed description will be given in the section dealing with the kinetochore at metaphase.

One consistency in electron microscope studies is the failure to detect any microtubular connections to the kinetochore before nuclear envelope dissolution in most organisms. From several earlier observations of prophase chromosome movement prior to nuclear envelope breakdown (cf. Mazia, 1961, p. 202-203), one

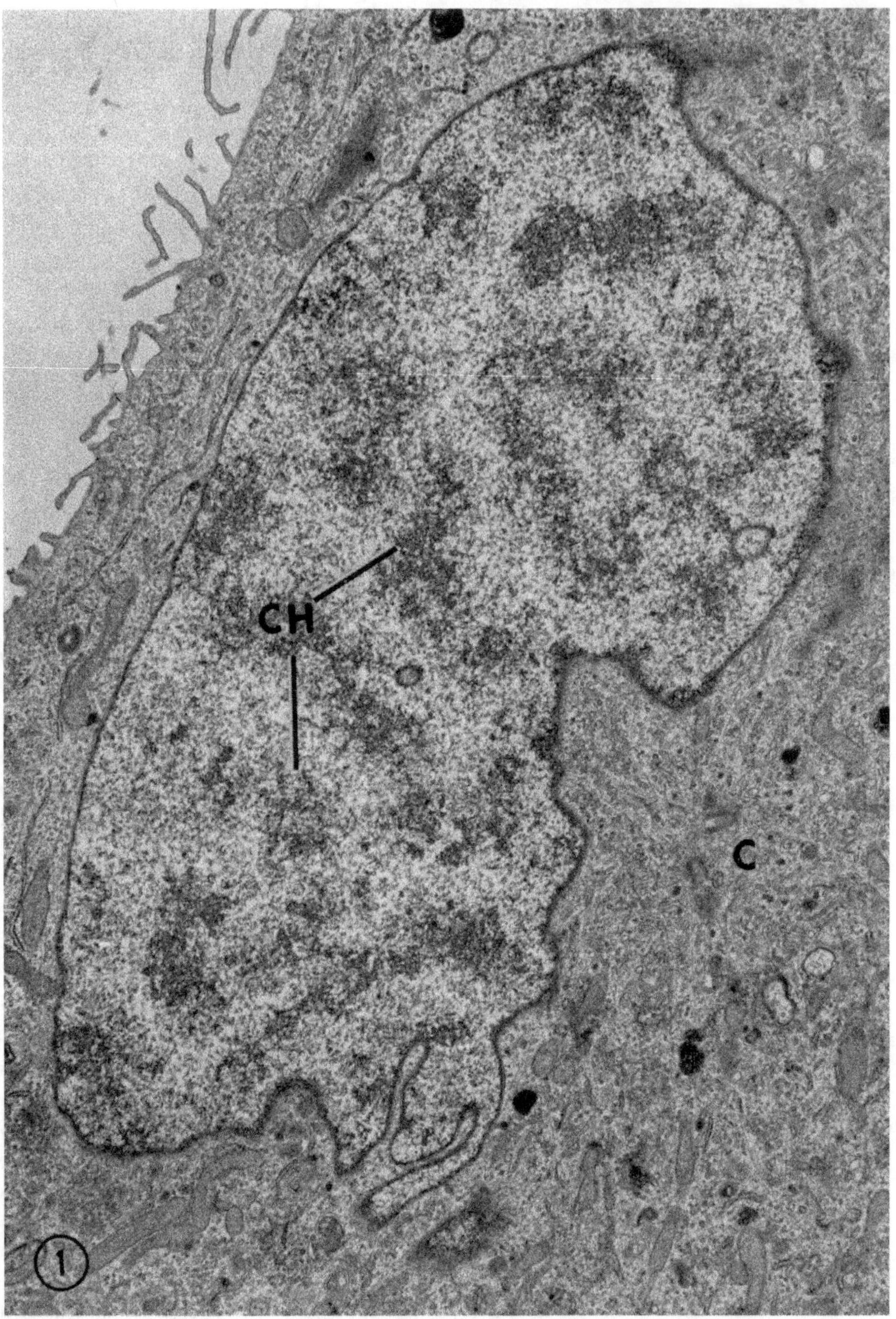

FIG. 1. Prophase nucleus of a rat kangaroo fibroblast with condensing chromosomes (CH). Note complete complement of centrioles (C) in the cytoplasm. × 9,000.

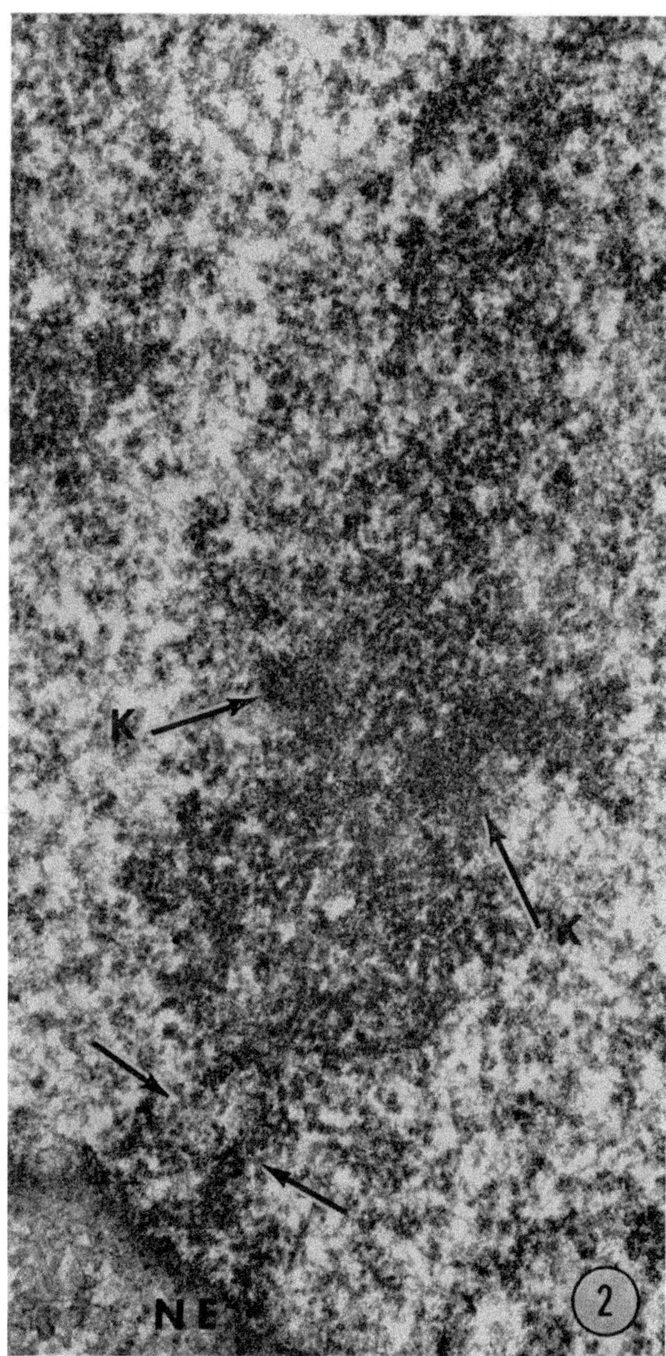

Fig. 2. Higher magnification of early prophase chromosome showing very early appearance of sister kinetochores (K). Unmarked arrows indicate "stalks" or points of attachment to the nuclear envelope (NE). × 60,000.

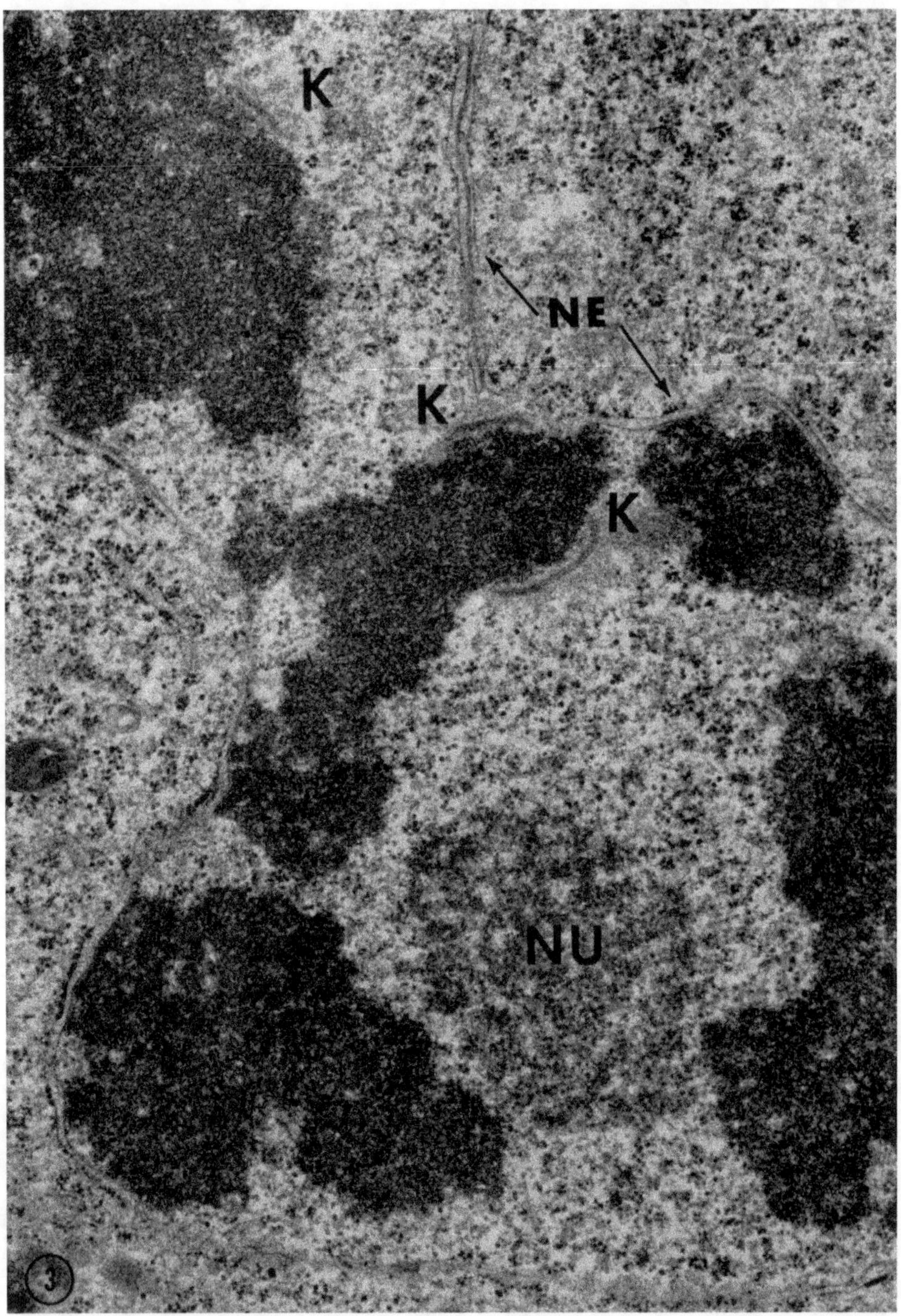

FIG. 3. Late prophase of Chinese hamster fibroblast showing fully formed kinetochores (K). The nuclear envelope (NE) is almost completely intact. The granular mass within the nucleus is the nucleolus (NU). × 40,000.

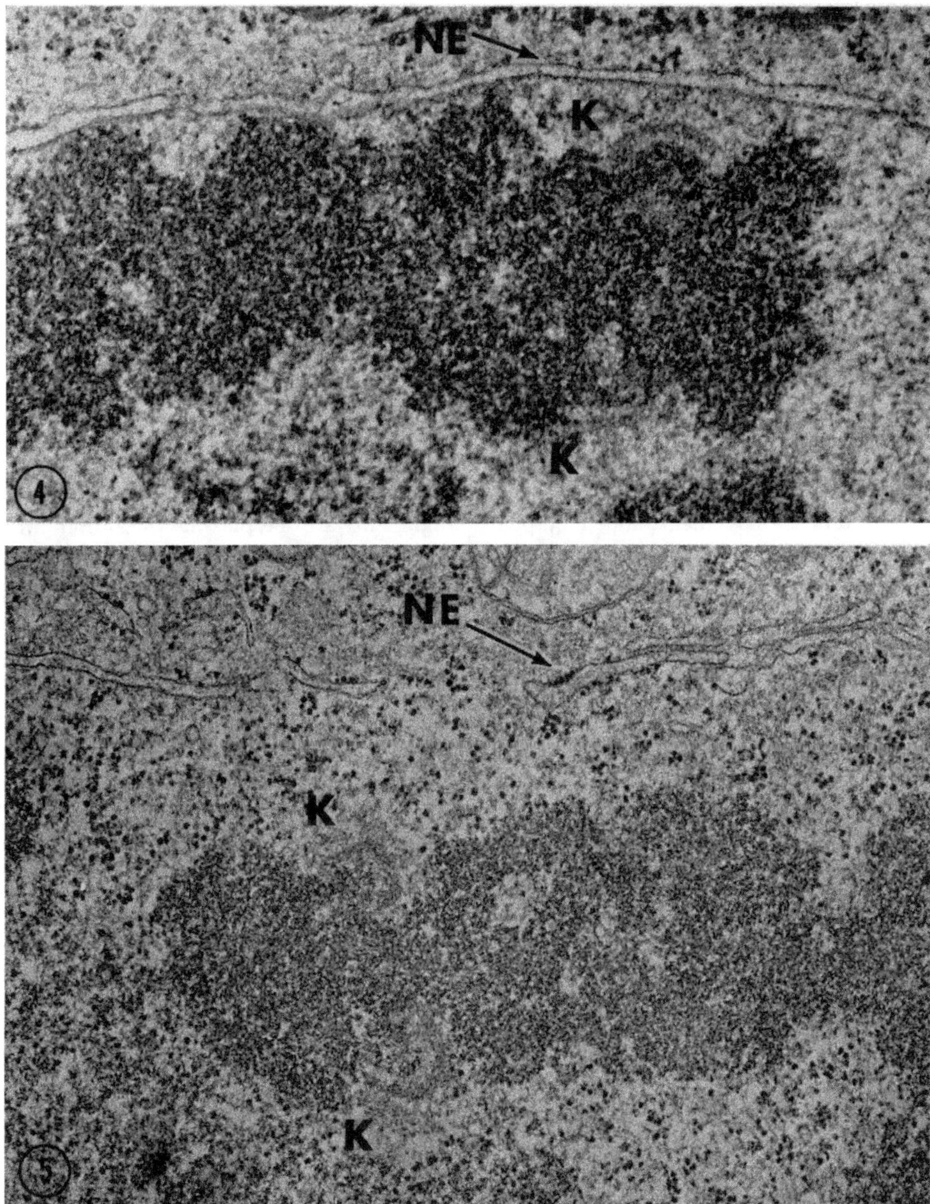

FIG. 4. Detailed view of sister kinetochores (K) on prophase chromosomes. The nuclear envelope (NE) is still intact. × 60,600.

FIG. 5. Late prophase chromosomes with both sister kinetochores (K). Both appear to be symmetrical at this stage. The nuclear envelope (NE) is breaking down. Chinese hamster fibroblast. × 33,000.

might have expected some microtubular association with the kinetochore at prophase. In each case, however, observations were made with the light microscope. It would be interesting to study meiotic prophase in such organisms as *Loxa* (Schrader, 1947) or *Anisolabis* (Schrader, 1953) with the electron microscope to see if prophase chromosome movement reported in these organisms is brought about by the usual microtubular attachments to the kinetochores. One would suspect that the nuclear envelope in such cases might not be completely intact. It is obvious in mammals that microtubules become associated with kinetochores only after the nuclear envelope begins to break down. Similar observations have been reported by Harris (1965) in studies of sea urchin mitosis.

Prometaphase

The disruption of the nuclear envelope signals the onset of prometaphase. During this period the chromosomes undergo directed movement (metakinesis) and eventually become aligned onto a metaphase plate. The detailed problems of prometaphase movement and metaphase alignment are perplexing ones and will be discussed in a future volume in this series. Here we will be concerned with the structure and microtubular association of kinetochores prior to metaphase alignment.

It is clearly evident that the kinetochore is functional in metakinesis. Microtubules can be seen in association with the kinetochore before the nuclear envelope has completely disassociated (Fig. 6b). In some sections of kangaroo cells the chromosomes appeared to undergo a "false anaphase movement" at the beginning of prometaphase. That is, one sister kinetochore of each chromosome was polarized to a particular region of the cell (Figs. 6a and 6b). For this to occur, one sister would have to be more active for a brief period than the other. In other words, functional asynchrony between sister and kinetochores should occur at prometaphase. Thus far we have not been able to detect the differences in fine structure or microtubular association of sister kinetochores in prometaphase, which might be expected from such behavior.

Luykx (1965b) reported that kinetochores of *Urechis* eggs may go through a "trial and error" period during meiotic prometaphase in which microtubular connections form between kinetochores and poles at random. Such connections are often broken and formed anew up until metaphase. Such behavior might account for the "oscillatory" movement of chromosomes during prometaphase which is seen in tipulid spermatocytes (Dietz, 1956), in lobster spermatocytes (Schrader, 1947), in *Haemanthus* endosperm (Bajer, 1958), and in the "prometaphase stretch" of mantids (Hughes-Schrader, 1947; Staiger, 1950).

Metaphase

Once the chromosomes have reached the metaphase plate, the sister kinetochores are symmetrically arranged on opposite sides of the chromosomes. In mammals, a bundle of microtubules, which is made up of some 10 to 20 individual microtubules, is attached to each sister kinetochore. The number of microtubules

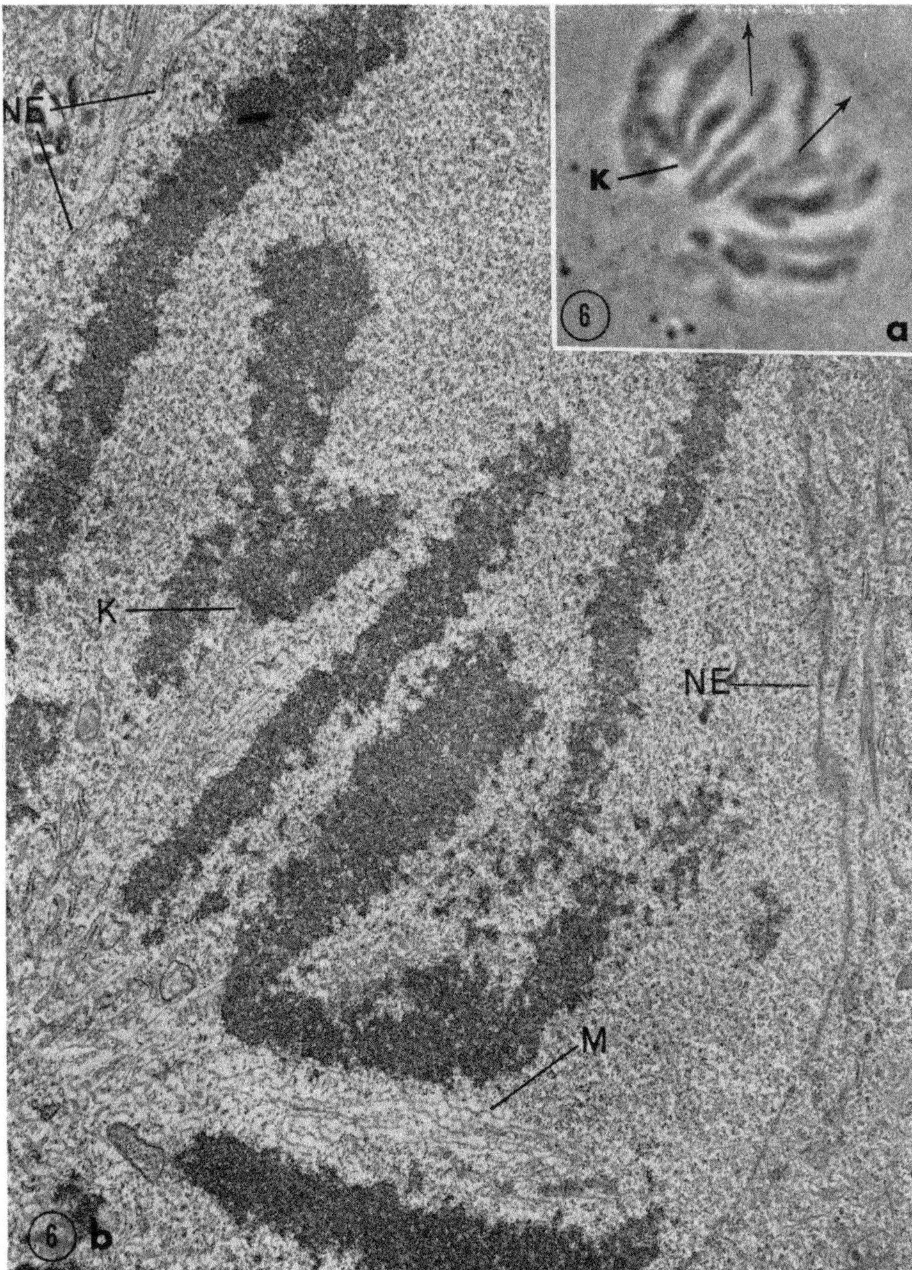

FIG. 6. Comparative light and electron micrographs of the same prometaphase fibroblast of the rat kangaroo. The chromosomes appear to be polarized to one area of the cell. Microtubules (M) can be seen although nuclear envelope (NE) is still partially intact. The kinetochores (K) are indicated in both micrographs. Arrows in 6a show faint outline of the nuclear envelope. Figure 6a reduced 5 percent from × 1,800; 6b reduced 5 percent from × 10,500.

associated with a kinetochore may be a function of its size. Harris and Bajer (1965) report some 50 to 100 microtubules attached to a single kinetochore in *Haemanthus* metaphase chromosomes. Figures 24a and 24b are light and electron micrographs of *Haemanthus* chromosomes; 20 to 30 microtubules can be seen at the kinetochore in a single section. In mammalian tissue culture cells a "sheet" of microtubules is associated with each kinetochore; the microtubules frequently lie in a flat plane, one above the other (Fig. 7). Such an arrangement may result

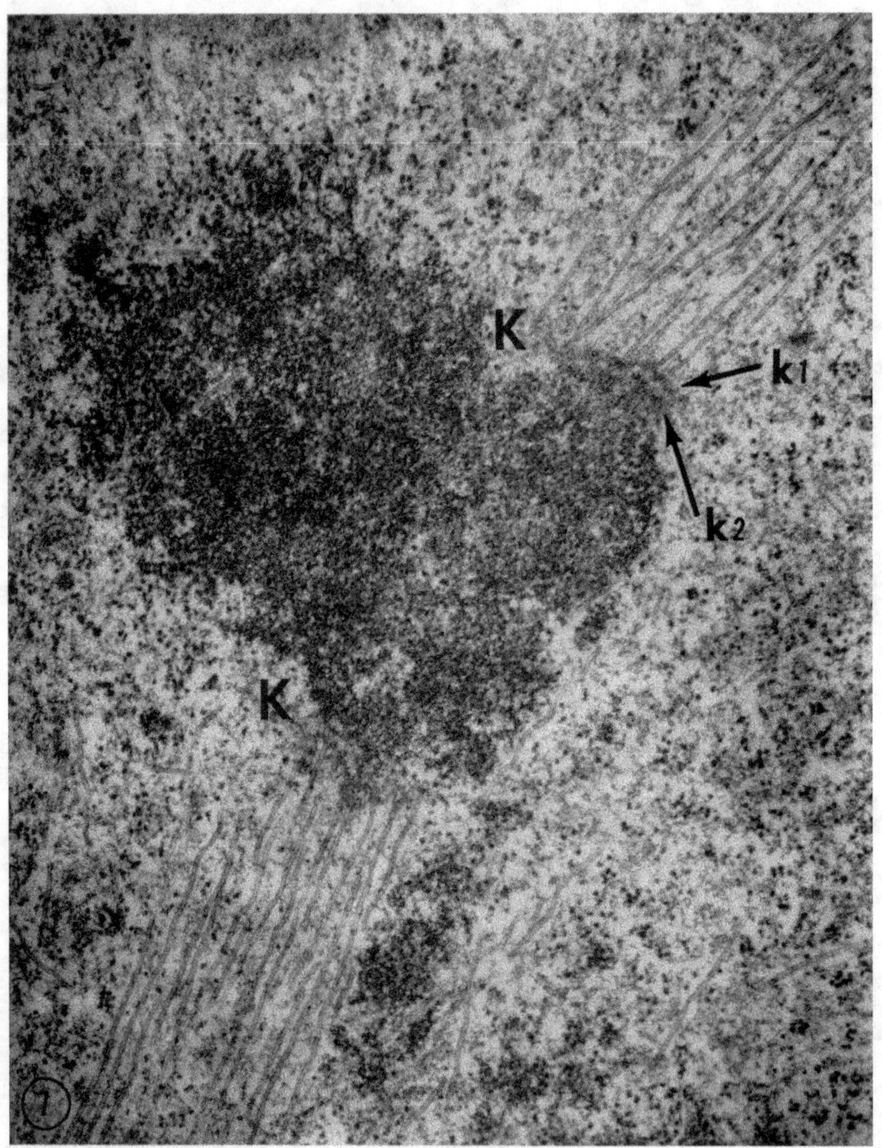

FIG. 7. Small subtelocentric chromosome of the rat kangaroo fibroblast at metaphase. Note "sheet" of microtubules attached to both sister kinetochores (K). Portions of both kinetochore filaments (K_1 and K_2) are seen on one daughter kinetochore. × 40,000.

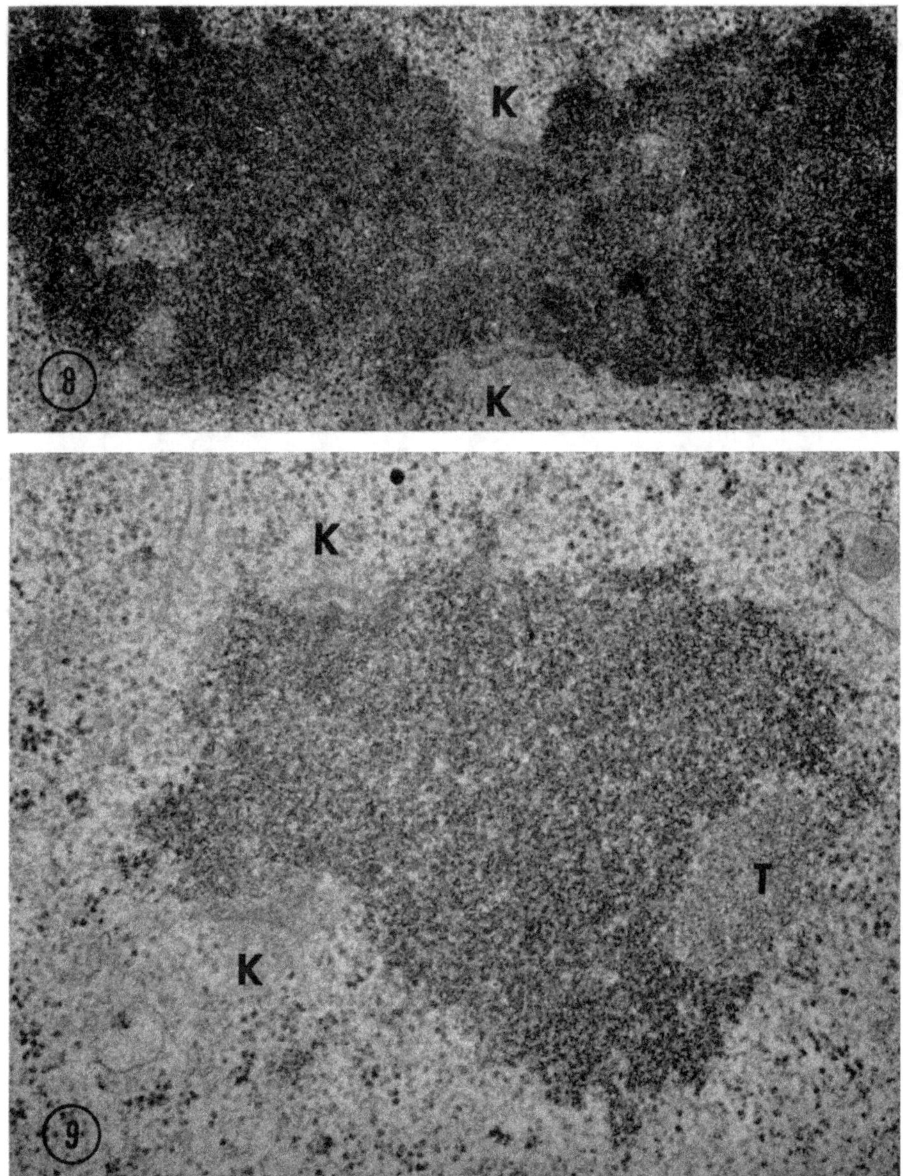

FIG. 8. Metacentric chromosome showing kinetochore filaments of both kinetochores (K). × 43,000.

FIG. 9. Small submetacentric (or subtelocentric) chromosomes with kinetochores (K) near one end of chromosome. The light-staining fibrillar mass on the opposite end (T) is the terminal nucleolus organizer. × 50,000.

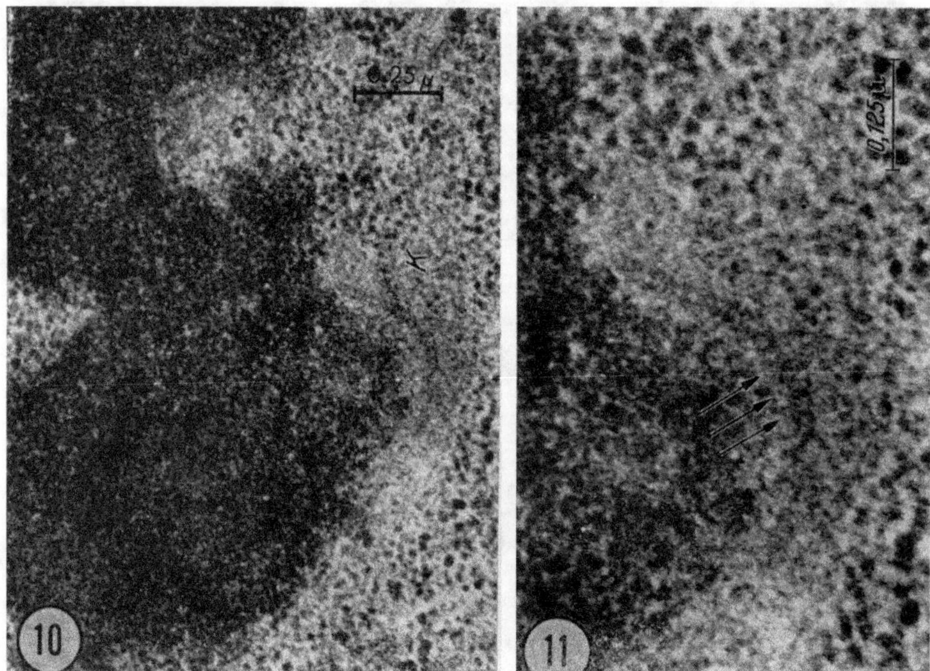

FIG. 10. Kinetochore (K) of Chinese hamster cell showing the double nature of the axial element. × 58,000.

FIG. 11. Higher magnification of the axial element shown in Figure 10. Points indicated by the arrows suggest a cohelical arrangement of the two elements. Chinese hamster fibroblast. × 124,000.

from the flattened conditions of tissue culture cells. Harris and Bajer (1965) found that in flattened *Haemanthus* cells microtubules were arranged parallel to each other. In unflattened cells, however, the chromosomal filaments were confined to cable-like bundles. Jokelainen (1967) has made an interesting observation regarding the relationship of microtubules to metaphase kinetochore in fetal rat cells. He described three types of relationships: (1) two sister kinetochores parallel to one another and located on opposite sides of a kinetochore constriction with microtubules bilaterally arched toward each pole, (2) preferential orientation of constriction to one pole with microtubules extending in a straight line to that pole, and (3) both sister kinetochores connected to opposite poles by straight microtubules. It is clear from cinemicrographic studies that chromosomes move erratically and oscillate as they approach the metaphase plate; they appear much like nervous thoroughbreds being led to the starting gate. Such behavior might account for Jokelainen's observations regarding the various orientations of the metaphase kinetochore.

We have been able to detect little change in the ultrastructure of the kinetochore from late prophase to metaphase. Therefore, the description to follow is that of a fully developed kinetochore. From our observations of normal and Colcemid-inhibited mitosis, we conclude that kinetochores become structurally mature in late prophase or prometaphase regardless of whether microtubules are associated with

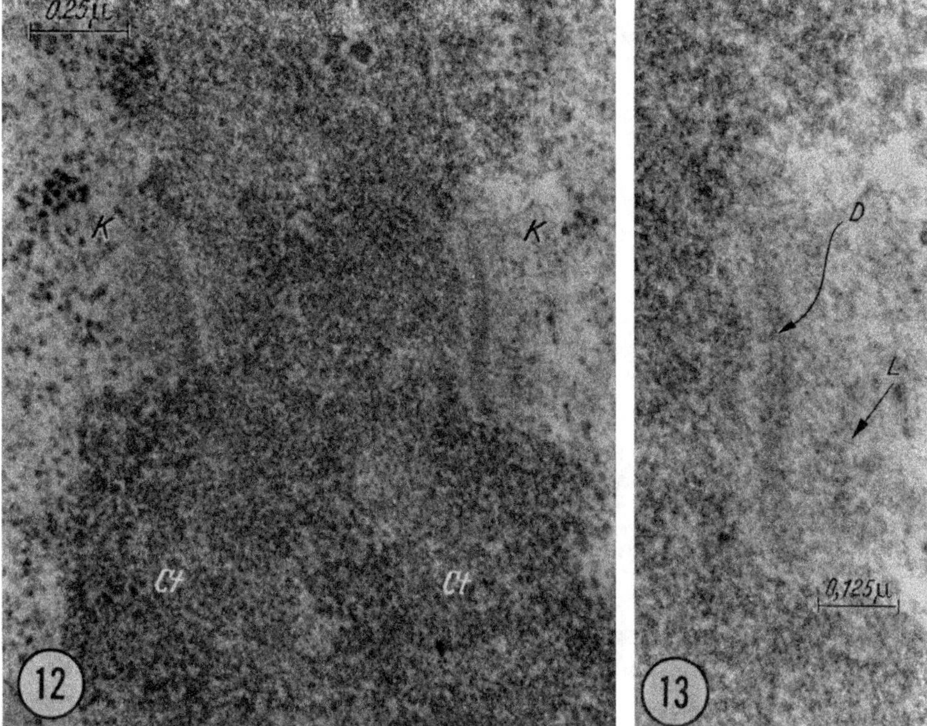

FIG. 12. Symmetrical kinetochore (K) on each chromatid (Ct) of a metacentric chromosome. × 40,000.

FIG. 13. Higher magnification of a kinetochore shown in Figure 8. A fibrous "loop" is indicated in the less-dense zone (L). (D) axial element. × 90,000.

them or not. We will return to this point later in our discussion of mitotic inhibitors.

The metaphase kinetochore is contained in an obvious constriction (primary constriction) on the chromosome. Typically the kinetochore may be located in the middle of the chromosome (metacentric) (Fig. 8), near one end of the chromosome (submetacentric or subtelocentric) (Fig. 9), or at the very end of the chromosome (telocentric). There is yet much debate as to whether a chromosome can be truly telocentric. That is, can it have only a single arm with the kinetochore occupying the very tip, or is there in telocentrics a small arm beyond the kinetochore which is below the resolution of the light microscope? There are strong arguments both in favor of, and against, the existence of true telocentrics (cf., John and Hewitt, 1966). The question, therefore, must remain open for further investigation.

Figures 10 to 17 depict the ultrastructure of the metaphase kinetochore in mammalian cells. In longitudinal profile, each sister kinetochore is made of two distinct elements termed kinetochore filaments. Although each sister kinetochore contains two kinetochore filaments, both are rarely seen in a single section. In a few instances we have been able to see both filaments (Figs. 16 and 17). The existence of two kinetochore filaments is best demonstrated by stacking transparencies of serial sections taken through the kinetochore region. We have examined numerous sets

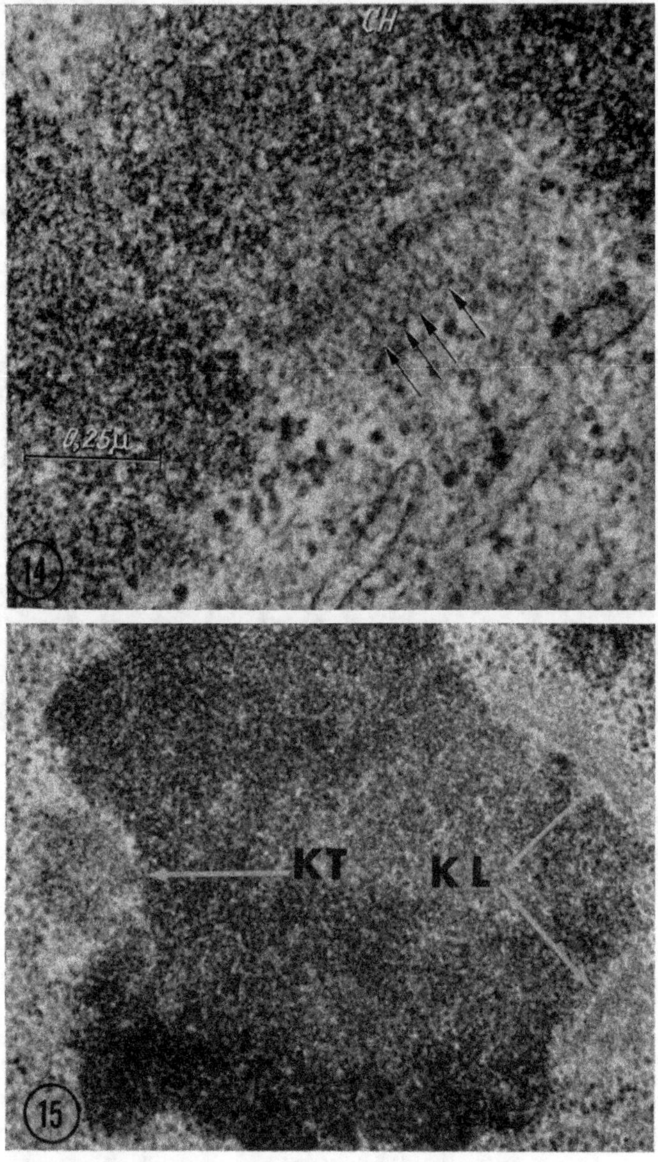

Fig. 14. The lateral fibrils of the less-dense zone are indicated by the arrows. Chinese hamster fibroblast. × 80,000.

Fig. 15. Section showing two aspects of the kinetochore. In one, (KL), kinetochore filaments are in the longitudinal section; in the other, (KT), the kinetochore filament is seen in the transverse section. Chinese hamster fibroblast. × 28,000.

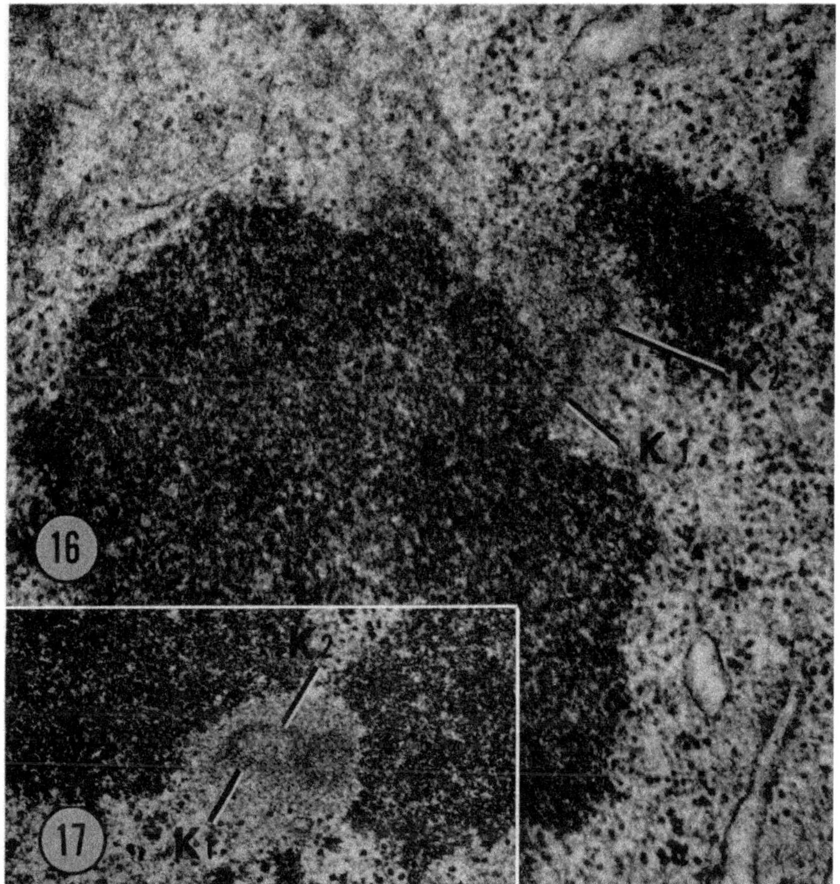

FIG. 16. A section showing both kinetochore filaments (K_1 and K_2). × 34,000.

FIG. 17. Two kinetochore filaments (K_1 and K_2) in a single section. Note clarity of lateral fibrils. Chinese hamster fibroblast. × 34,000.

of serial sections which were prepared in this manner, two of which are shown in Figures 18 and 19. In every case, each kinetochore was composed of two adjacent kinetochore filaments. The pair of filaments, arranged one above the other, extended approximately parallel to the long axis of the chromosome. In some sections they appeared to curve around the surface of the chromosome.

Each kinetochore filament consists of two components, an axial element and lateral fibrils. Each axial element consists of a pair of fibrils 50 to 80 Å wide, which appear parallel in some sections and twisted together in a cohelical manner in others (Fig. 11). Each of these fibrous structures is separated by a distance of 50 to 80 Å. The axial element originates at a condensed site on the chromatid, extends 0.2 to 0.5 μ along the chromatid through the primary constriction, and reinserts into another condensed region. Thus far it has not been possible to trace the axial elements after they enter the condensed regions of the chromosome. However, the elements are presumed to be continuous with the coarser fibrils in the arm.

The zone around the axial elements is made of fibrils (lateral fibrils) which are similar in dimension to those of the axial element (Figs. 13 and 14). They are, however, less densely stained in our material, and lie at right angles to the axial elements. In some sections (Figs. 13 and 14), they appear as loops which extend in a plane perpendicular to the long axis of the axial elements for a distance of 600 to 1,000 Å. When a kinetochore filament is sectioned transversely, the lateral fibrils can be seen to form the entire zone which completely surrounds the central axial elements (Fig. 15).

A simplified sketch of our interpretation of the mammalian metaphase kinetochore is shown in Figure 20. This model differs somewhat from that of other investigators who view the kinetochore as a "patch" or "plate" (Nebel and Coulon, 1962; Jokelainen, 1967). Jokelainen (1967) depicts the kinetochore of fetal rat cells as consisting of some three or four layers, when viewed in cross section. A face view is diagrammed as a circular profile. A somewhat similar model has been presented by Nebel and Coulon (1962). These investigators have demonstrated essentially the same structural components of the kinetochore that we have discussed in the previous paragraph. We differ primarily in our interpretation of how the components are arranged on the chromosome. However, it now appears certain that the kinetochore has a precise and definable ultrastructure in most organisms. Early electron microscope studies, which described the attachment sites of microtubules on the chromosomes as amorphous regions, simply do not hold up under current methods of chromosome preparation and improved electron microscopy.

Thus far, little has been said about the attachment of chromosomal microtubules to the kinetochore. From our studies, the 200 Å to 240 Å microtubules appear to be embedded in the fibrous loops and end at the axial elements (Figs. 7 and 30). By means of serial sections, it can be shown that each kinetochore filament has six to eight microtubules associated with it. Some investigators report the continuation of chromosomal microtubules (chromosomal spindle filaments) through the kinetochore into the chromosome proper (Barnicot and Huxley, 1965; Jokelainen, 1967). In our studies of mammalian cells, we have obtained little evidence to support this observation. However, we have observed continuous microtubules (continuous spindle filaments) which extend from pole to pole and penetrate the chromosomes at various locations, including the immediate vicinity of the kinetochore (Figs. 21a to 21c). In fact, chromosomal and continuous microtubules often are in close physical proximity in mammalian cells. Our data suggest, therefore, that chromosomal microtubules end at the axial elements of the kinetochore filaments, while continuous microtubules may penetrate the surrounding zone and continue through the body of the chromosome toward the opposite pole.

It is not unusual to observe microtubules extending through the condensed arms of chromosomes at all stages of mitosis. In every instance, those microtubules penetrating the chromosomes are surrounded on all sides by a clear zone (Figs. 21b and 21c). In other words, an obvious tunnel exists in the chromosome through which the microtubules pass. The penetration of microtubules through the chromosome has also been reported in HeLa cells (Robbins and Gonatas, 1964), amebas (Roth and Daniels, 1962), cranefly spermatocyte chromosomes (Behnke and Forer,

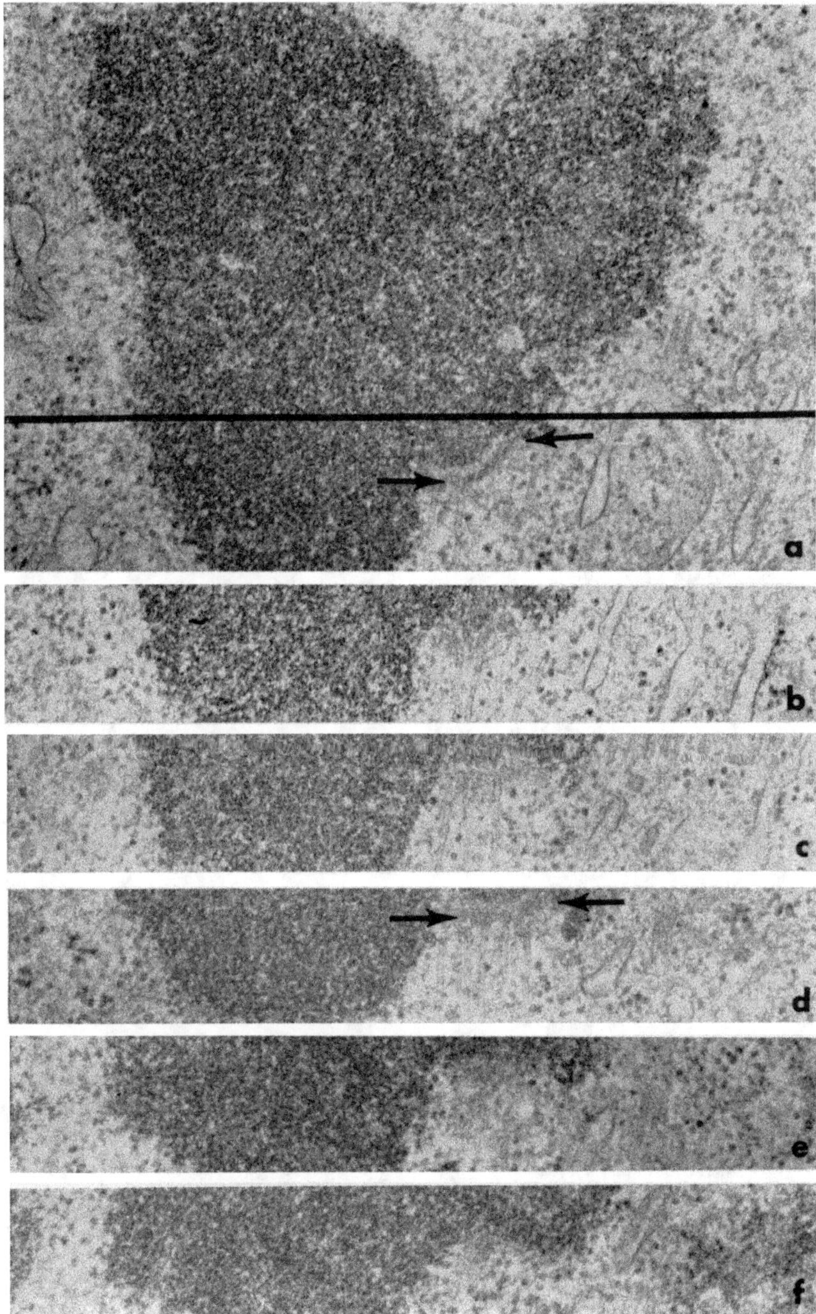

FIG. 18. Six serial sections through one sister kinetochore at metaphase. Two kinetochore filaments can be clearly demonstrated. The arrows in (a) delineate one filament which is sectioned completely through in (b). The second filament begins in (d) and disappears in (f). Note a total of 13 to 14 microtubules associated with both kinetochore filaments. × 55,000.

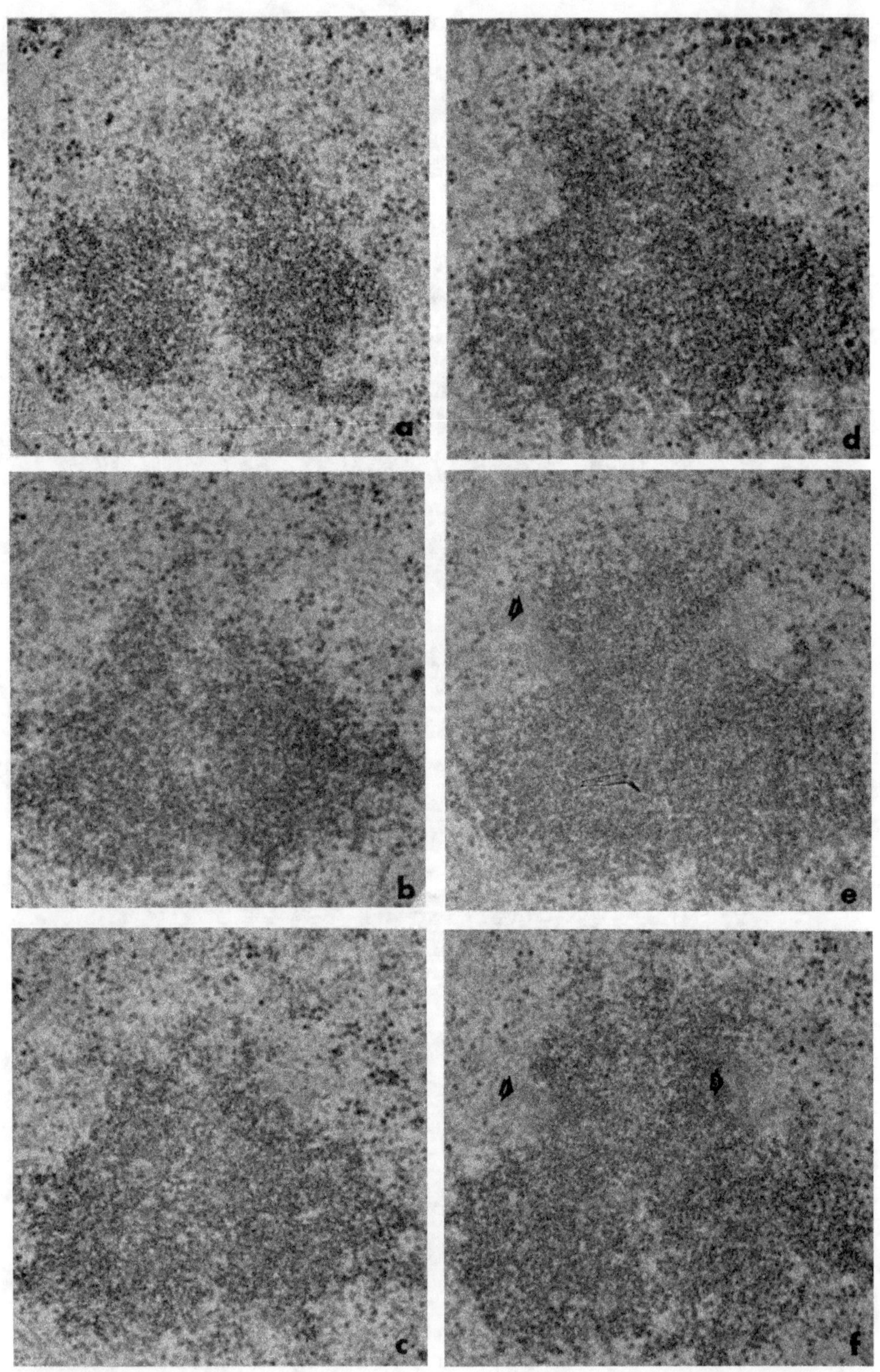

FIG. 19. Twelve consecutive serial sections through a small metaphase chromosome. Arrows 1 and 2 trace the kinetochore filaments in one kinetochore and arrows 3 and 4 delineate the

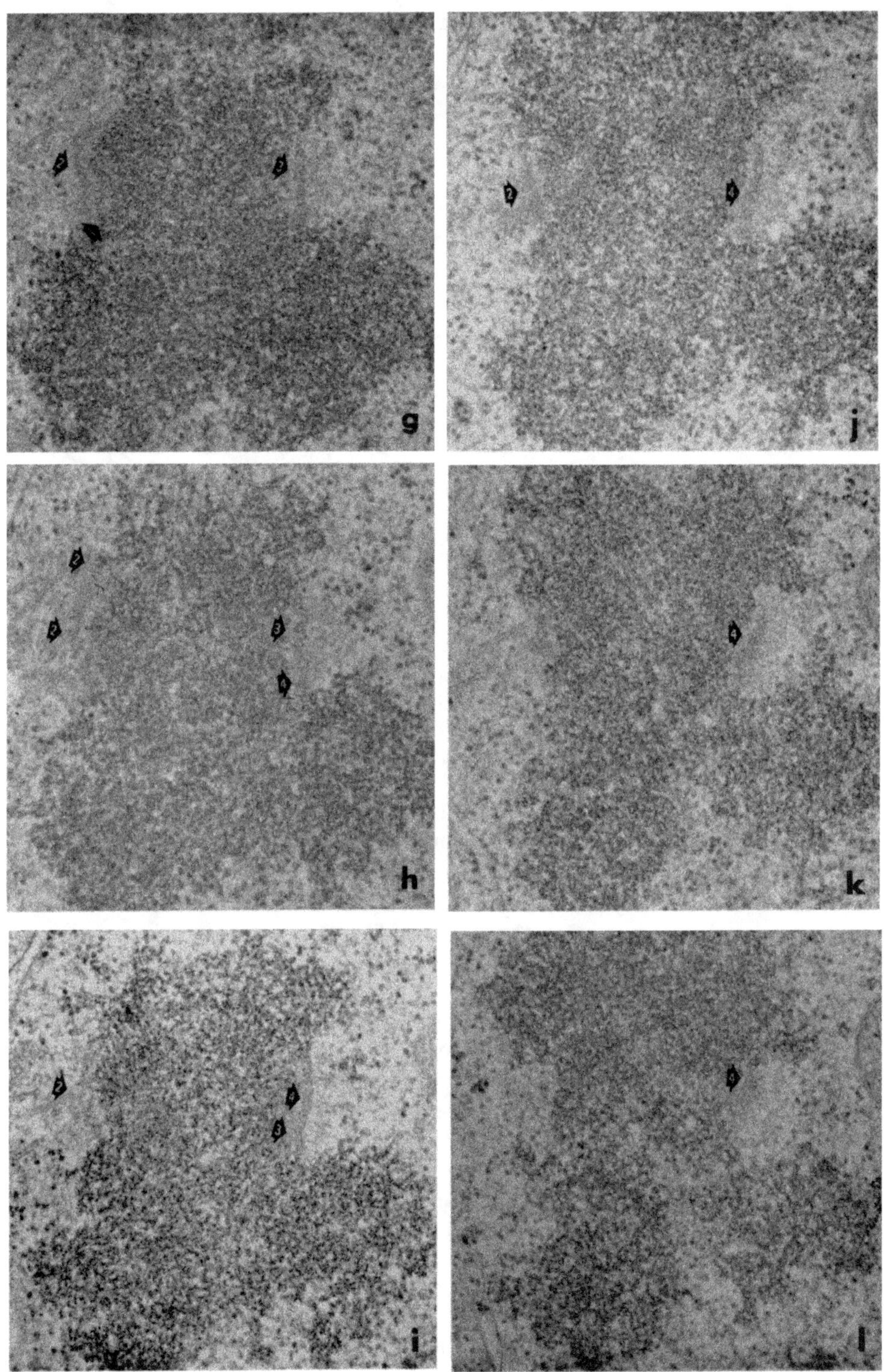

filaments in the other. The absence of any structural continuity between sister kinetochores is clearly demonstrated. × 55,000.

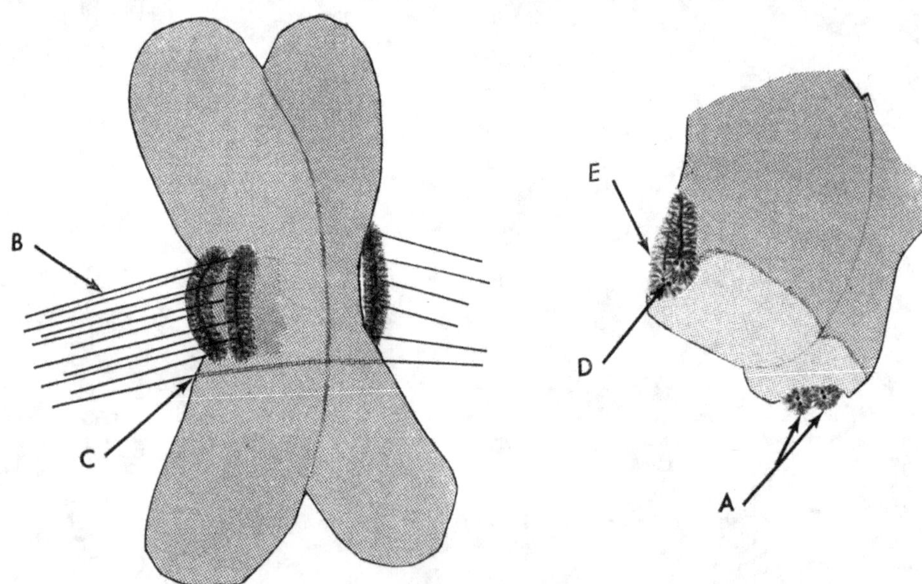

FIG. 20. Diagrammatic representation of metaphase kinetochore structure (A) Two kinetochore filaments of one sister kinetochore. (B) Chromosomal microtubules attached to both kinetochore filaments. (C) Continuous microtubules (pole-to-pole) extending through chromosome arms. (D) Axial elements consisting of a pair of 50 to 80 Å fibrils. (E) Lateral fibrils, 50 to 80 Å, which extend out from the axial elements.

1966), and the bug *Rhodinus* (Buck, 1967). *Rhodinus* is characterized by the absence of a localized kinetochore; chromosomal microtubules are attached to the chromosome throughout its length (diffuse kinetochore). The structural aspects of the diffuse kinetochore will be reviewed in a later section.

Anaphase and Telophase

Anaphase is signaled by the separation and movement of chromatids toward opposite poles. At this time, the kinetochore filaments are stretched in the direction of movement (Fig. 22). Both kinetochore filaments of each kinetochore are frequently seen at approximately the same plane during anaphase. Such behavior indicates that both filaments are being stretched out from the chromosome to about the same degree. Stretching of the kinetochore away from the chromosome has been described in numerous studies. Lima-de-Faria (1958) observed this phenomenon in *Tradescantia virginia,* and Bajer (1965) found it in *Haemanthus.* In *Tradescantia,* the kinetochore undergoes a transition from a "cylinder shape" in metaphase to a more "conic appearance" in telophase. In our electron micrographs, the cylinders of Lima-de-Faria are apparently the kinetochore filaments which are stretched out from the chromosomes when they are pulled toward the poles. However, we have never really observed that the filaments become bent into cones. In fact, as shown in Figures 22 and 23, the leading surface of the kinetochore filament advancing to the poles is usually oriented perpendicular to the chromosomal microtubules.

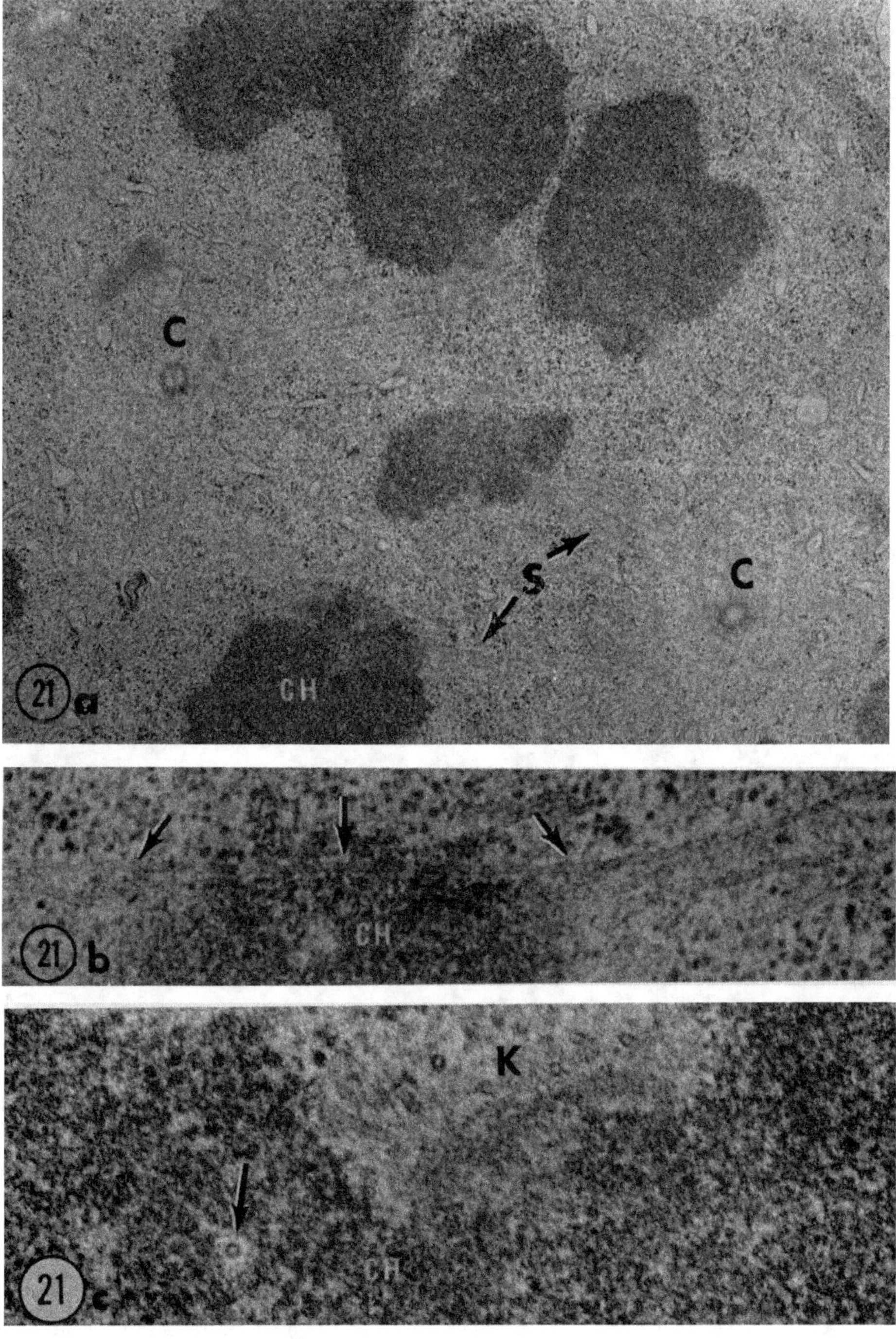

FIG. 21. (a) Survey of metaphase showing spindle tubules (S) and centrioles (C) at both poles. × 20,000. (b) Higher magnification of chromosome (CH) shown in (a). Arrows indicate a microtubule which extends through the chromosome (continuous spindle tubule). × 80,000. (c) Cross section of microtubule (arrow) extending through the condensed chromosome near the kinetochore (K). Note clear area around microtubule. × 90,000.

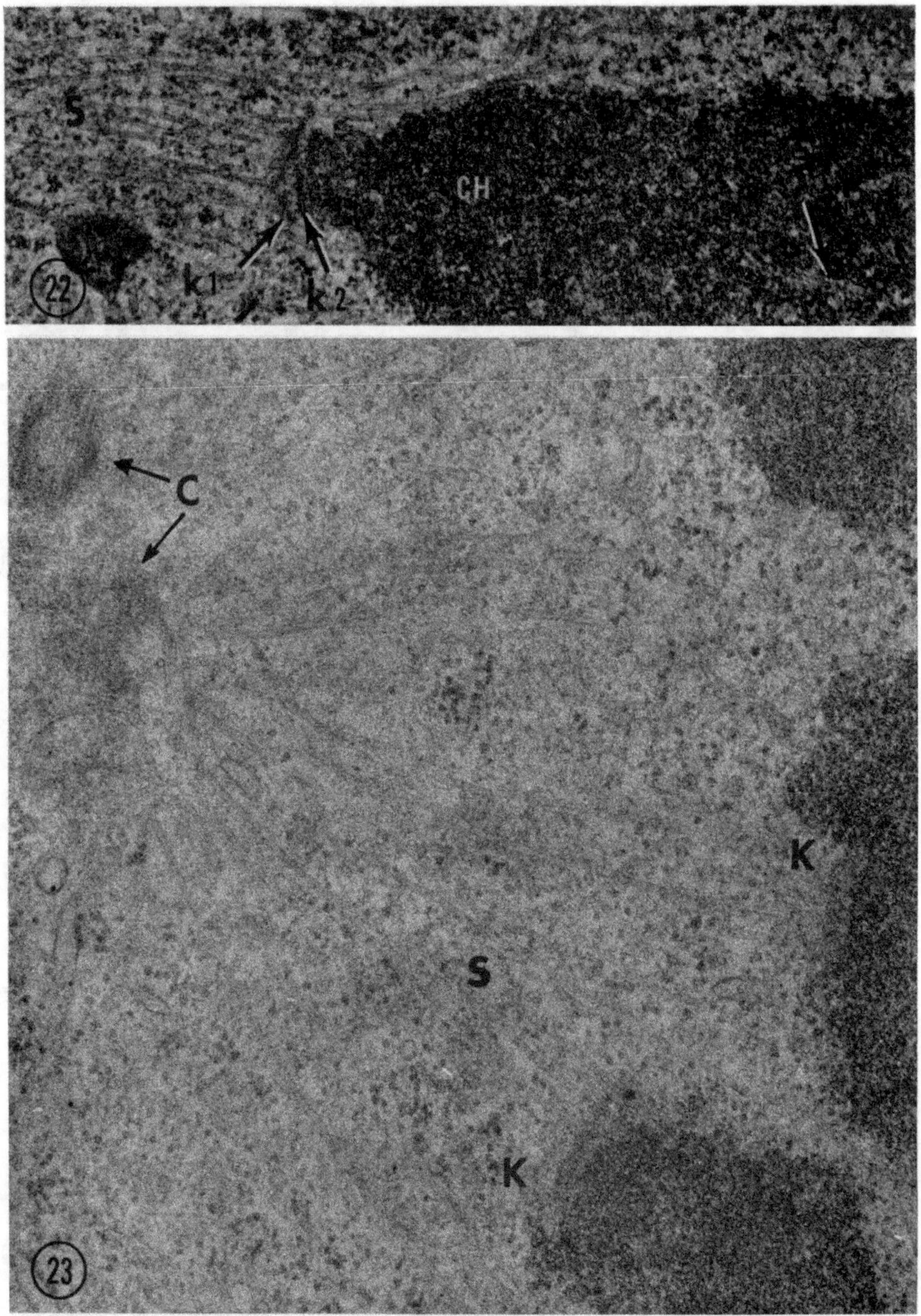

FIG. 22. "Stretched" kinetochore in early anaphase showing both kinetochore filaments (K_1 and K_2). Arrow points to microtubule within chromosome. (CH) chromosomes, (S) spindle tubules. × 40,000.

FIG. 23. Anaphase. Spindle tubules (S) converge on a single member of a centriole (C) pair. (K) kinetochore. × 44,600.

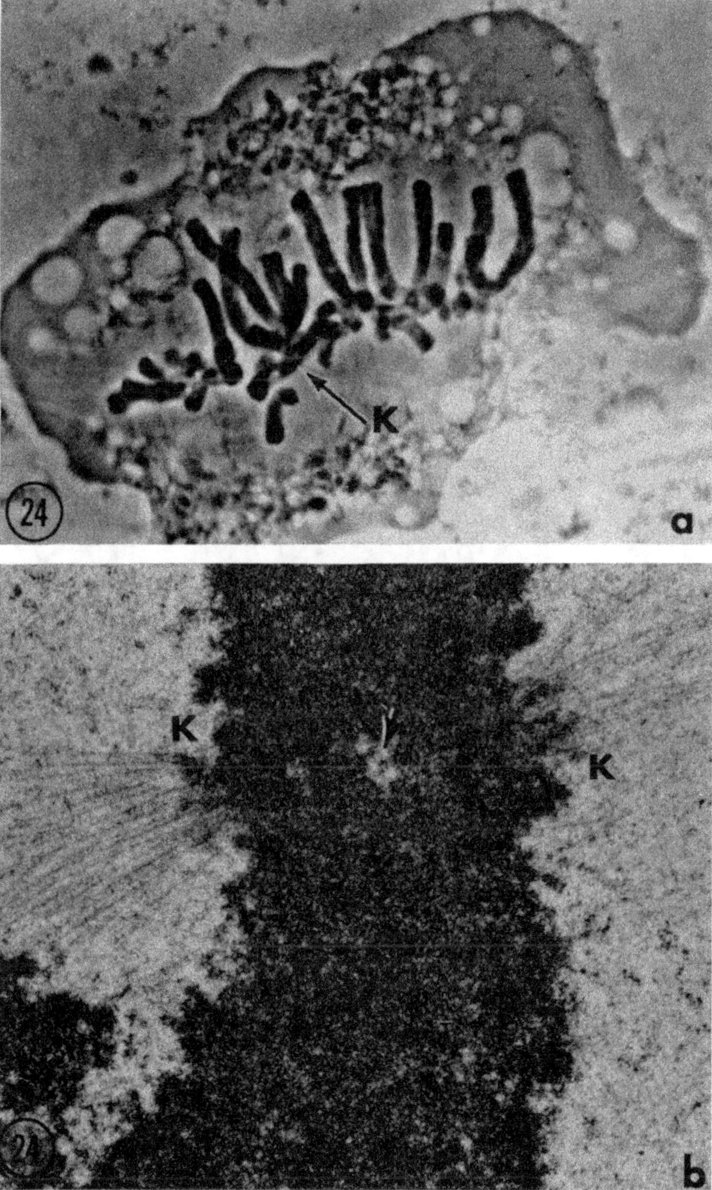

FIG. 24. (a) Light micrograph of chromosomes from *Haemanthus* endosperm. Arrow points to the kinetochore region. (b) Kinetochore in the region indicated by arrow in (a). Note numerous microtubules which "converge" on the kinetochore. Small arrow in center indicates separation of chromatids. (Both micrographs courtesy Dr. Andrew Bajer.) × 15,000.

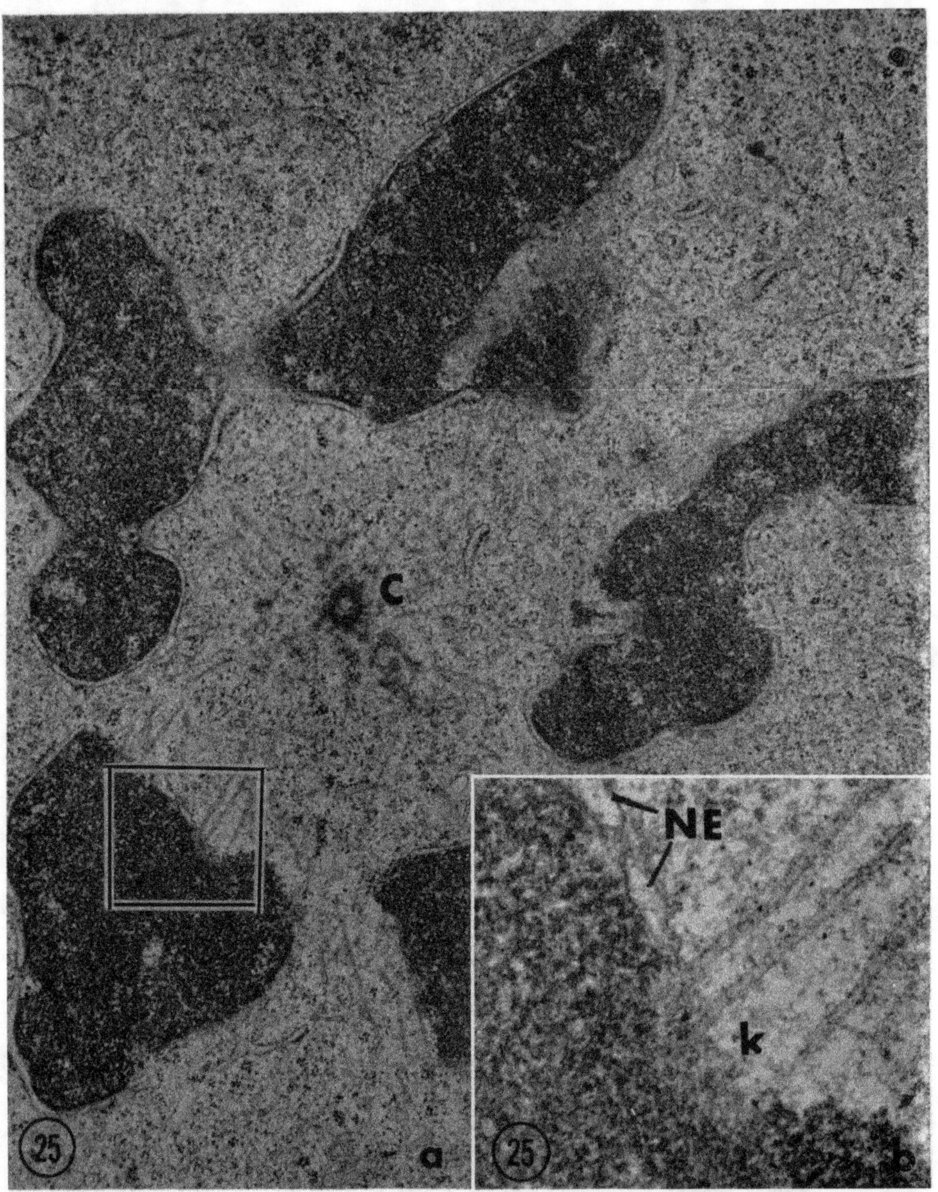

FIG. 25. (a) Telophase (polar view). Chromosomes near one centriole (C). The nuclear envelope has almost completely formed around the chromosomes except at the kinetochores. × 19,000. (b) Higher magnification of area within the square of (a). Note microtubules which terminate in kinetochore (K). Nuclear enevelope (NE) ends just above the kinetochore. × 76,000.

In electron micrographs, we examine sections of kinetochores which have not been distorted or flattened by squashing. Therefore, structures seen in light-microscope squash preparations are likely to differ somewhat from those seen in electron micrographs. This does not mean that all descriptions of kinetochores of light-microscope squash preparations are artifacts. Indeed, the simplicity of the squash preparation has made it the most useful single technique for chromosome studies and has contributed greatly to our knowledge of kinetochore structure and behavior.

A logical solution to the problem of comparing structures seen with the light microscope to those seen with the electron microscope is a technique which permits the observation of the same chromosome with both high-resolution light optics and electron microscopy. Such a procedure has recently been developed in our laboratory (Brinkley et al., 1967a). Figures 5a and 5b show, however, that while the kinetochore region can easily be seen with a phase microscope, little structural detail is evident until we section and examine with the electron microscope. Figures 24a and 24b show a similar preparation of *Haemanthus* chromosomes. More structures can be seen in the kinetochore regions with light microscopy if we treat cells in hypotonic solution, fix them in acetic acid alcohol, and make a conventional squash preparation. The fine structure in such preparations, however, is destroyed (Brinkley, unpublished data).

As the chromosomes approach the pole in telophase, the advancing regions fuse to form a semicircle around the centrioles (Fig. 25a). At this time, membranous vesicles become aligned along the arms of the chromosomes and produce the nuclear envelope. The kinetochore and associated microtubules remain evident until late telophase when formation of the nuclear envelope is almost complete (Figs. 25a and 25b). The kinetochore apparently disappears by becoming as electron-dense as the remainder of the chromosome. Perhaps this is achieved by further condensation of the kinetochore filaments to a degree equal to that of the remainder of the chromosome arms. When the nuclear envelope is completely formed around the telophase chromosomes, the kinetochore and the chromosomal microtubules are no longer evident.

THE KINETOCHORE IN MEIOSIS

The structural aspects of meiosis in most animal cells are, for the most part, like those of mitosis. That is, a division apparatus forms which is composed of essentially the same elements as those in the mitotic apparatus, namely centrioles at each pole, continuous microtubules which extend from pole to pole, and chromosomal microtubules which extend from the kinetochore to the poles. The primary difference, of course, is the orientation of chromosomes on the metaphase plate during the first division. This process is initiated by the pairing of homologous chromosomes in meiotic prophase. As the paired homologs (bivalents) arrive at the metaphase plate, they behave individually like chromatids in mitosis. The two sister kinetochores of each homolog are functionally single and are oriented to the same pole. Anaphase separation results in the reductional division of chromosomes followed by a second division which produces four nuclei. Each of these contains essentially half the number of chromosomes (haploid number) of the somatic (diploid number) cell. The second division is essentially a mitotic division. In some organisms, such as the coccids (Hughes-Schrader, 1955), the above sequence of kinetochore orientation is reversed.

The difference between the behavior of chromosomes in meiosis and mitosis is governed largely by the activity and orientation of the kinetochores. The question arises, therefore, as to whether the kinetochores of meiotic chromosomes are structurally different from those of the mitotic chromosomes. Darlington (1937) proposed, in his "precocity theory of meiosis", that meiotic chromosomes pass through metaphase I with undivided kinetochores. That is, each homologous chromosome contains a single kinetochore. Numerous investigators have opposed Darlington's theory on the basis that a pair of kinetochores can be seen on each chromosome in anaphase I (Iwata, 1940; Lima-de-Faria, 1956; Schrader, 1939). More recently, the problems of doubleness of the meiotic kinetochore at the first division has been approached at the electron microscopic level.

Ultrastructure of Meiotic Kinetochores

The ultrastructural basis of meiosis has not been studied as thoroughly as the ultrastructural basis of mitosis. The reason for this is obvious; it is considerably more difficult to obtain all the division stages of meiotic tissue in thin sections. Where electron microscope studies have been done, the doubleness of the kinetochore in meiosis I is still in question. Nebel and Coulon (1962) reported that the metaphase I kinetochore of pigeon chromosomes appeared undivided. The kinetochore could be detected as early as late pachytene when it appeared as a single spherical mass whose electron density was considerably greater than that of the remainder of the chromosome. By metaphase it displayed a complex "laminated" structure consisting of several distinct layers.

By swelling anaphase I chromosomes with hypotonic solution prior to a brief osmium fixation, Luykx (1965a) convincingly demonstrated the existence of two distinct kinetochores on anaphase I chromosomes of *Urechis* eggs. Using this method of fixation, he found that the kinetochore was considerably more electron-dense than the rest of the chromosome. Measurements of the individual kinetochores on anaphase I chromosomes gave values which corresponded to sister kinetochores in mitosis and second-division meiosis. Luykx's data strongly suggest that each homolog (half-bivalent) contains the usual pair of sister kinetochores seen in mitosis, but is oriented to the same pole. The ultrastructural organization of the meiotic kinetochores of *Urechis* eggs is much like that generally seen in mitotic chromosomes. Luykx interprets the organization of the kinetochore in meiosis as circular in face view and consisting of two dense layers separated by a light space in transverse section. Each of the dense layers, according to our interpretations, would correspond to a kinetochore filament (see p. 131 and p. 133).

Recently, one of us (Brinkley) began a study of the ultrastructure of meiotic chromosomes in grasshopper testes in collaboration with Dr. Bruce Nicklas at Duke

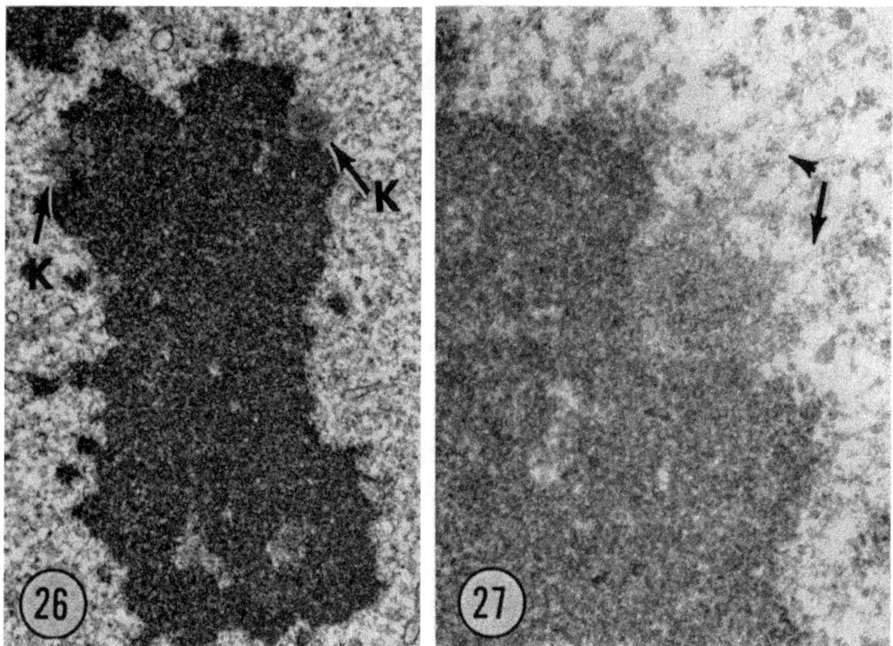

FIG. 26. Metaphase I bivalent of grasshopper showing kinetochore (K) as a light staining spherical mass contained in a small "cup" on each chromosome. × 23,000.

FIG. 27. Higher magnification of one kinetochore shown in Figure 26. Microtubules (arrows) appear to extend into a mass of 50 to 80 Å filaments which are less densely stained than those in the chromosome. × 70,000.

University. Nicklas has modified the flat-bedding method of Brinkley et al. (1967a) to permit preselection of meiotic stages with the phase microscope prior to sectioning for electron microscopy. Figures 26 and 27 are electron micrographs of a metaphase I bivalent of *Melanoplus differentialis*. The kinetochore is a lightly staining mass which is approximately 0.25 μ in diameter and is located in a small cup-like constriction near the tip of the chromosome. The latter may correspond to Schrader's (1953) "commissural cup." At higher magnification, the kinetochore appears as a mass of 50 to 80 Å fibrils. Because of the preliminary nature of our studies at this stage, we cannot comment on the differences between meiotic and mitotic kinetochores or interpret their three-dimensional organization at metaphase I.

EFFECTS OF MITOTIC INHIBITION ON KINETOCHORE BEHAVIOR

Compounds which inhibit mitosis by interfering with normal spindle formation, such as colchicine, Colcemid, vinblastine, and vincristine, have been used in cytological studies for years (cf., Eigsti and Dustin, 1955; Biesele, 1958). The ultrastructure of cells arrested with vincristine and vinblastine has been studied by George et al. (1965), and the fine structure of colchicine-arrested cells has been examined by Robbins and Gonatas (1964). Stubblefield and Klevecz (1965) found that Chinese hamster tissue culture cells arrested with Colcemid (a derivative of colchicine) recovered when the drug was removed, and the entire population of arrested cells completed division and proceeded synchronously into the next interphase cycle. This method of synchrony is especially well suited for accumulating large quantities of cells for biochemical studies. Recently Brinkley et al. (1967b) studied the ultrastructural basis of Colcemid inhibition and reversal on Chinese hamster cells *in vitro*. Their study indicated that at a concentration sufficient to inhibit mitosis (0.06 μg/ml) Colcemid did not completely prevent the formation of chromosomal microtubules. With the disruption of the nuclear membrane, the chromosomes migrated around the centriole complex, which remained unseparated during Colcemid treatment (Figs. 28 and 29). As long as the cells were exposed to the drug (up to 18 hours), microtubules formed from only one of the sister kinetochores, the one nearer the centrioles. The other sister kinetochore, although structurally normal, remained totally inactive and free of microtubules (Fig. 30). This behavior resulted in the formation of a "chromosomal sphere" with microtubules radiating from the inner kinetochore to the centriole, with complete suppression of those sister kinetochores on the outside of the sphere. How can this selectivity in kinetochore function be explained when both sisters apparently have developed to the same degree of maturity and are exposed to the same cytoplasmic environment? The answer apparently is related to the existence of only a single pole in Colcemid-arrested cells and its spatial arrangement with the outside daughter kinetochores. For the outer kinetochores to form microtubules toward the centrioles which are located in the center of the chromosomal sphere, microtubules would have to be assembled on the inner surface of the kinetochore and extend through the chromosome. The fact that this behavior has not been observed supports the early hypothesis by Östergren (1951) which states that each sister kinetochore has a kinetic and an akinetic side. The kinetic side is always distant to the chromosome body and, through some interaction with the poles, can assemble microtubules. The akinetic side is that which faces the chromosome proper. Thus, during Colcemid inhibition, the existence of only a single pole results in the orientation of chromosomes such that only one sister kinetochore of each chromosome has a kinetic side facing a pole.

Immediately upon removal of the Colcemid from the culture media, the centriole pairs migrate to opposite poles (Figs. 31a to c and 32). Concomitant with centriole separation is the appearance of microtubules from both sister kinetochores to each of the centriole pairs. It is only after reversal of Colcemid inhibition that a bipolar spindle is produced and chromosomes are moved to a metaphase plate (Fig. 33). Thus Colcemid does not arrest cells at metaphase, as has been commonly reported, but at some intermediate stage between prophase and metaphase. The primary function impaired by Colcemid poisoning is the migration of centriole pairs to opposite poles, an event which apparently requires microtubules.

The mechanism of action of colchicine, and presumably Colcemid, is to inhibit microtubular polymerization by binding to the active sites of the protein subunits (Taylor, 1965; Borisy, 1967; Borisy and Taylor, 1967a). The completeness of inhibition seems to depend largely upon concentration of inhibitor. Thus at a higher concentration (0.3 μg/ml) microtubules were not seen in sections of arrested Chinese

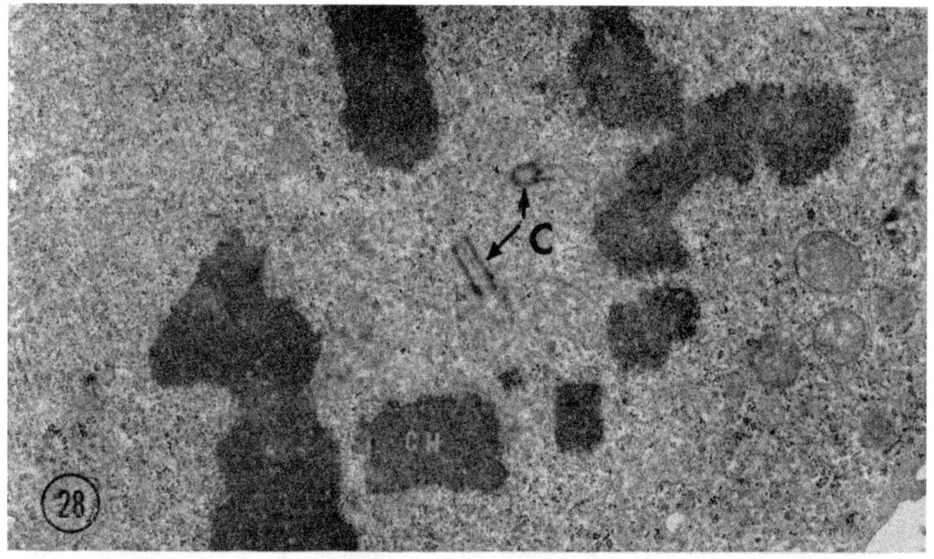

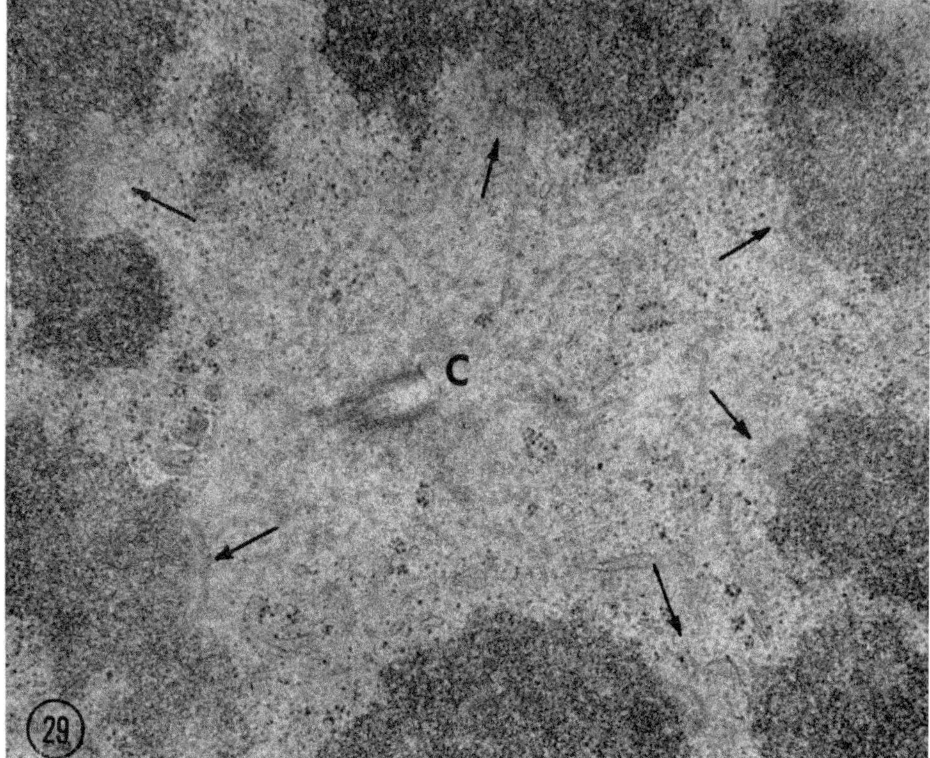

FIG. 28. Cell arrested with Colcemid. Chromosomes (CH) are drawn around the centrioles (C). Two-hour Colcemid treatment. × 22,000.

FIG. 29. Higher magnification of Colcemid arrested chromosomes. Arrows indicate kinetochores (K) surrounding a centriole (C). Microtubules can be seen extending from the kinetochores toward the centriole. × 30,000.

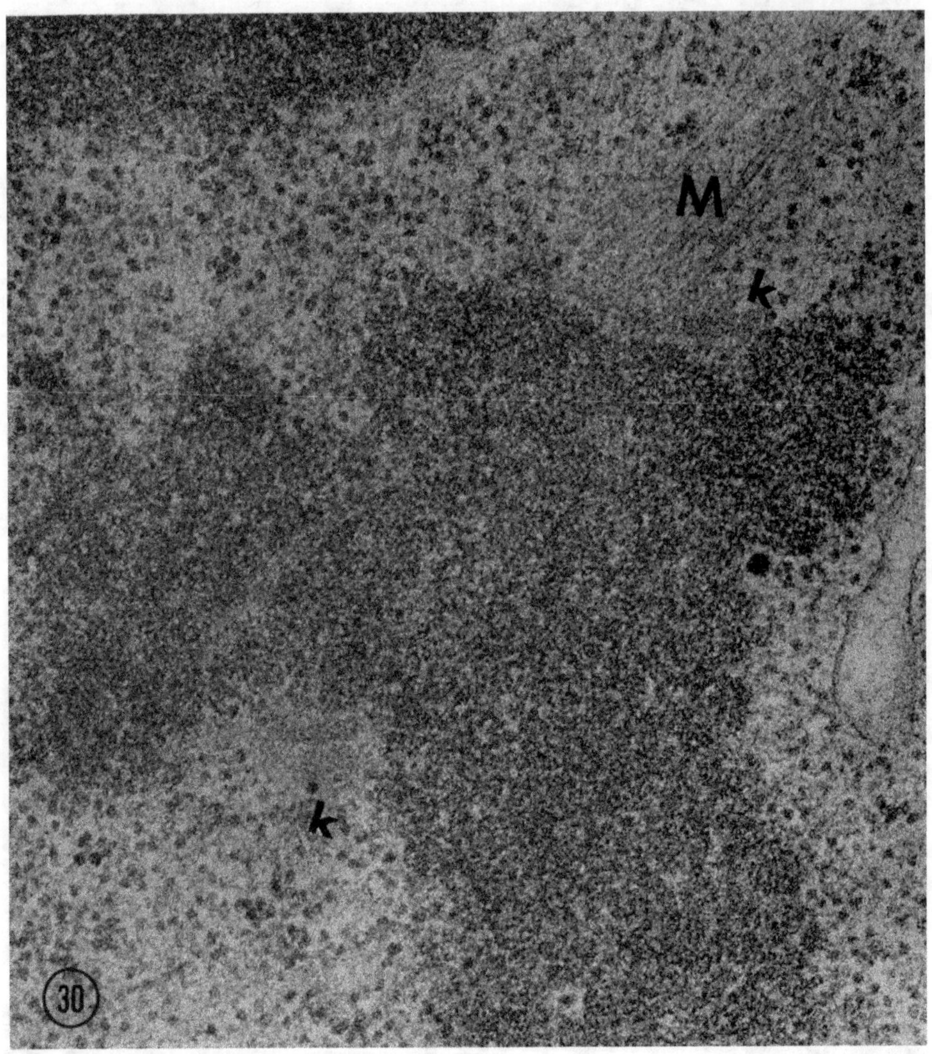

FIG. 30. Chromosome from cell which has been treated with Colcemid for 2 hr. Note microtubules (M) which extend from only one kinetchore (nearest the centrioles). The sister kinetochore although structurally similar, has no microtubules. × 51,000.

hamster cells (Brinkley and Stubblefield, unpublished data). At 0.06 μg/ml, Colcemid differentially blocks microtubules which are necessary for centriole migration, but not kinetochore activity. Thus, it is a useful tool for studying the interaction of these two components and their role in the organization of the mitotic apparatus.

One point of interest regarding the recovery of cells from Colcemid inhibition is that microtubule growth with respect to the "suppressed" kinetochore can be observed in this period. Using serial section analysis, we have found that microtubules appear to grow (increase in length) from the kinetochore to the poles im-

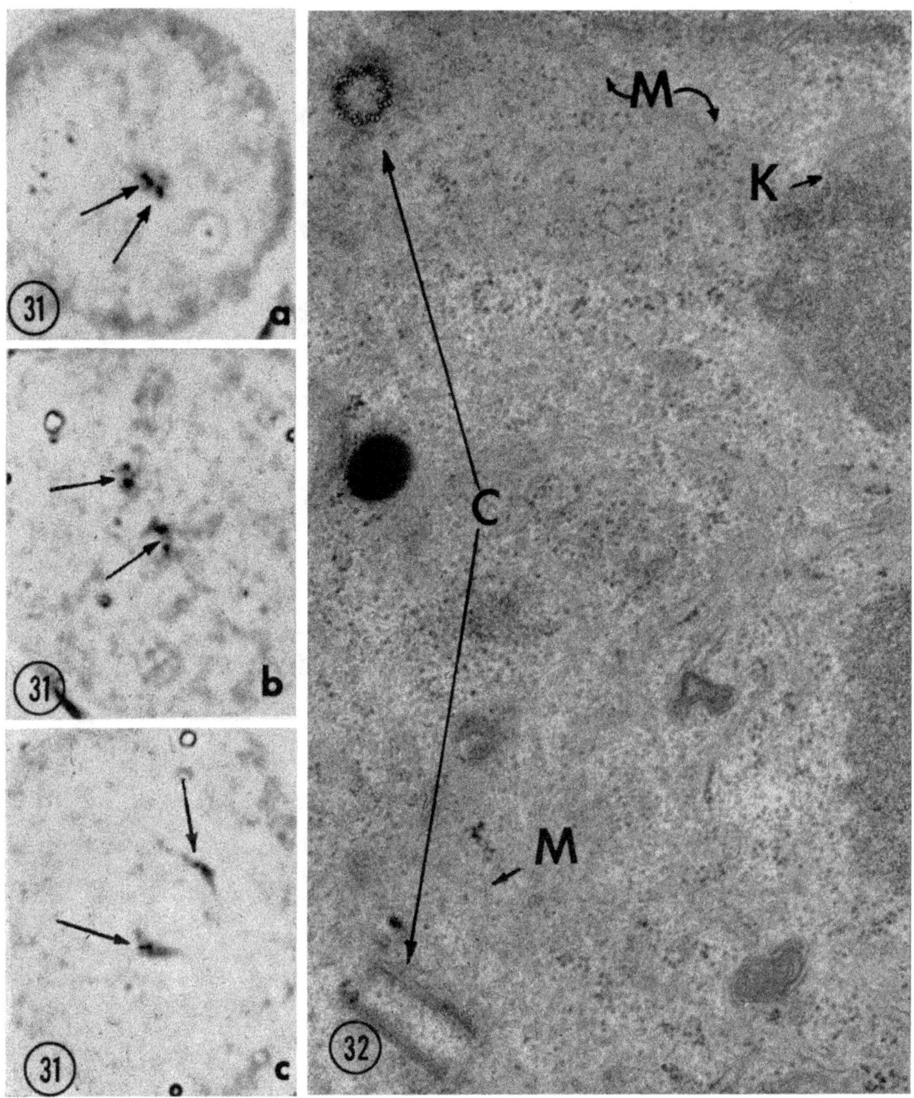

FIG. 31. Stages in the migration of centrioles (arrows) following the removal of Colcemid. Phase contrast. (See Brinkley et al., 1967b, for procedure.) (a) 2 hr Colcemid treatment. (b) 15 min after reversal. (c) 20 min after reversal. Reduced 10 percent from × 1,840.

FIG. 32. Longitudinal and transverse sections of centrioles (C) during stages of reversal. Note microtubules (M) which extend toward both kinetochores (K) of a chromosome in the upper right-hand corner of the micrograph. Five minutes after removal of Colcemid. Reduced 10 percent from × 43,500.

mediately after Colcemid is removed (Brinkley and Stubblefield, unpublished data). The problem of whether the spindle microtubules develop from the pole to the kinetochore or in the reverse direction has not been clearly resolved in earlier studies. Went (1966) proposed that, depending upon the material, both patterns of

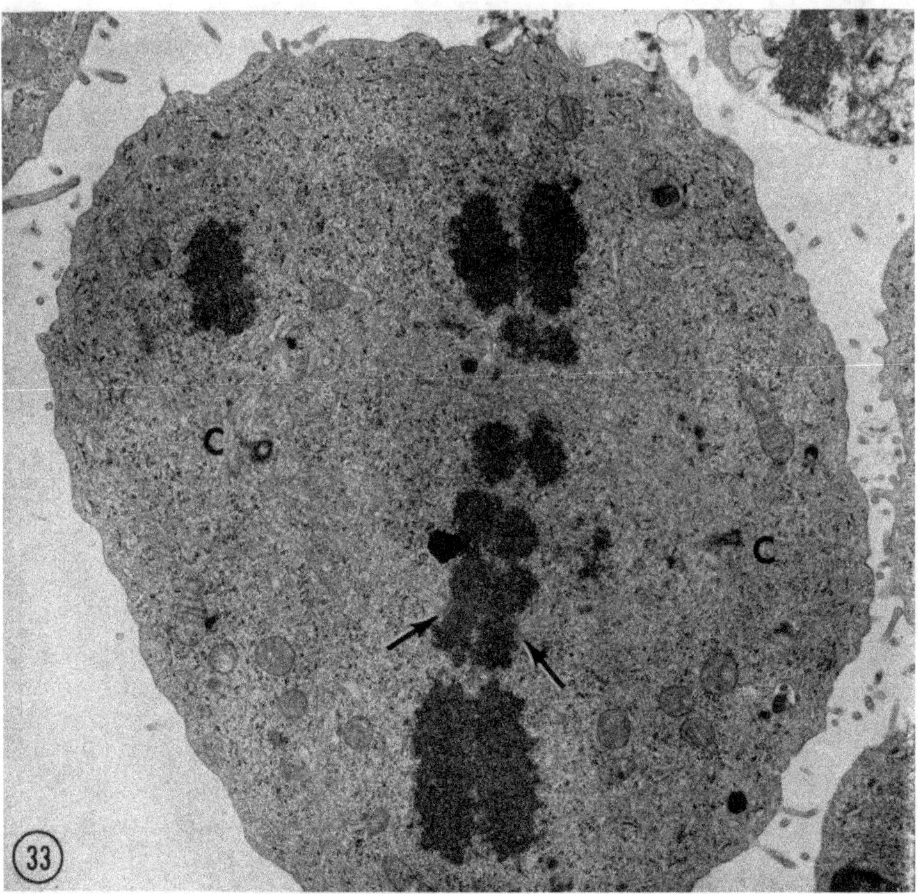

FIG. 33. Typical metaphase cell with chromosomes aligned on metaphase plate. Spindle tubules can be seen extending from kinetochores (K) to centrioles (C) located at each pole. One chromosome is still lagging away from the plate. Thirty minutes after removal of Colcemid. Reduced 12 percent from × 32,600.

development may prevail. Studies of Cleveland (1957) indicate that in flagellate protozoa the kinetochore-associated tubules grow from the astral rays, which also produce the continuous (pole-to-pole) microtubules. Our data regarding the growth of microtubules from the kinetochore to the poles after Colcemid reversal support the birefringence studies of Inoué (1964) and Forer (1965) which argue that microtubules are assembled at the kinetochore and grow toward the poles.

THE KINETOCHORE AS A SPECIALIZED "GENE"

The kinetochore may be confined to a specific site on the chromosome as in most metazoan organisms (localized kinetochore) or, as in the case of hemipteran insects (Schrader, 1935; Hughes-Schrader, 1942) and some plants (Brown, 1954), may be diffusely distributed along the chromosome arms (diffuse kinetochore). Regardless of its location, the kinetochore is much like a classical "gene" in that it is activated at a specific time in the division cycle and is suppressed when its function is completed at the end of cell division. Furthermore, it remains a stable and integral part of the chromosome and must be replicated at some time during the period for chromosome replication. In most cases, it is divided equationally in mitosis, but undergoes reduction division in meiosis (cf. Rhoades, 1961). Evidence for the existence of DNA in the kinetochore has been demonstrated by the Feulgen staining procedure in the plants, *Secale cereale* and *Tradescantia bracteata* (Lima-de-Faria, 1950, 1958) and the newt, *Triturus viridescens* (Gall, 1954). Efforts to localize nucleic acids in the kinetochore by means of ultrastructural cytochemistry have not met with complete success. This is largely because of the lack of specificity of staining procedures and the poor morphological preservation which follows DNase and RNase digestion. Luykx (1965a) reported that the kinetochore of *Urechis* eggs was insensitive to DNase digestion. His results are inconclusive, however, since his criterion for removal of DNA was based on a lack of or weak Feulgen staining at the light microscope level.

Further evidence for the presence of "genes" in the kinetochore was suggested by the work of Auerbach (1947), who observed that treatment of wild-type male *Drosophila melanogaster* with a mustard gas compound produced an X chromosome which exhibited abnormal kinetochore behavior. The treated X had a tendency to follow the untreated X chromosome to the same cell in meiosis (nondisjunction). The treated X also tended to be lost on the spindle during meiosis (see discussion by Cuevas-Sosa, 1967).

One perpetual problem in localizing markers of any type in the kinetochore region is that of resolution. One must be careful to avoid confusing the chromatin on either side of the kinetochore with the kinetochore itself. There is evidence in the literature, however, that the so-called "heterochromatin" adjacent to the kinetochore may, in fact, affect the activity of the kinetochore itself (Lindsley and Novitski, 1958). Stern (1936) concluded that somatic crossing over in *D. melanogaster* occurred most frequently adjacent to the kinetochore. Crouse (1960), working with a series of translocations between autosomes and the X chromosome, demonstrated in *Sciara coprophila* that heterochromatin adjacent to the kinetochore was responsible for determining which chromosome underwent precocious anaphase movement with nondisjunction at the second meiotic division and eventual elimination. Her study, like that of Lindsley and Novitski (1958), suggested that kinetochore behavior was determined by the adjacent heterochromatin.

Sister chromatid exchanges in the kinetochore region have been demonstrated by Cuevas-Sosa (1967) in human lymphocytes and in a human neoplastic cell line (H. Ep. 2). Cells pulse-labeled with ^{3}H-uridine for 8 hours and examined at metaphase in the second generation showed some chromosomes with obvious exchanges at the kinetochore. Out of 6,050 Colcemid-arrested chromosomes observed, 97 contained sister chromatid exchanges in the kinetochore region. Efforts were made to rule out twisting of chromatids in the kinetochore as reported by Prescott and Bender (1963) by scoring only those chromosomes in which the chromatids were clearly separated and flattened.

Chiasma formation within the kinetochore has been proposed by John and Hewitt (1966) as the mechanism for formation of ditactic bivalents in some species of grasshopper. In so-called "true telocentric chromosomes" meiotic bivalents occasionally form by end-to-end association at the kinetochore region. The authors propose that such a configuration is maintained by a chiasma between homologous kinetochore filaments. Their model, however, is based upon the existence of true telocentric chromosomes (see p. 131).

THE KINETOCHORE AND THE LAMPBRUSH ORGANIZATION OF METAZOAN CHROMOSOMES

The Kinetochore and the Nucleolus Organizer

If the kinetochore can in fact be thought of as a "special" genetic locus, its large size and activity in mitosis suggest similarity to another mitotically active region, the nucleolus organizer. The latter is a giant locus which contains redundant linear information for transcription of ribosomal RNA (cf., Wallace and Birnstiel, 1966). Our only information regarding the similarity of the kinetochore with such loci is morphological and comes from ultrastructural studies of the nucleolus organizer in Chinese hamster *(Cricetulus griseus)* and Tasmanian rat kangaroo *(Potorous tridactylis)* tissue culture cells (Brinkley and Hsu, 1966; Hsu et al., 1967). The Tasmanian rat kangaroo cell strains PtK_1 and PtK_2, are female and male aneuploid lines which contain 11 and 12 chromosomes respectively. A prominent secondary constriction exists on the long arm of the X chromosomes near the kinetochore. Hsu et al. (1967) have demonstrated that the naturally-occurring secondary constrictions of the rat kangaroo are nucleolus organizers and, in the female, share a single large nucleolus (Fig. 34).

The ultrastructure of the nucleolus organizer region is somewhat similar to that of the kinetochore. Both have the same staining capacity and are primarily composed of 50 to 80 Å fibrils. In prophase, the nucleolus organizer region is greatly extended (Fig. 35). At this time, the region contains a pair of dense axial "cores" surrounded by 50 to 80 Å fibrils and 150 to 200 Å granules. In prophase of Chinese hamster cells the nucleolus is still incorporating tritiated uridine. However, as metaphase is approached the secondary constriction narrows and RNA synthesis ceases (Hsu et al., 1965; Arrighi, 1967). Although we have not yet studied RNA synthetic activity during mitosis in rat kangaroo cells, the same behavior is likely. As metaphase is reached, the nucleolus organizer region becomes more condensed than the kinetochore (Fig. 36), but still remains less condensed than the rest of the chromosome arm. This organization is maintained throughout anaphase. As telophase is reached, the nuclear envelope is formed and the nucleolus organizer begins to synthesize RNA again (McClure and Hsu, unpublished data). At this time electron-dense granules reappear along the periphery of the 50 to 80 Å fibrillar loops in the secondary constriction (Fig. 37).

In summary, the fine structure of the nucleolus organizer is similar to that of the kinetochore in the following ways: (1) it contains the same lampbrush or-

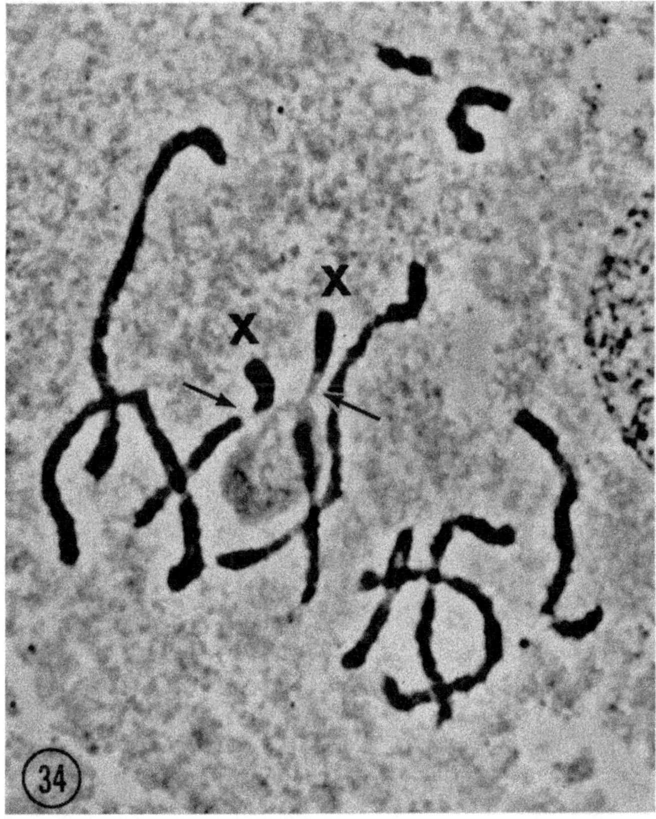

FIG. 34. Prophase chromosomes of rat kangaroo cell line PtK_1. Acetic orcein, phase contrast. Note large nucleolus attached to secondary constriction (arrows) of the X chromosomes. (Photograph courtesy of Dr. T. C. Hsu.) × 2,500.

ganization with dense axial cores and 50 to 80 Å lateral loops in prophase while RNA synthesis is assumed to be occurring; (2) during midmitosis, when metabolic activity is shut off, it becomes more condensed but retains the same filaments and axial core, although they are less extended; (3) when activity is resumed in telophase the nucleolus organizer again "opens up" displaying the lampbrush type organization. In every case we were careful to identify the secondary constriction by light microscopy before proceeding with thin sectioning and ultrastructural analysis according to the method of Brinkley et al. (1967a).

The Kinetochore and Polytene Chromosome "Puffs"

The oldest and perhaps best-known example of morphological manifestation of genetic activity is that of "puff" formation in giant polytene chromosomes of Diptera. Studies in several laboratories have demonstrated that the finest chromatin threads in the puff region are filaments of approximately 100 Å in diameter which appear to be arranged into lampbrush loops (Beermann and Bahr, 1954). As a

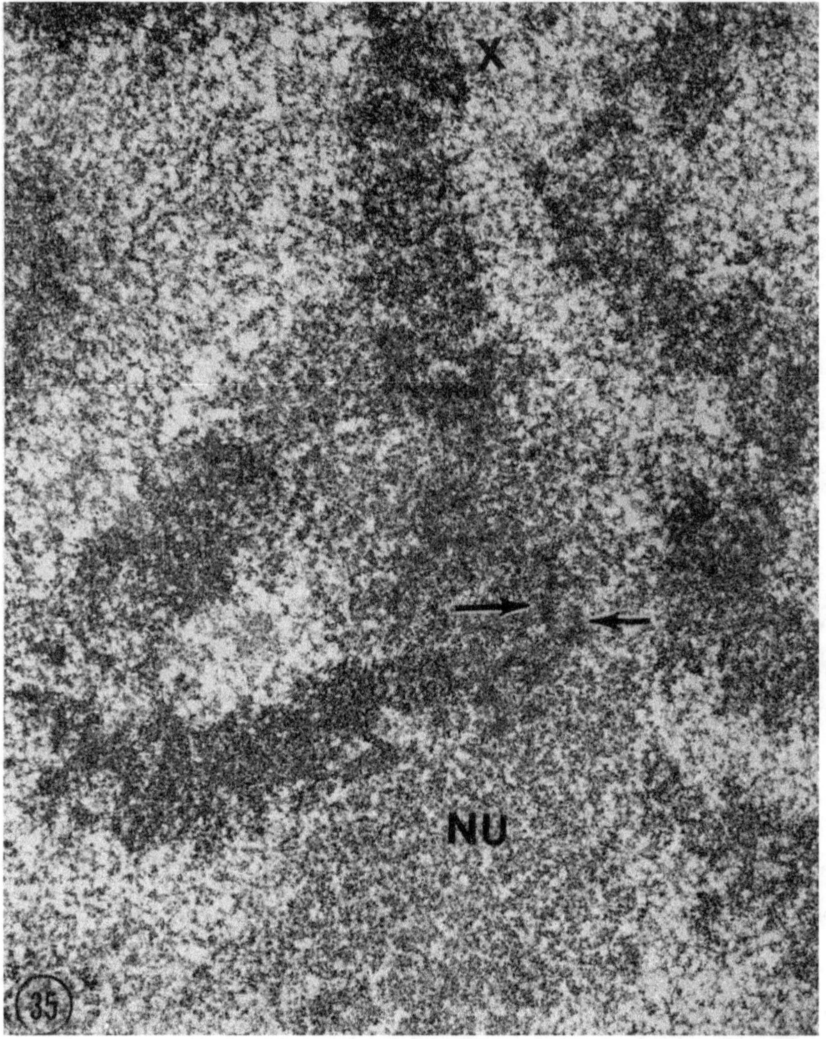

FIG. 35. Electron micrograph of X chromosome of the rat kangaroo. The nucleolus (NU) can be seen associated with the secondary constriction. Arrows indicate dense "axial core" through secondary constriction. × 22,850.

puff develops from a condensed band the coarse fibrils of the bands unfold into fine threads which loop out in all directions. RNA synthesis occurs in the puff and appears to exist structurally in the form of 300 Å "Balbioni ring" particles along the loops (Stevens, 1964; Swift, 1965). Presumably each of the thin threads is a DNA double helix which transcribes the RNA. When RNA synthesis stops the chromatin threads recondense, the puff disappears, and the dense band reappears (cf., Swift, 1965). The similarity in morphology of the kinetochore to salivary gland chromosome puffs is implicit, although in the case of the puff we are dealing with structures which are several orders of magnitude larger than those of the

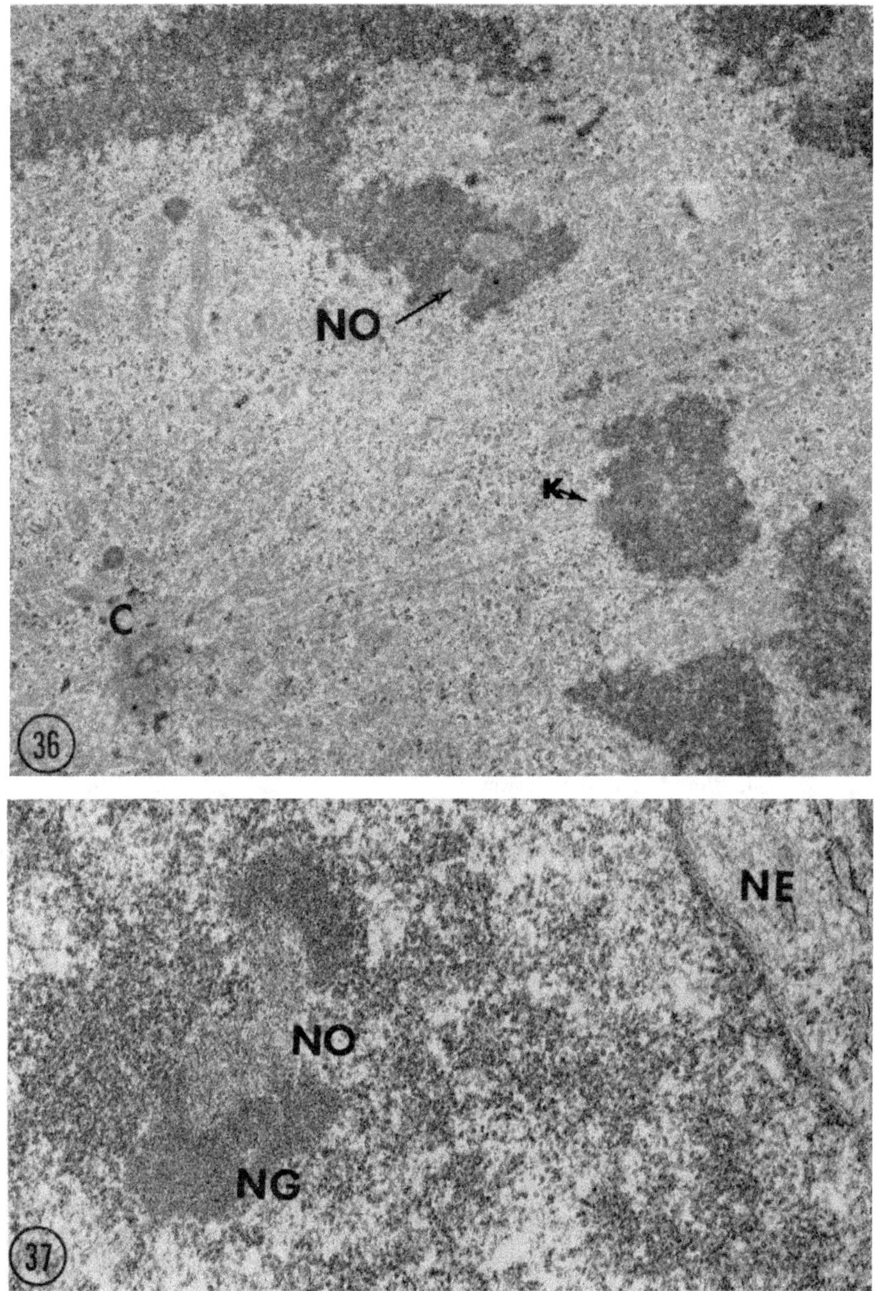

FIG. 36. Metaphase in rat kangaroo showing nucleolar organizer (NO), kinetochore (K), and centriole (C). × 10,000.

FIG. 37. Nucleolar organizer (NO) in telophase after reformation of nuclear envelope (NE). At this stage dense nucleolar granules (NG) which appear along the periphery of the nuclear organizer are presumed to be RNA. × 35,000.

kinetochore. The organization and basic components, namely, chromatin threads organized into lampbrush loops, are essentially the same. The puff, like the kinetochore, appears and disappears in a precise and definable pattern, which is apparently regulated by factors and events in the immediate protoplasmic environment.

The Kinetochore and Lampbrush Chromosomes

The arrangement of microfibrils into a lampbrush-like organization is best demonstrated in the oocytes of salamanders (e.g., *Triturus*) during meiotic prophase (Gall, 1963b; Callan and Lloyd, 1960). During the extended growth of the primary oocyte, the diplotene chromosomes undergo pronounced elongation and morphological change. They may reach a length of 1 mm, which is several hundred times longer than the corresponding somatic chromosomes. Each homolog contains a pair of axial threads on which Feulgen positive chromomeres are located. From the chromomeres lateral loops extend for a distance of 10 to 15 μ (cf., Moses, 1964). Careful digestion with deoxyribonuclease by Gall (1963a) has demonstrated that both the lateral loops and interchromomeric threads contain DNA. Gall found that breaks in the lateral loops exhibited two-event kinetics while those in the interchromomeric threads followed four-event kinetics. This strongly suggested that each homolog contained a pair of DNA duplexes while each loop was a single DNA duplex. Presumably at the chromomere each duplex loops out in opposite directions. RNA synthesis takes place along the periphery of the loops.

The kinetochore can be recognized in lampbrush chromosomes of the axolotl (Callan, 1966). Being inactive, however, the axial threads are condensed into a series of chromomeres which are void of lateral loops. The kinetochore extends for a region of 10 μ in these chromosomes. When the kinetochore becomes active in metaphase it would likely form lampbrush loops much like those described in mammalian kinetochores. The remainder of the chromosome being metabolically quiescent at that time would be condensed.

The Kinetochore and the Synaptinemal Complex

An organization more closely allied to that of the kinetochore is seen in the so-called synaptinemal complex. Moses (1958) and Fawcett (1956) were the first to describe this component in thin sections of paired homologous chromosomes during meiotic prophase. Since the structures seen were related to synaptic pairing, Moses (1958) appropriately termed it the synaptinemal complex. Basically it consists of a tripartite structure containing a pair of axial elements, each approximately 100 Å in diameter, separated by a less dense zone some 120 mμ to 150 mμ wide. A central pairing line midway between the axial fibrils can often be seen. The axial fibrils are in close association with laterally displaced microfibrils 70 to 100 Å in diameter. The latter are thought to loop out at right angles to the axial elements (Moses, 1964). The presence of nucleic acids, presumably DNA, in the lateral fibrils and axial element has been demonstrated at the ultrastructural level by Coleman and Moses (1964).

The synaptinemal complex, like the polytene puff and the meiotic chromosomes of amphibian oocytes, is another case of lampbrush-type organization associated with metabolic activity along the chromosome. Studies by Utakoji (1966) have shown that ^{3}H-uridine is incorporated throughout the length of the chromosome during pachytene in Chinese hamster spermatocytes. The label in every case appeared to be associated with the lateral loops. The nucleoli of the same cells, however, failed to show any label. The function of chromosomal RNA synthesis in primary spermatocytes is thus far unknown.

Status of the Kinetochore in Chromosome Organization

On the basis of light microscope studies, two general hypotheses exist regarding the structural relationship of the kinetochore to the remainder of the chromosome (cf., Schrader, 1953). One maintains that the kinetochore is, for the most part, a product "exuded" by the chromosome. Sharp (1934) and Schrader (1939) proposed that the so-called spindle spherule, tiny globules at the end of chromosomal spindle filaments, were products of the "genonema." (The latter term was given to threads which extended from one end of the chromosome to the other.) A somewhat similar hypothesis was put forth by Matsuura (1941). Nebel (1939) and Darlington (1939) proposed that the kinetochore was a liquid body containing parallel-oriented micelles. The other hypothesis maintained that the kinetochore is an integral part of the chromosome, which is structurally different because of the timing of condensation (Carothers, 1936). The latter perhaps is more in keeping with current concepts of the structure and function of kinetochore and chromosome.

Studies of chromosome structure show that the elements seen in the metaphase kinetochore fit well into the current scheme of chromosome organization. Light microscope studies clearly show that each chromatid is double, consisting of two half-chromatids. Further division of the half-chromatids has also been reported (cf., Steffensen, 1959; Cole, 1967). The strandedness of the chromatid in relation to the DNA–histone complex is still debatable. However, from studies of lampbrush chromosomes (Gall, 1963b), polytene chromosomes (Swift, 1965), and the synaptinemal complex (Moses, 1964), we can assume that only a few DNA–histone strands (perhaps two to four) make up the metaphase chromosome. The degree of condensation along the chromosome and the obvious "multistrandedness" of the condensed regions is therefore a result of the superimposition of coiled strands in this region.

Each chromatid contains a kinetochore (sister kinetochore) which, as we have shown, is divided into two separate kinetochore filaments. The most plausible interpretation, therefore, is that each half-chromatid is continuous with a single kinetochore filament. The kinetochore filament may in fact represent an "active locus" on the half-chromatid. The same site becomes condensed and inactive after mitosis or meiosis, whereas other loci suppressed during division become active in the next interphase. Structurally, therefore, we envision the arrangement of the 50 to 80 Å fibrils of the kinetochore filaments in much the same way as Moses (1964) views the microfibrils of the synaptinemal complex. Reports of DNA in the kinetochore have come from Feulgen staining at the light microscope level. Until it can be demonstrated unequivocally that axial elements and lateral loops of the kinetochore contain DNA, we must admit that homologies between the lampbrush structures of the kinetochore and those of lampbrush chromosomes, polytene chromosome puffs, the synaptinemal complex, and nucleolus organizers are based largely on similarity in morphological pattern. Nevertheless, to view the kinetochore as an active locus of the metaphase chromosome provides a more functional insight regarding its structure and arrangement on the chromosome.

The existence of two kinetochore elements on each chromatid supports the contention that the chromosome is a "multineme" structure and is not composed of a single DNA strand extending from one end of the chromosome to the other. If each 50 to 80 Å fibril of the axial element is a superhelix containing a single Watson-Crick DNA duplex, the chromatid would contain at least four parallel strands of DNA or a total of eight strands in the metaphase chromosome.

The problem of control and regulation of the kinetochore locus is even further from our grasp. What stimulates its activity in prometaphase and inactivates it in telophase? At this time too little is known about the regulation of conventional genes to speculate on how the cell may regulate a locus whose status as a gene is still questionable.

VARIATION IN KINETOCHORE STRUCTURE AND BEHAVIOR

The Diffuse Kinetochore

The descriptions in this paper are directed primarily to so-called localized kinetochores which are confined to a specific site on the chromosome. There are, however, notable exceptions to this pattern. The chromosomes of hemipteran insects are all characterized by the presence of a "diffuse kinetochore" which, as the term implies, is distributed throughout the length of the chromosome (Schrader, 1935; Hughes-Schrader, 1942). A similar arrangement is also found in the chromosomes of the wood rush, *Luzula* (Brown, 1954).

Experimental evidence for the diffuse nature of the kinetochore of coccid chromosomes was first obtained by Hughes-Schrader and Ris (1941) and Hughes-Schrader and Schrader (1961). These investigators found that when such chromosomes were fragmented by irradiation each fragment behaved as if it contained its own kinetochore. That is, each broken segment moved in a directed manner on the spindle. Generally, "acentric" fragments from chromosomes with localized kinetochores may exhibit some movement after irradiation, but not in a directed manner like the fragments of coccid chromosomes. The results from irradiated coccid chromosomes therefore suggested that the kinetochore "substance" was distributed throughout the length of the chromosome.

Further evidence along these lines is seen from the manner in which mitotic chromosomes of *Steatoccus* and *Luzula* move to the pole in anaphase. Instead of a particular region leading the rest of the chromosome, the entire chromosome moves to the pole with its axis perpendicular to that of the spindle (reviewed by Mazia, 1961).

The diffuse kinetochore as such is not seen in light microscope preparations. Recently, however, Buck (1967) examined the fine structure of the kinetochore in the bug *Rhodnius prolixus*. His study indicated that the kinetochore existed in the form of less-dense material along the surface of the chromosome. For the most part, the material of the kinetochore was structurally similar to that seen in localized kinetochores. Further investigations may show that the structural elements in the diffuse kinetochore are the same as those seen in localized kinetochores. In the former, they are obviously continuous throughout most of the chromosome length.

If the kinetochore is in fact a genetic locus, it is difficult to imagine how it could exist throughout the length of the chromosome, as in the case of the diffuse kinetochore. One possible explanation is that many chromosomes with localized kinetochores fuse together, forming fewer giant "polycentric" chromosomes (Schrader, 1935; Vaarama, 1954). Obviously, the answer must await further cytological investigation.

Neocentric Activity

While spindle attachment and chromosome movement generally involve a well-defined chromosomal locus, other sites may also undergo directed movement. Rhoades (1952) has described so-called "neocentric" behavior of chromosomes in several strains of maize. In such cases, the ends of chromosomes may undergo abrupt poleward movement at a time when the kinetochores are still at the metaphase plate. Such behavior strongly suggests that regions other than the usual site of spindle attachment can function as a kinetochore. As Mazia (1961) indicates, however, neocentric movement obviously plays a rather insignificant role in mitosis since in every case the kinetochore eventually takes over the role of anaphase movement. Neocentric behavior may be important in meiosis, however, in the mechanism of coorientation of the kinetochore (Bajer, personal communication). Nevertheless, it would be interesting to know the structural basis of such behavior. Cases of neocentric activity in *Haemanthus* chromosomes (Östergren and Bajer, 1960) are now being investigated at the ultrastructural level (Bajer, personal communication).

THE CENTRIOLE

The centriole has been studied by light microscopy for almost a century, but only in the last decade has its ultrastructure been defined. Currently, many of the details of its architecture are still uncertain, but enough is known to make some reasonable attempts to correlate structure and function. Although our own experience is largely limited to a single species, the Chinese hamster centriole, nothing in the literature concerning centrioles and basal bodies of other species suggests that they are basically different. The centriole appears to be a very stable structure, presenting an almost identical ultrastructure in fungi (Aldrich, 1967; Renaud and Swift, 1964), algae (Ringo, 1967), protozoa (Gibbons and Grimstone, 1960), echinoderms (Harris, 1962), molluscs (Gall, 1961), amphibians (Reese, 1965), birds (Doolin and Birge, 1966), mammals (Sorokin, 1962; Stubblefield and Brinkley, 1967). Every flagellated metazoan cell seems to contain a basal body of essentially the same design as the mammalian centriole. Thus, although we describe only a single species, we anticipate that this description, and the resulting theory of centriole function, will hold true in most, if not all, cases. These details of centriole structure were first presented by Stubblefield and Brinkley (1967), and that work should be consulted for more information.

Structure

The centriole is a minute cylindrical structure 0.5 to 0.7 μ long and 0.20 to 0.25 μ in diameter. Its most striking feature is perhaps the "pinwheel" pattern of its wall elements when viewed in cross section (Fig. 38). The centriole wall consists of microtubules fused together in groups of three, each group termed a triplet. Since the three microtubules of a triplet almost lie in one plane, each triplet is actually a blade which runs the full length of the centriole. Characteristically, there are nine of these triplet blades comprising the wall. The angle of the blade is not constant, but changes from one end of the centriole to the other in some situations. The two ends of the centriole are different, containing different structures (see below), and have been defined functionally in basal bodies. The distal end is capable of generating the shaft of a cilium or flagellum (Gibbons and Grimstone, 1960; Gall, 1961), while the opposite proximal end is always associated with the formation of the daughter, or procentriole, when reproduction occurs. When viewed from the proximal end, the triplet blades slant outward in a counterclockwise direction, as in Figure 38.

In favorable sections, a "cartwheel" structure is visible in the proximal end of the centriole, with delicate spokes radiating from a central hub. The cartwheel is seen in only the end section of centrioles (Fig. 39), but occurs repeatedly throughout much of the length of basal bodies (Gibbons and Grimstone, 1960). The spokes of the cartwheel, which are apparently segmented along their length and at their distal ends, terminate in a structure projecting from the triplet blade.

In the distal end of the centriole a more amorphous structure is found with eight fold, rather than nine fold, symmetry as in the cartwheel (Stubblefield and Brinkley, 1967). This structure is seen only in the distal end section of the centriole and has not yet been detected in basal bodies (see montage, Fig. 40b). At the centriole distal end, transition fibers are also found radiating peripherally from the triplet blades (Fig. 40). These occur also in basal bodies and are thought to function in ciliogenesis in the attachment of the ciliary sheath to the basal body (Fawcett, 1961).

The lumen of the centriole usually appears empty, except for a small membrane-bound vesicle termed the central vesicle or nucleoid; this structure seems to be a constant feature of Chinese hamster centrioles, but may not appear in some species. It is usually about 600 Å in diameter (Fig. 41).

With careful fixation and photography, a second structure is demonstrable in the centriole lumen. A large helix winds about eight to ten turns through the centriole just inside the triplet blades. In favorable cross sections, only part of a helix turn is visible in the thickness of a single section (Fig. 38). The structure is

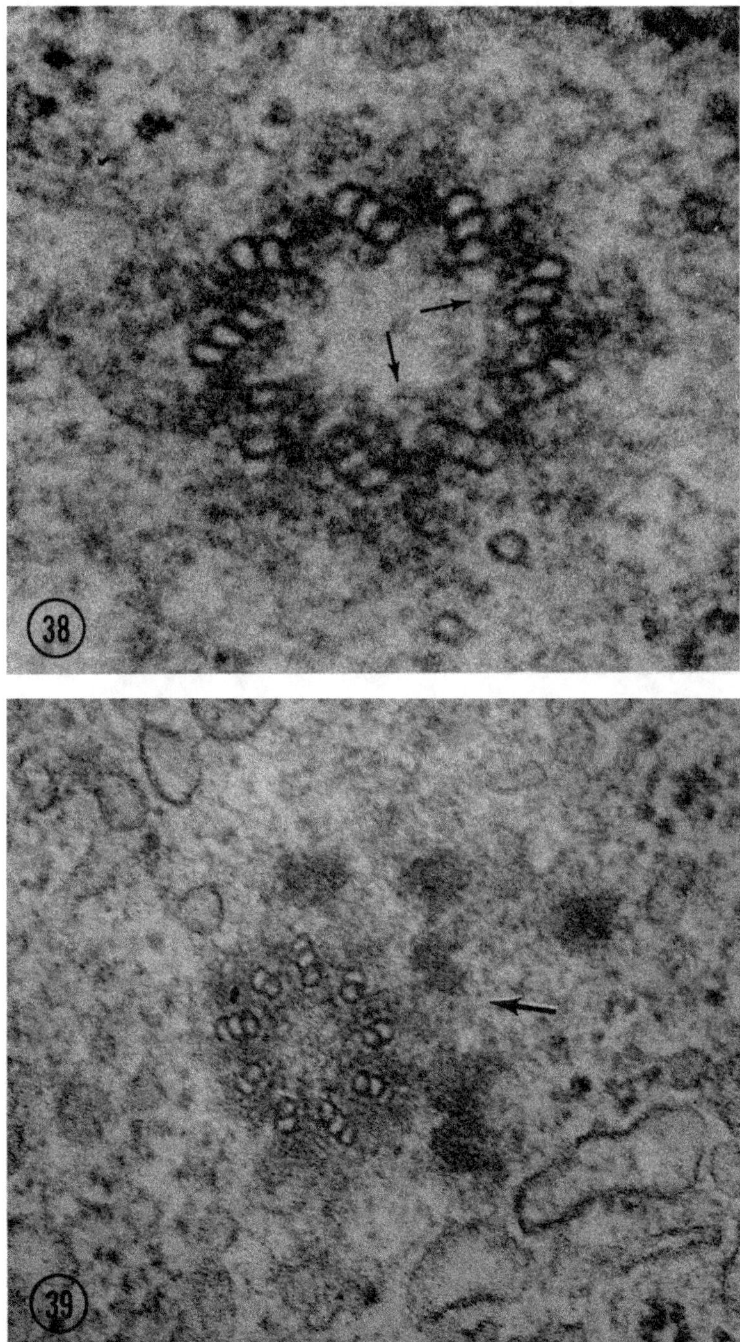

FIG. 38. Cross section through a centriole of the Chinese hamster. The arrows point to the ends of a segment of the helix in the lumen. ×225,000.

FIG. 39. Cross section through the proximal end of a centriole showing the cartwheel and pericentriolar satellites. One satellite (arrow) is attached to two microtubules, one inserting directly into the centriole wall. × 152,000.

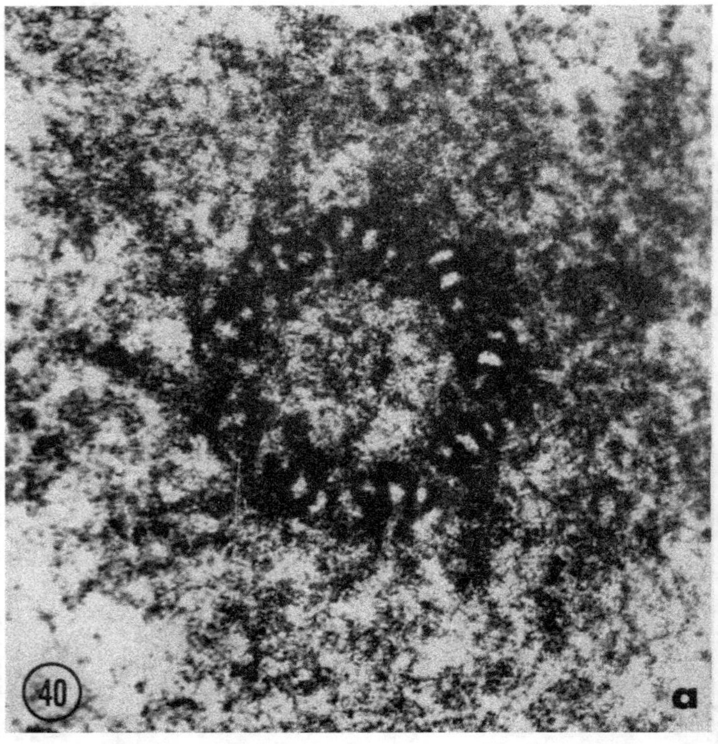

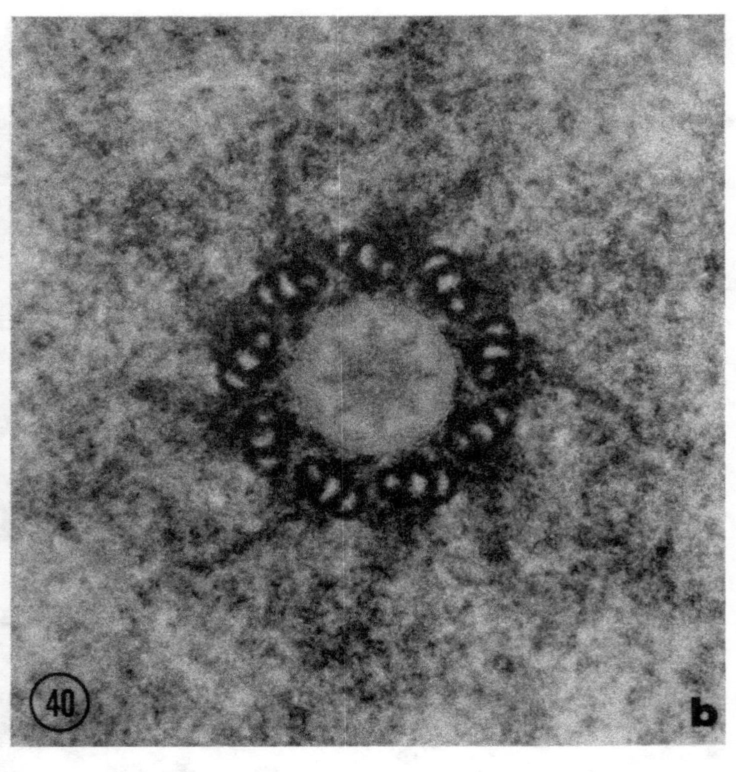

FIG. 40. Cross section through the distal end of a centriole; transition fibers radiate from the triplets on the outside. In (a) the original electron micrograph is shown, while (b) reveals more detail in the photograph after image enhancement by Markham rotation (Markham et al., 1963) (a montage with stellate structure showing eight fold symmetry inserted in the lumen of the nine triplets). The stellate structure reinforces best as an eight fold structure. (From Stubblefield and Brinkley. 1967. *In* Warren, K. B., The Origin and Fate of Cell Organelles. Courtesy of Academic Press, Inc.) × 229,000.

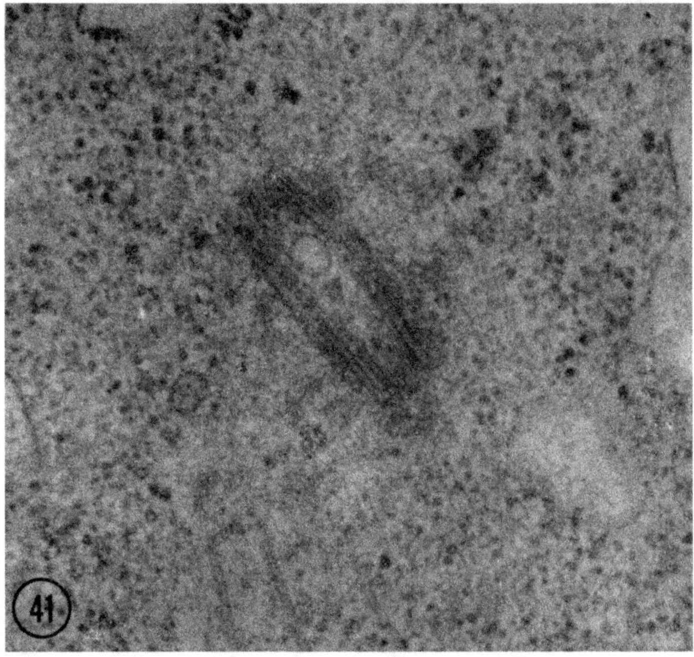

FIG. 41. Oblique section of a centriole showing a prominent central vesicle. × 70,000.

visible in some longitudinal centriole sections and has also been seen more prominently in centrioles undergoing reproduction (Fig. 45) (see also Stubblefield, 1968). The helix is about 1,300 Å in diameter with a calculated stretched length of about 5 μ. The helix filament is itself only about 60 to 70 Å in diameter. The helix appears to be centered in the centriole lumen by its contact with dense projections from the inner microtubule of each triplet blade. Evidence was presented earlier (Stubblefield and Brinkley, 1967) that these projections were RNase digestible. Pericentriolar satellites, densely-staining bodies occurring near centrioles, sometimes appear to be linked by fine filaments to dense material occupying the spaces between adjacent triplet blades (Fig. 42). Recent experiments suggest that the satellites and the dense wall material are also almost completely removed by RNase treatment (Figs. 42 and 43).

Reproduction

In an earlier report (Stubblefield, 1968) the centriole reproduction cycle in Chinese hamster fibroblasts was presented. Centriole reproduction is somehow geared to the reproductive cycle of the cell as a whole, so that daughter centrioles appear while the cell nucleus is involved in DNA synthesis. This also was observed to be the case in HeLa cells by Robbins et al. (1968). Centrioles reproduce by a generative mechanism in which a single daughter is assembled alongside the parent. There is some evidence (Stubblefield, 1968) that each part of the parent structure

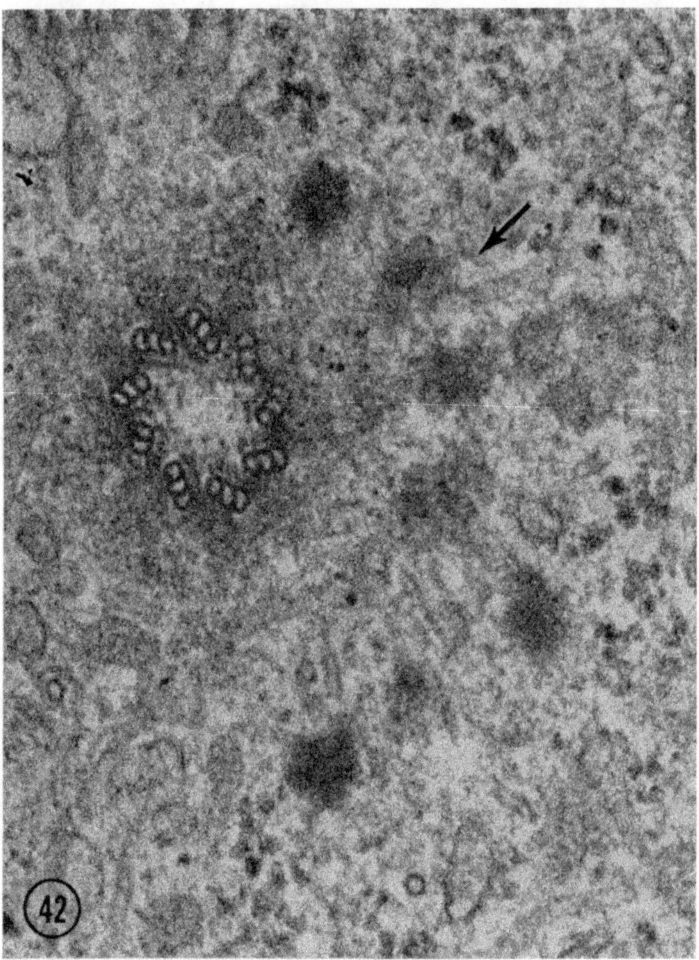

FIG. 42. Centriole with numerous pericentriolar satellites. Many appear to be attached to the centriole wall by filaments. Note the microtubule terminating near one satellite (arrow). × 108,000.

is duplicated, and the copy then transferred bodily to the site nearby where the daughter is being assembled.

The daughter, or procentriole, is first visible about 6 hours before mitosis as a short cylinder oriented perpendicular to the parent cylinder and near the proximal end (Fig. 44). The daughter forms at right angles to the parent because the microtubules generated by the parent are used to form the wall of the procentriole; these microtubules all are oriented perpendicular to the long axis of the parent (Stubblefield and Brinkley, 1967). By the time of mitosis, most of the centriole substructures (helix, nucleoid, cartwheel, etc.) have been added, and the cylinder lengthened to about half the normal length (Fig. 45). Usually, the immature centriole does not function in spindle formation, but is simply carried to the spindle pole attached to its parent. This was documented by Murray et al. (1965). However, in cells treated with Colcemid for several hours to block mitosis, we have often

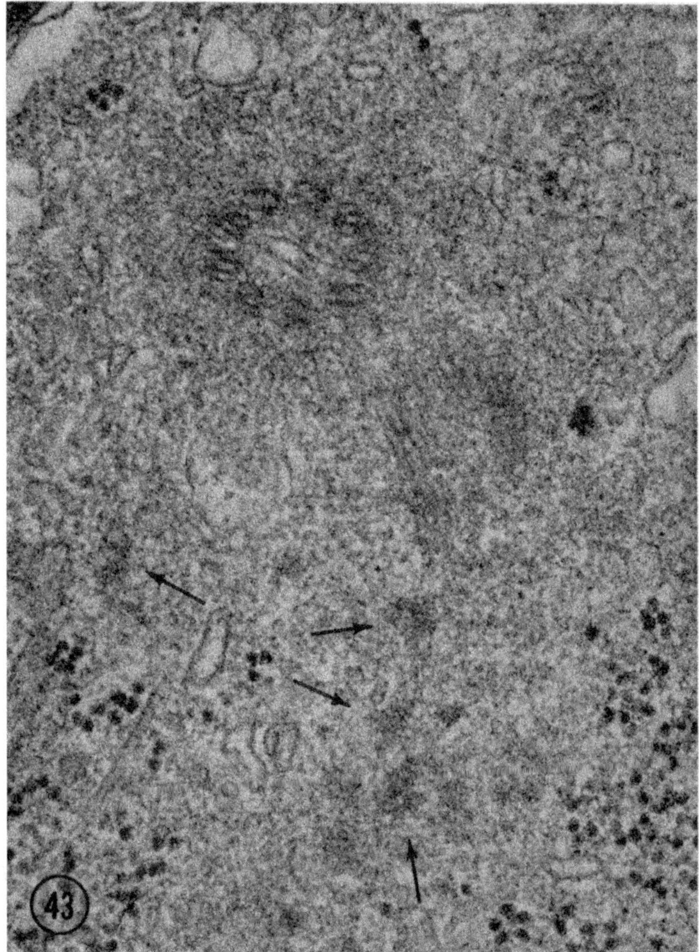

FIG. 43. Centriole similar to that shown in Figure 42, but treated for 30 min with ribonuclease (1 mg/ml at 37°C) immediately after glutaraldehyde fixation. Note the loss of density in the pericentriolar satellites (arrows) and in the centriole wall. (The lack of detail in the triplets is due to the angle of the section and not the enzyme digestion.) × 91,000.

observed cells with functioning new daughter centrioles; they apparently can mature in the arrested metaphase cell (Stubblefield, 1968). In some cases the daughter will separate from the parent and establish its own spindle pole, leading to a multipolar spindle. In untreated cells the daughter does not mature until several hours after mitosis. Thus each cell normally receives two centrioles (one mature and one immature) at telophase, and these begin to reproduce in synchrony with chromosome replication as the cell prepares for the next mitosis.

Function

What role does the centriole play in the mitotic mechanism? It may be argued that since higher plants achieve mitosis without centrioles (Mazia, 1961), they

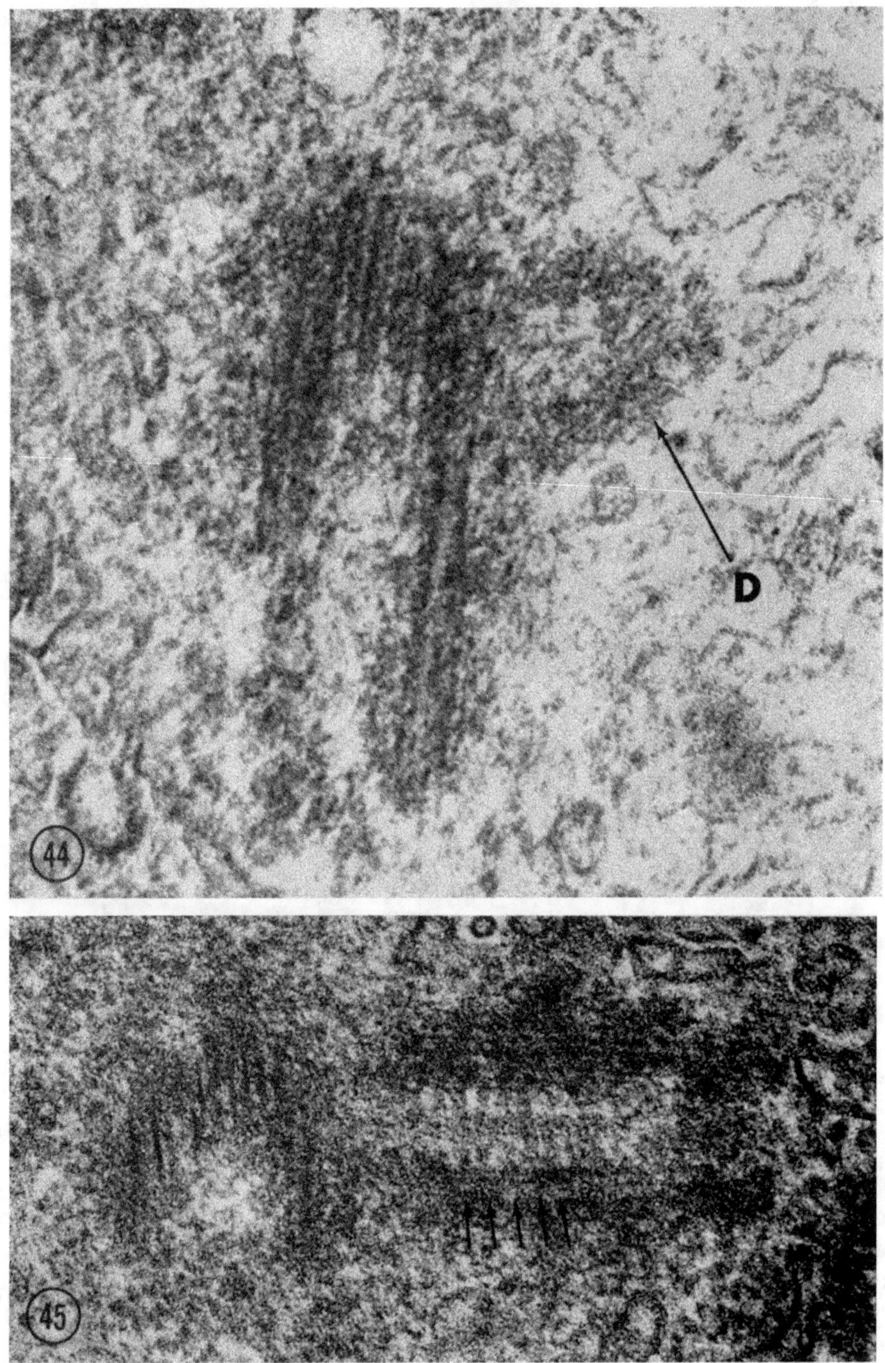

FIG. 44. Centriole with attached daughter (D) procentriole forming at right angles to the parent. Reduced 5 percent from × 90,000.

FIG. 45. Parent and daughter centrioles of a cell in prophase. The procentriole is now longer than the one in Figure 44, but not yet full length. In the lumen of the daughter several turns of the helix can be clearly seen (arrows). Reduced 5 percent from × 82,000.

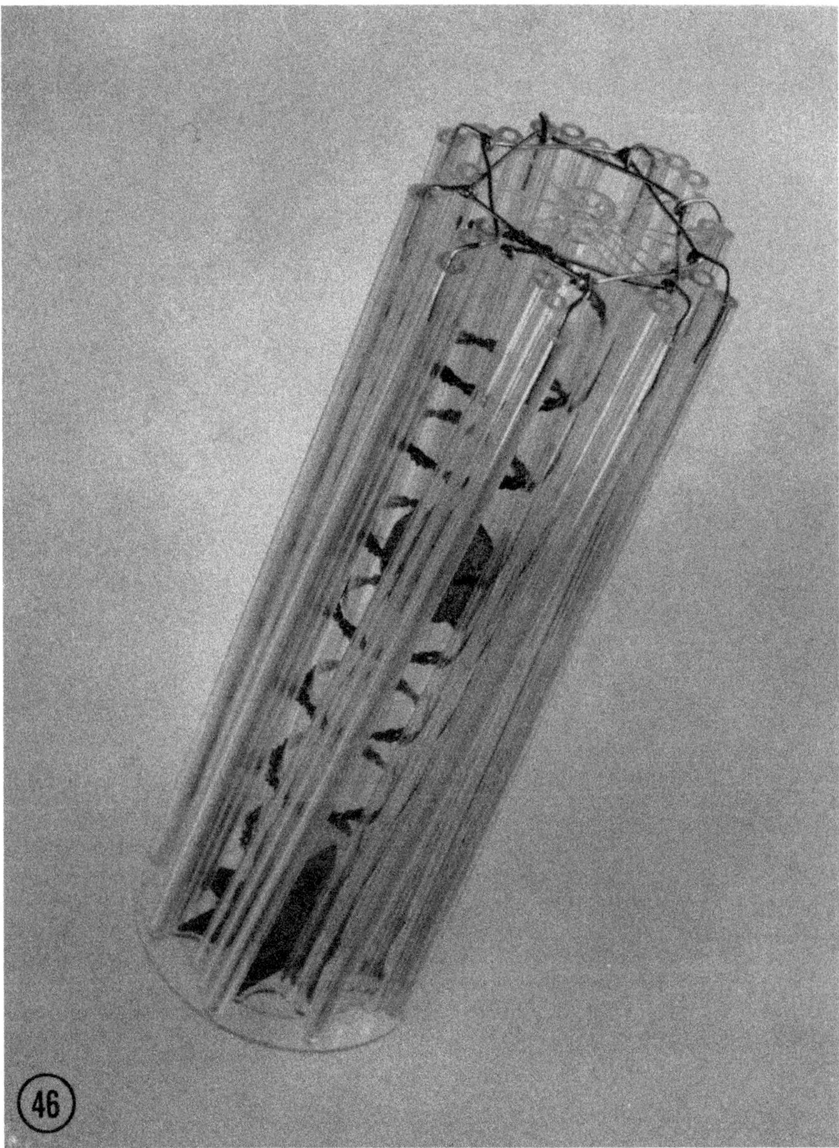

FIG. 46. Plastic model of a centriole constructed according to the description given in the text. The model has moving parts as diagrammed in Figure 47.

play no important role at all. From this viewpoint, the spindle mechanism in animal cells achieves the partitioning of centrioles simply to assure future cell generations of basal bodies for ciliogenesis. However, our own studies of Colcemid effects in mammalian fibroblasts suggest that the centriole participates actively in the mitotic mechanism (Brinkley et al., 1967b). Three phenomena support this view: (1) Colcemid blocks centriole separation and effectively prevents the formation of the bipolar mitotic spindle; (2) the continuous microtubules of the spindle are formed

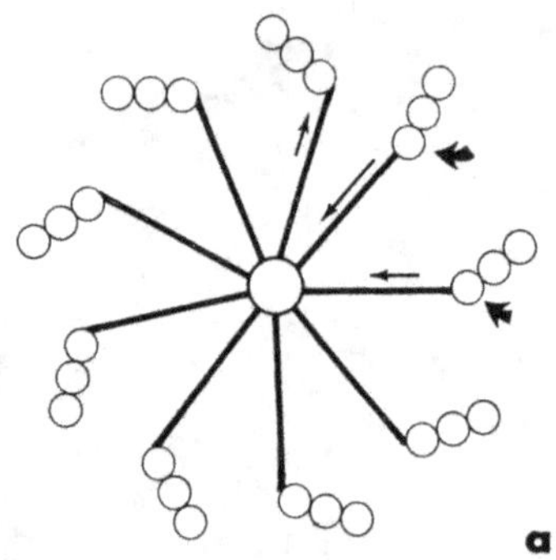

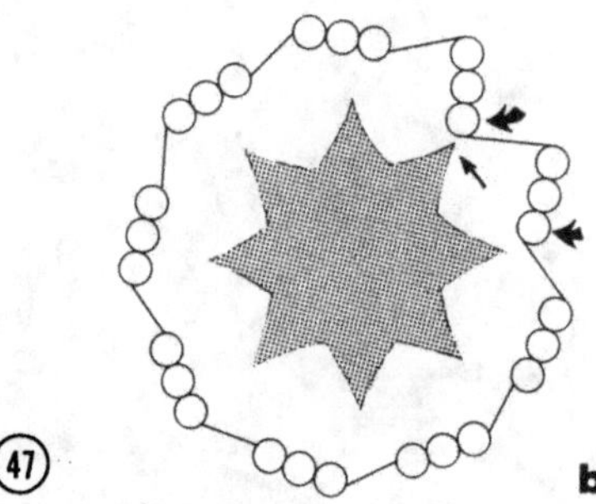

FIG. 47. Diagram of the centriole mechanism as proposed in the text. In the proximal end (a) sequential contraction of the "spoke" fibers (clockwise sequence) pivots the triplets along the outer microtubule and pulls the inner microtubule toward the hub of the cartwheel (arrows). At the distal end of the centriole (b) the pivoted triplets (large arrows) interact sequentially with the points of the octagonal structure in the lumen, each interaction moving the structure one seventy-second of a turn counterclockwise (small arrow), assuming the interaction to be some form of attraction. Mutual repulsion would drive the central structure in a clockwise direction. At the end of the stroke the triplet is returned to its original position by elastic interconnections in the distal end.

as the centrioles separate in recovery from Colcemid treatment; and (3) too many functioning centrioles in a cell often result in multipolar mitosis and an obvious disruption of the normal process (Stubblefield, 1968). Our studies thus suggest that the centrioles determine whether and where the chromosomes will migrate in anaphase. By whatever means the division mechanism may operate in acentriolar cells, it seems clear to us that centrioles are important in mitosis of cells that have them.

Assuming, then, that centrioles are an active, and not a passive, part of the mitotic apparatus, how may we correlate their precise, symmetrical ultrastructure with their role in mitosis? To begin with we must keep in mind that centrioles serve a dual purpose; part of centriole ultrastructure may be dictated by the biomechanical considerations of ciliogenesis and ciliary function. Two lines of evidence will be pursued here; one is related to the nature of the overall centriole structure itself, while the other is related to the structure of microtubules.

If a plastic mock-up of the centriole is constructed according to the descriptions given earlier in this section, some interesting possibilities appear. Three clues suggested to us that the centriole was a kinetic structure. The first was the homology between centrioles and basal bodies; since a cilium beats, perhaps the basal body has moving parts. A second consideration was the observation that the angle of the triplet blade in the centriole wall is not constant; perhaps it is hinged and the pitch of the blade can be changed. Thirdly, the finding of a structure with eight fold symmetry surrounded by nine triplet blades suggests rotational interaction. These ideas led to the model shown in Figure 46, which presents the centriole as a *rotary engine* and a classical machine of molecular dimensions.

How would such a machine function? The power would derive from the cartwheel structure situated in the proximal end. If each triplet blade is hinged along its outer microtubule, then contraction of the spoke fibers of the cartwheel would pull the triplet blade into a position of maximum angle in relation to the centriole cylinder surface, as in Figure 47. Sequential contraction of the spokes would lead to sequential interaction of the triplet blade with the eight teeth on the octagonal distal end structure and cause its rotation. In effect, the latter structure is a cam, the triplet blades are connecting rods, and the cartwheel spokes are pistons. The purpose of such a mechanism would be to rotate the helix in the centriole lumen, for reasons which will become apparent later. The kinetic parts of the centriole are diagrammed in Figure 47 according to this hypothesis.

THE ULTRASTRUCTURE OF MICROTUBULES

As was mentioned earlier, we feel that the available evidence strongly supports the conclusion that part of the microtubules of the mammalian mitotic apparatus are somehow generated by the centriole. Two kinds of spindle fibers have been found in many species (Mazia, 1961), those connecting the chromosomes to the mitotic poles (chromosomal microtubules), and others which run from pole to pole (continuous microtubules). In Chinese hamster fibroblasts the pole-to-pole microtubules are generated by the centrioles.

In order to understand how centrioles generate microtubules, we must first attempt a molecular description of the microtubule. Ultrastructure studies of microtubules have been presented by Gall (1966), Ledbetter and Porter (1964), and Moor (1967) using several species and techniques. Probably the least subject to artifact and easiest to interpret are the studies of Moor, who examined yeast microtubules using the technique of freeze-etching. It is tempting to correlate all of the evidence into a single model of microtubule structure, but there is no a priori reason for believing that microtubules of all species must be identical, or even that all microtubules of a single cell are alike in ultrastructure.

Microtubules of the Chinese hamster are generally very similar to those of other species. In cross section they appear to be hollow tubes of 200 to 250 Å in outer diameter and 120 to 150 Å inner diameter (Fig. 48). They are composed

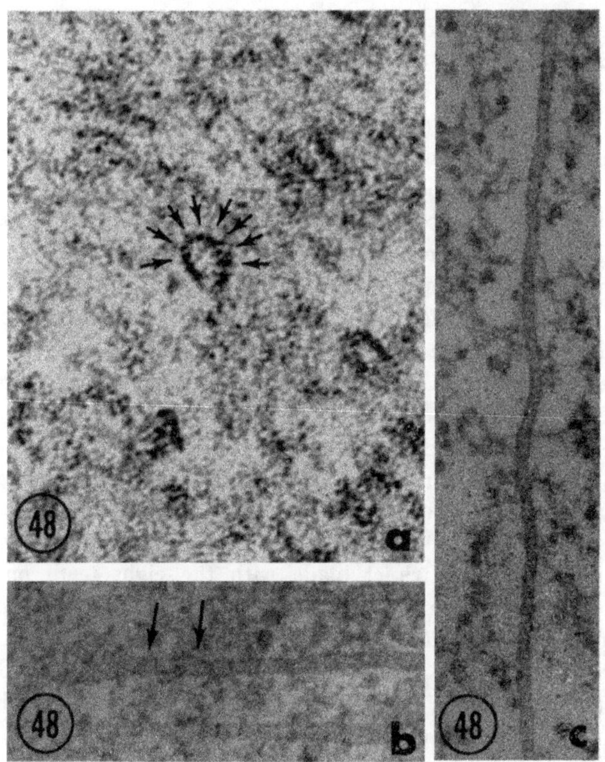

FIG. 48. Thin sections through microtubules of the Chinese hamster revealing some periodicity. In cross section (a), individual subunits can be seen around more than half of the circumference; the spacing suggests 13 units to the complete circle. Longitudinal sections (b) sometimes reveal a periodicity of about 80 Å (arrows), and in some cases exhibit a gross helical configuration (c). (a) × 225,000, (b) × 98,000, (c) × 74,000.

of subunits, about 40 Å in diameter, arranged in rows parallel to the long axis of the tubule (Fig. 49), and numbering probably 13 around the cross-sectional view (Fig. 48). The studies of Moor (1967), and also our own, indicate that the microtubule is helical (Fig. 49). Earlier studies of Zimmerman (1960) indicate that isolated spindles are ribonucleoprotein in nature, with an RNA content of about 6 percent. Based on these observations, we propose the model shown in Figure 50. The position of the RNA in this structure is only hypothetical, however, and is assumed by analogy to the known structure of tobacco mosaic virus (TMV). In the case of microtubules the pitch of the helix requires a double helix of protein subunits (see Moor, 1967), rather than a single one as in TMV, so it seems likely that the RNA occupies only one of the two intersubunit helical "grooves." Although this model seems to fit our own observations best, one might expect variations of this structure in different species. Some microtubules may be completely analogous to TMV in structure, ie., single helix, while others may be double or even triple helix, in nature. The number of protein subunits seen in cross section could in different species conceivably vary from 10 up to perhaps 20 and the structure still be generated in the same basic way, as described next.

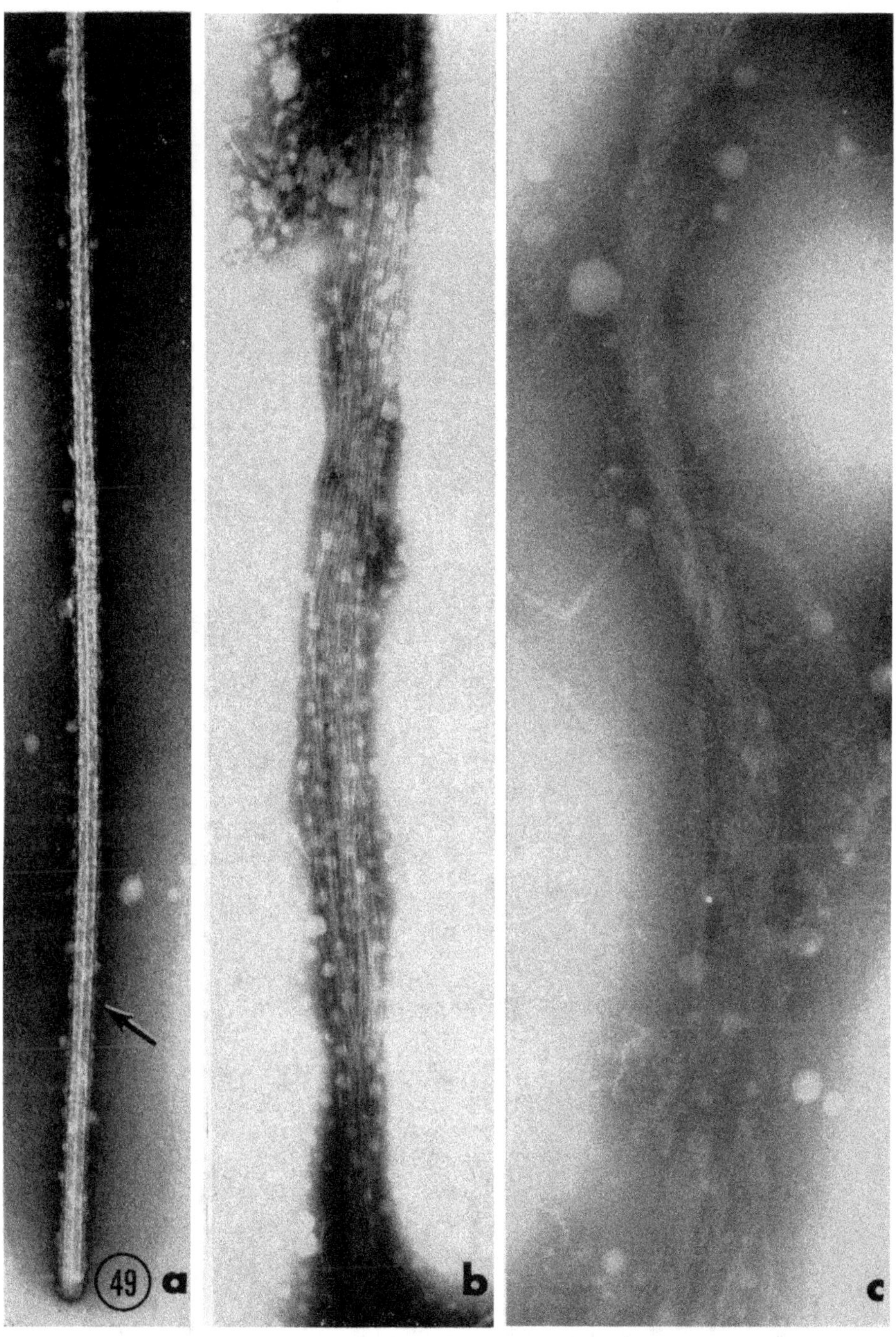

FIG. 49. Microtubules negatively stained with phosphotungstic acid. Spindles were isolated from Chinese hamster mitotic cells in 12 percent hexylene glycol (pH 6.5), fixed in 3 percent glutaraldehyde, and sonicated briefly. An intact microtubule is shown in (a); individual subunits are visible (arrow). In (b) a disrupted microtubule is composed of numerous fine filaments which are apparently the linear columns of subunits in the original structure. However, other microtubules dissociate into fibers that have a definite helical pattern (c). (a) × 90,000, (b) × 91,000, (c) × 115,000.

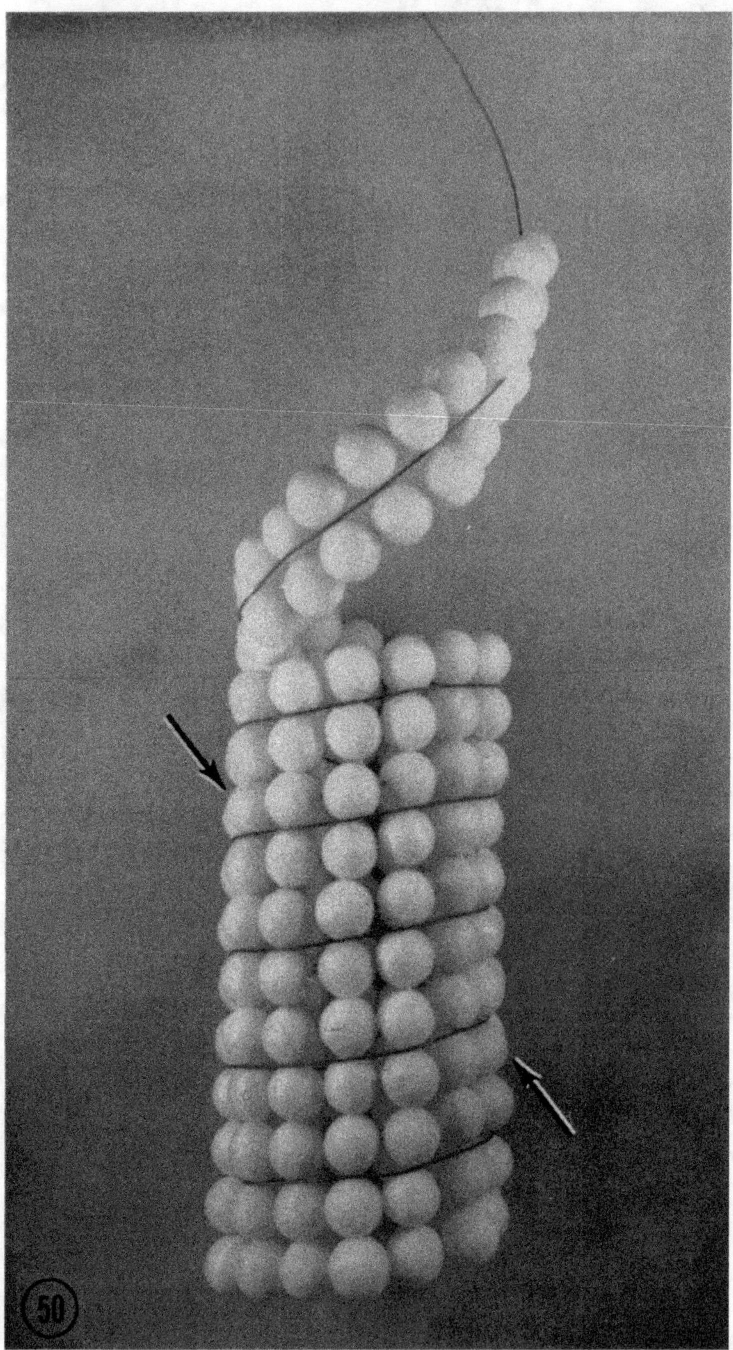

FIG. 50. Hypothetical model of microtubule based on evidence presented in the text. The model demonstrates the theoretical assembly as a double helix of protein subunits attached to an RNA fiber. Disruption of the structure along a secondary helix (arrows) would yield the configuration shown in Figure 49c.

CENTRIOLE AND KINETOCHORE INTERACTIONS

In discussing kinetochore structure, we have already indicated that part of the microtubules of the mitotic apparatus (chromosomal microtubules) end in direct contact with the axial element of the kinetochore. The studies of Inoué (1964) indicate that these microtubules are probably generated by the kinetochore. If so, then a tentative concept of their formation is relatively simple to imagine. Assume (as was discussed earlier, pp. 150–152) that the kinetochore is simply a chromosomal gene segment that is active in mitosis; then the RNA produced by the kinetochore could simply self-assemble with the protein subunits abundantly present in the cytoplasm (Went, 1966, p. 75) to make a microtubule.

The idea of interpreting the kinetochore as an active gene segment is very intriguing, for here we may possibly have a direct morphological view of genetics in action at the molecular level. Carrying the molecular interpretation a bit further, if each microtubule inserts into the kinetochore at one end of an RNA transcribing segment of the chromosome, then we are faced with the question of how a relatively short DNA segment can produce such a long RNA to run the length of a microtubule. There are two possibilities. On the one hand, we may have misinterpreted microtubule architecture, i.e., the RNA may be in short pieces rather than in one continuous length. On the other hand, it would be possible to explain how the kinetochore could generate a long RNA strand, if the DNA which gives rise to microtubule RNA is not in the axial element but rather in fibrous loops around it. If these loops are actually circular DNA, then they could cycle endlessly through an RNA polymerase attaching them to the axial element to produce long lengths of RNA. This idea is depicted in Figure 51.

Microtubule Generation by the Centriole

With these concepts in mind, let us return to the question of centriole structure and microtubule formation. How are the continuous microtubules of the spindle generated? First, we may ask if spindle microtubules ever directly contact the centriole as they do in the case of the kinetochore. Two modes of association have been observed. We have often seen microtubules inserting directly into the centriole wall between triplet blades as in Figure 39. Other workers have published photomicrographs illustrating this as well (Gall, 1961, Fig. 16). In other cases microtubules are frequently seen to end in contact with pericentriolar satellites as in Figure 39 and 42. This also has been observed by others (Szollosi, 1964; Murray et al., 1965). However, if these satellites are largely RNA, as was indicated earlier, and if they arise from the wall of the centriole, as Figure 39 and 42 suggest, then these two origins are easily reconciled; and a general theory of microtubule formation may be formulated.

If microtubule assembly by the centriole requires RNA synthesis, as it probably does at the kinetochore, then the centriole should contain DNA. No direct evidence is available to indicate that DNA is present in centrioles; however, there

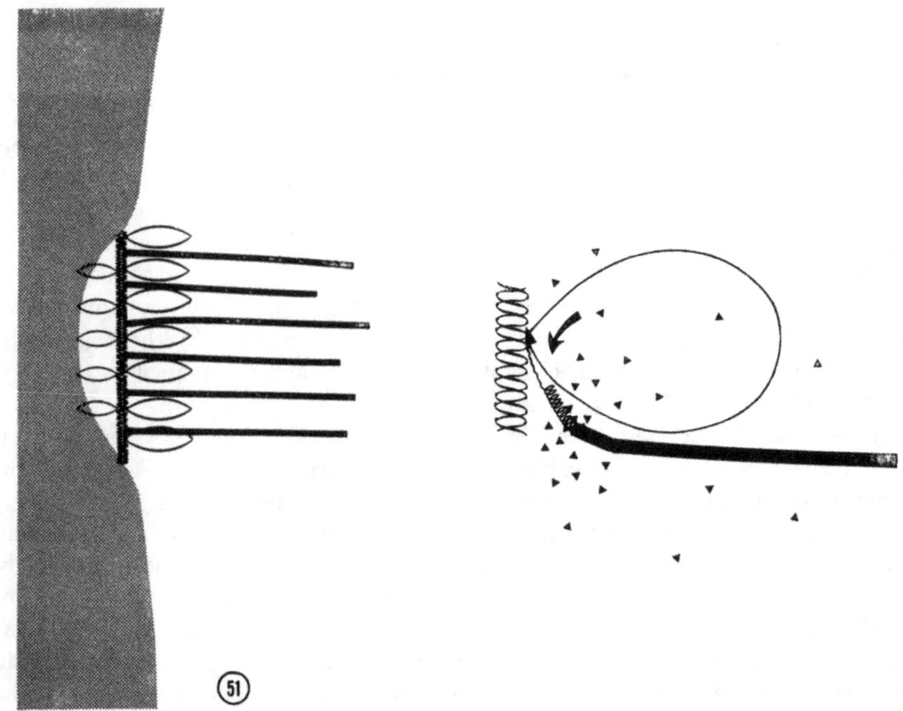

FIG. 51. Diagram of kinetochore mechanism for microtubule formation. Transcription of microtubule RNA from circular DNA in kinetochore lateral loops is proposed as a likely explanation combining ultrastructure appearance and kinetochore function.

is some indication that basal bodies of some ciliates do contain DNA (Randall and Disbrey, 1965; Smith-Sonneborn and Plaut, 1967). If the large helix of the centriole lumen were DNA, then the relationship of ultrastructure to function in the centriole becomes clear. The radial engine design of the centriole provides a mechanism to turn the DNA helix in the centriole lumen. As it turns, transcription of the DNA into RNA would occur at each point where the helix contacts a triplet blade, and the RNA strands would emerge from the centriole wall between the blades all along the centriole length. These RNA strands would either generate microtubules directly by a self assembly process or would be stored as pericentriolar satellites for later assembly into microtubules.

We may generalize the concept of microtubule assembly during mitosis as follows: the release of free RNA into the cytoplasm by any structure may result in microtubule formation, provided that the protein subunits that form the microtubule walls are freely available. Polysome formation in mitotic cells appears to be suppressed (Steward and Shaeffer, 1967), and the reason might well be to prevent ribosomes from competing with microtubule subunits for positions on kinetochore or centriole RNA. As the RNA is generated the protein subunits would probably first attach to the RNA and then assemble into the helical configuration described earlier. As in TMV self-assembly, we may assume that the length of microtubule formed would depend on the length of RNA generated (Anderer, 1963). The kinetochore might be expected to generate long microtubules of unspecified length, but centriole architecture dictates microtubules of a length specified by the length of DNA helix in the centriole lumen. There is some evidence to indicate that this arrangement may indeed make good sense, and studies are in progress to test these ideas further.

First, let us consider the microtubules produced by the centriole. One of the early events of mitosis is the separation of the centriole pairs (parent and daughter to each pole) and the concurrent production of the continuous spindle (Mazia, 1961). We have shown in earlier studies (Brinkley et al., 1967b) that this process is blocked by Colcemid, and it now seems patently clear that microtubule formation by the centrioles results in their separation, ie., they propel themselves to the opposite mitotic poles by pushing against each other through the spindle which they generate. If each centriole can generate microtubles, how do the microtubules from opposite centriole pairs interact? One possibility is suggested in a study of the Flemming body, a structure produced during cytokinesis. The Flemming body is the last connecting link between the two daughter cells, and it is always found to be packed with microtubules, presumably those of the continuous (pole-to-pole) spindle. These microtubules enter the Flemming body from either end singly, but are seen to clearly overlap in the middle of the structure (Fig. 52). The finding of microtubule doublets in the equator of the spindle was first reported by Krishan and Buck (1965). This observation implies that when the centrioles were close together their microtubules interacted to form fused doublets. As the centrioles push away from each other, this region of doubleness would remain equidistant from the poles and thus define the equator of the continuous spindle.

The pole-to-pole distance in Chinese hamster fibroblasts is characteristically about 8 μ. Thus, each centriole must make microtubules 4 μ long. If the microtubule length is specified by the RNA length, as we have suggested, then the centriole must be capable of producing about 30 μ of RNA, considering the helical microtubule assembly postulated earlier. This presents a superficial difficulty, however, since the helix in the centriole lumen, presumably the DNA from which the RNA is transcribed, is only about 5 μ long. The thickness of the helix strand, about 60 to 70 Å, suggests a supercoiled structure, however, which may provide sufficient stretched length for theoretical agreement. Thus, the initial spindle length would be specified by centriole architecture.

A careful measurement of the movements of the centrioles and chromosomes in mitosis of Chinese hamster fibroblasts (Fig. 53) reveals a second motion which must be accounted for. Not only do the chromosomes move toward the poles, but the poles themselves separate further in a process called spindle elongation (Mazia, 1961). It would appear that the continuous microtubules must somehow be lengthened in this process, but just how a new length of RNA could be spliced in is not readily understandable. It would seem more likely that the initial spindle length corresponds to only part of the helix length in the centriole, and that the remainder of the helix DNA is transcribed in making the RNA for microtubule lengthening in spindle elongation. If this were the case then the implied stretched length of the helix fiber must be increased to about 50 μ.

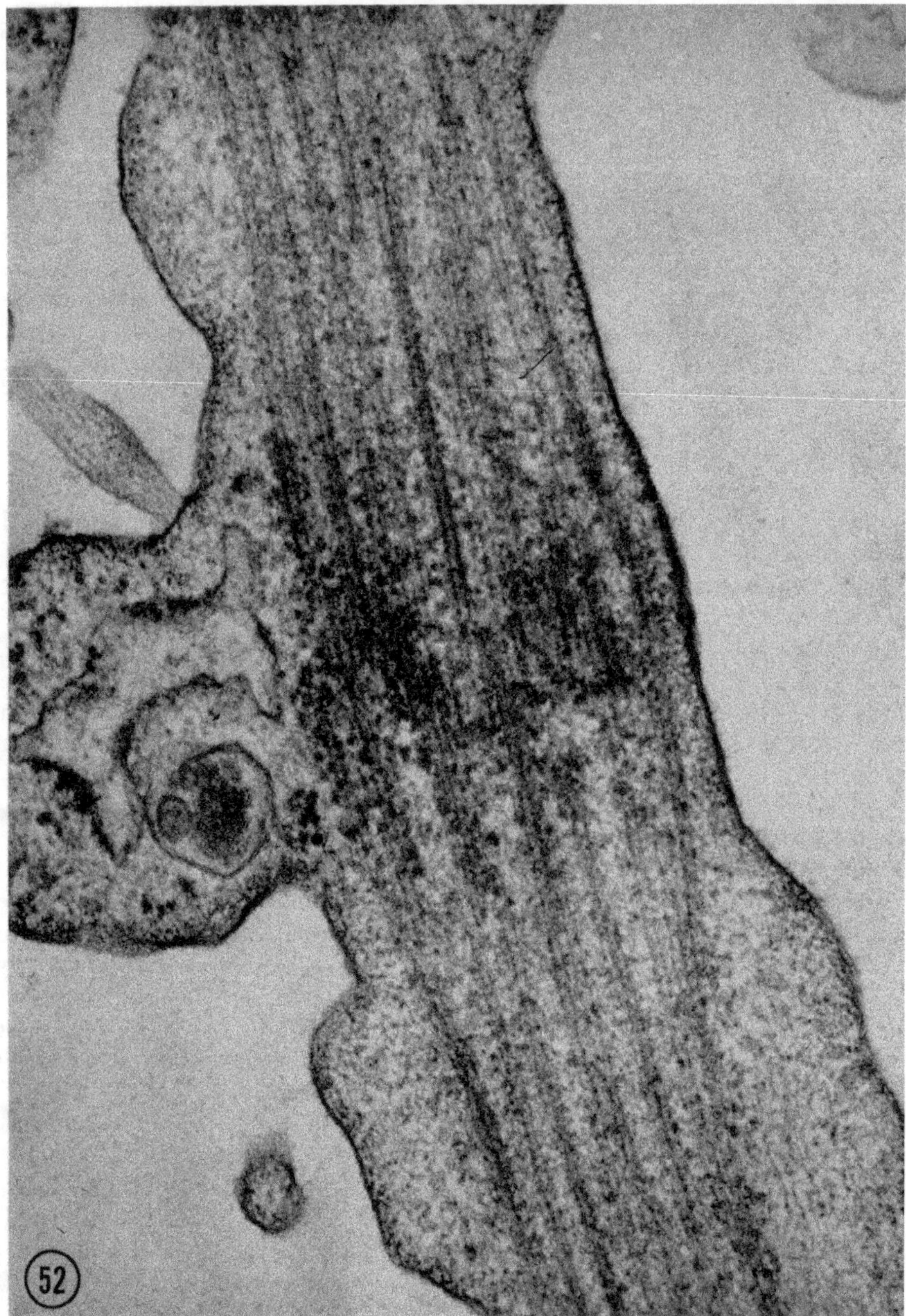

Fig. 52. Flemming body of the Chinese hamster. Microtubules, remnants of the continuous spindle, enter the structure at each end and appear to interdigitate in the middle. × 79,000.

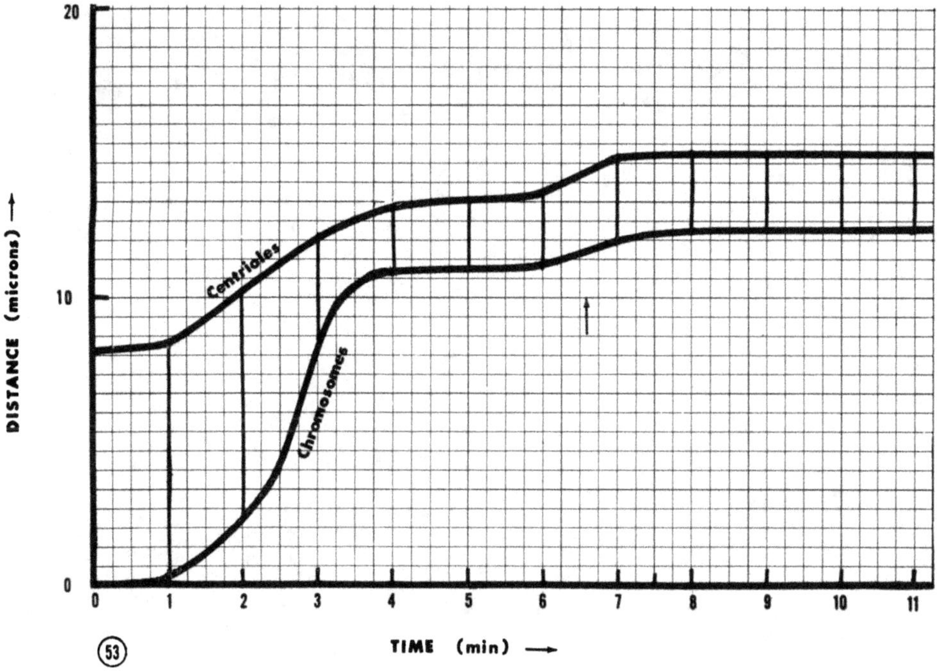

FIG. 53. Movements of the mitotic centers and chromosomes in a mitotic Chinese hamster fibroblast. The data were extracted from a time-lapse motion picture of a cell already in metaphase at the beginning of the sequence. At anaphase, as the chromosome kinetochores separate (lower line), the centrioles also move further apart (upper line) as the chromosome-to-pole distance decreases (one-half of the distance between the two lines). Cytokinesis (arrow) further increases the spindle length, probably as a result of bundling of the pole-to-pole microtubules.

Microtubule Formation by the Kinetochore

On the other hand, microtubule formation by the kinetochore is somewhat different. The studies of Inoué (1964) suggest that the chromosomal fibers of the spindle are in a dynamic equilibrium, under continual assembly at the kinetochore and continual disassembly at the poles. Careful studies of serial sections of mitotic spindles have convinced us that while the continuous microtubules are frequently seen to be attached directly to the centriole wall, the chromosomal microtubules simply end near the pole with no specific attachment (Fig. 54). It thus appears that kinetochore microtubules have free distal ends, while centriole microtubules do not.

The consequences of free ends on the kinetochore microtubules are twofold. In the first place, degradation of these microtubules at their free ends and assembly at the kinetochore end would account for the apparent dynamic equilibrium suggested by Inoué. Secondly, microtubules under such an equilibrium state would almost certainly be rotating at a high rate of speed. This latter concept follows both

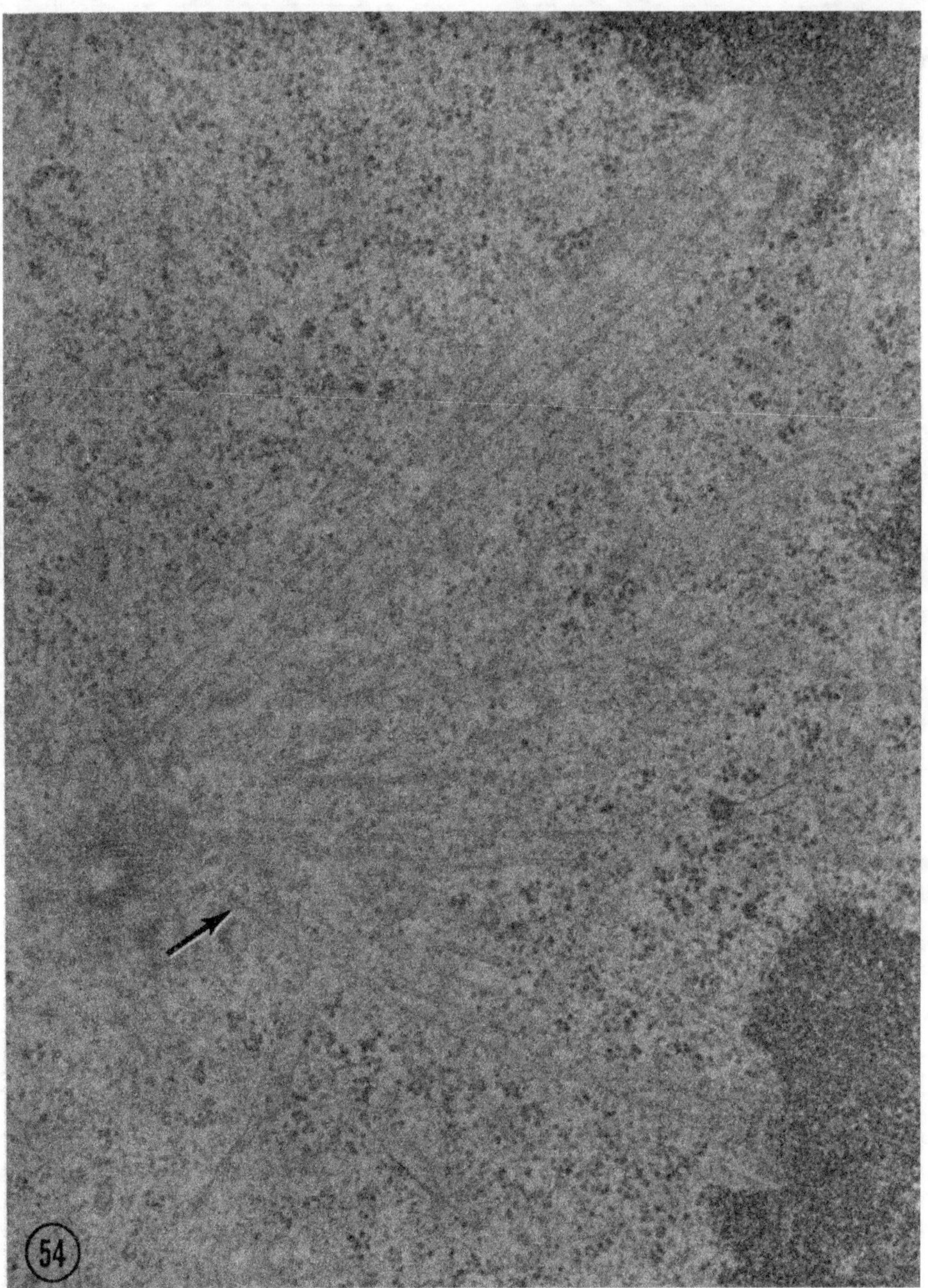

FIG. 54. Section of a mitotic center showing termination of chromosomal microtubules near the centriole, but not directly attached. × 43,000.

from considerations of nucleic acid transcription and a helical assembly of the microtubule itself. Preliminary calculations show that the rapid rotation of the 10 to 20 helical microtubules attached at each kinetochore (as in Fig. 7) should provide ample motive force for anaphase travel of the chromosome (Jarosch, 1964). Since the centriole microtubules appear to overlap and fuse at the equator, they are not free to rotate, but this should not hinder microtubule assembly at the poles, since the single-stranded RNA linking the microtubule to the centriole should be an effective swivel to allow helical assembly of the microtubule if it is not free to rotate.

It is, of course, impossible to correlate the concepts developed in this review with all of the various findings of many investigators over the years. A few unusual and interesting phenomena are so consistent with our concepts, however, that they require mentioning. The elegant studies of Nicklas (1967) and Nicklas and Staehly (1967), in which chromosomes removed from a meiotic spindle by micromanipulation can be seen to reorient and integrate with the division mechanism in such a way that normal anaphase is possible, are consistent with our interpretations. Chromosomal microtubules can be broken off of the kinetochore, and they can be regenerated and realligned with the spindle. Apparently no specific attachment of the chromosomal microtubules to the poles is required.

Indeed, the spindle pole is not needed in order for the basic mechanism to operate, i.e., the rapidly rotating chromosomal microtubules can pull the chromosomes unaided by the centrioles. The poles serve to focus the activities of all kinetochores so that the chromosomes are brought into close contact as the nucleus reforms in telophase. Thus, the problem of acentriolar cells (as in higher plants) is readily resolved. It would appear that the continuous microtubules always present in such cases serve to coordinate the migration of all chromatids to opposite ends of the cell, and that the uniting of all chromosomes into a single nucleus is solved by some other mechanism when divergent spindles are employed. A related observation, that colchicine readily induces polyploidy in plant but not animal cells, can be correlated with the lack of centrioles in plant cells. In animal cells, chromosome reduplication leads to a multipolar spindle because of concurrent centriole reproduction and maturation, and the resulting disorganization seldom allows the viable genome combinations that occur readily in plants. In fact, it seems that the centriole is something of a hazard in animal cells, for once the cell is committed to replication and mitosis, cell division must be completed on schedule or else a multipolar spindle might ruin the whole process (Stubblefield, 1968).

Finally, the concept of a diffuse kinetochore, described in an earlier section, is readily understandable in terms of many chromosomal "genes" making metaphase RNA and thus generating microtubules. Another variation of this same concept might occur in some organisms in which many chromosomal sites could potentially be kinetochores, but some control mechanism would activate only one such site on each chromosome during mitosis. Such an explanation would account for the finding of constant chromosome number with varying metaphase chromosome morphology in the genus *Peromyscus* (Hsu and Arrighi, 1966).

CONCLUSION

Much of the foregoing discussion about the mitotic mechanism is admittedly theoretical and hypothetical. However, the concepts are open to testing, and some preliminary information already in our hands supports our viewpoint. Probably the most critical dogma challenged by our presentation is the notion that mitotic cells do not synthesize any RNA (Crippa, 1966). It is true that the bulk of RNA synthesis is shut off during mitosis, but careful measurements of synchronized metaphase cells reveals ^{3}H-uridine incorporation into RNA at about 5 to 10 percent of the interphase rates. Preliminary attempts to localize this label by high-resolution autoradiography suggests that kinetochores are actively incorporating the isotope. More rigorous testing of this result is obviously required, but at least the early evidence is not disappointing.

The evidence that microtubules contain RNA is still disputed by some (Went, 1966, p. 70), but our views favor this finding. More solid evidence would certainly be desirable, and we anticipate that it will be forthcoming. Centrioles should contain DNA, but a finding to the contrary would not be fatal to the rest of our presentation, since microtubule RNA could possibly be made in the centriole from a parent RNA; we consider this finding unlikely. The kinetic assumptions of our hypothesis, i.e., centriole motors and rotating microtubules, will be more difficult to confirm, but it is hoped that some biophysical tools can be brought to bear on these predictions in the near future.

ACKNOWLEDGMENTS

This study was supported in part by Research Grant HD–2590 from the National Institutes of Health. The authors gratefully acknowledge the technical assistance of Miss Patricia Murphy, Miss Marion Gay, Mr. John Carnes, Mr. Wayne Wray, and Mr. Mike Lieppman. We wish to thank Mrs. Joy Herman for secretarial assistance, and in addition we especially thank Dr. T. C. Hsu, Drs. Margery, and Charles Shaw, University of Texas M. D. Anderson Hospital and Tumor Institute, Dr. Bruce Nicklas, Duke University, and Dr. Andrew Bajer, University of Oregon, for helpful suggestions and critical review of the manuscript. Appreciation is extended to Dr. Andrew Bajer and Dr. T. C. Hsu for permission to publish their excellent micrographs.

REFERENCES

Aldrich, H. C. 1967. The ultrastructure of meiosis in three species of *Physarum*. Mycologia, 59:127–148.

Anderer, F. A. 1963 Recent studies on the structure of tobacco mosaic virus. *In* Advances in Protein Chemistry, Anfinsen, C. B., Anson, M. L., and Edsall, J. T., eds., New York, Academic Press, Inc., Vol. 18, pp. 1–35,

Arrighi, F. E. 1967. Nucleolar RNA synthetic activity in Chinese hamster cells *in vitro* and the effects of Actinomycin D and Nogalamycin. J. Cell. Physiol., 69:45–52.

Auerbach, C. 1947. Abnormal segregation after chemical treatment of *Drosophila*. Genetics, 32:3–7.

Bajer, A. 1958. Cine-micrographic studies on mitosis in endosperm V. Formation of the metaphase plate. Exp. Cell Res., 15:370–383.

——— 1965. Behavior of chromosomal spindle fibres in living cells. Chromosoma, 16:381–390.

Barnicot, N. A., and H. E. Huxley. 1965. Electron microscope observations on mitotic chromosomes. Quart. J. Micr. Sci., 106:197–214.

Beermann, W., and G. Bahr. 1954. The submicroscopic structure of the Balbiani ring. Exp. Cell Res., 6:195–201.

Behnke, O., and A. Forer. 1966. Intranuclear microtubules. Science, 153:1536–1537.

Biesele, J. J. 1958. Mitotic Poisons and the Cancer Problem, Amsterdam, Elsevier Publishing Company.

Borisy, G. G. 1967. The mechanism of action of colchicine. Colchicine binding to sea urchin eggs and the mitotic apparatus. J. Cell Biol., 34:535–548.

——— and E. W. Taylor. 1967. The mechanism of action of colchicine. Binding of colchicine–^{3}H to cellular protein. J. Cell Biol., 34:525–533.

Brinkley, B. R. 1965. The fine structure of the nucleolus in mitotic divisions of Chinese hamster cells *in vitro* J. Cell Biol., 27:411–422.

——— and T. C. Hsu. 1966. Structure and behavior of specialized regions of mammalian chromosomes: kinetochores, telomeres, and nucleolar organizers. J. Cell Biol., 31:16A.

——— and E. Stubblefield. 1966. Fine structure of the kinetochore of Chinese hamster cells *in vitro*. Chromosoma, 19:28–43.

——— P. Murphy, and L. C. Richardson. 1967a. Procedure for embedding *in situ* selected cells cultured *in vitro*. J. Cell Biol., 35:279–283.

——— E. Stubblefield, and T. C. Hsu. 1967b. The effects of Colcemid inhibition and reversal on the fine structure of the mitotic apparatus of Chinese hamster cells *in vitro*. J. Ultrastruct. Res., 19:1–18.

Brown, S. W. 1954. Mitosis and meiosis in *Luzula campestris*. Univ. Calif. Publ. Bot., 27:231–278.

Buck, R. C. 1967. Mitosis and meiosis in *Rhodnius prolixus*: The fine structure of the spindle and diffuse kinetochore. J. Ultrastruct. Res., 18:489–501.

Callan, H. G. 1966. Chromosomes and nucleoli of the axolotl, *Ambystoma mexicanus*. J. Cell Sci., 1:85–108.

——— and L. Lloyd. 1960. Lampbrush chromosomes of crested newts *Triturus cristatus*. Phil. Trans., B243:135–219.

Carothers, E. E. 1936. Components of the mitotic spindle with special reference to the chromosomal and interzonal fibers in the Acrididae. Biol. Bull., 71:469–491.

Chang, J. P., and C. W. Gibley, Jr. 1968. Ultrastructure of tumor cells during mitosis. Cancer Res., 28:531–534.

Cleveland, L. R. 1957. Achromatic figure formation by multiple centrioles of *Barbulanympha*. J. Protozool., 4:241–248.

Cole, A. 1967. Chromosome structure. *In* Theoretical and Experimental Biophysics, Cole, A., ed., New York, Marcel Dekker, Inc., Vol. 1, pp. 305–375.

Coleman, J. R., and M. J. Moses. 1964. DNA and the fine structure of synaptic chromosomes in the domestic rooster (*Gallus domesticus*). J. Cell Biol., 23:63–78.

Crippa, M. 1966. The rate of ribonucleic acid synthesis during the cell cycle. Exp. Cell Res., 42:371–374.

Crouse, H. 1960. The controlling element in sex chromosome behavior in *Sciara*. Genetics, 45:1429–1443.

Cuevas-Sosa, A. 1967. Crossing-over and the centromere. Cytogenetics, 6:331–341.

Darlington, C. D. 1937. Recent Advances in Cytology, 2nd ed., New York, Blakiston Division of McGraw-Hill Book Company.

——— 1939. Misdivision and the genetics of the centromere. J. Genet., 37:341–365.

Deitz, R. 1956. Die Spermatocytenteilungen der Tipuliden II. Mitteilung graphische Analyse der Chromosomenbewegung während der Prometaphase I in Leben. Chromosoma, 8:183–211.

Doolin, P. F., and W. J. Birge. 1966. Ultrastructural organization of cilia and basal bodies of the epithelium of the choroid plexus in the chick embryo. J. Cell Biol., 29:333–346.

Eigsti, O. J., and P. Dustin. 1955. Colchicine in Agriculture, Medicine, Biology, and Chemistry, Ames, Iowa, Iowa State University Press.

Fawcett, D. W. 1956. The fine structure of chromosomes in meiotic prophase of vertebrate spermatocytes. J. Biophys. Biochem. Cytol., 2:403–406.

——— 1961. Cilia and Flagella. *In* The Cell, Brachet, J. and Mirsky, A. E., eds., New York, Academic Press, Inc., Vol. 2, pp. 217–298.

Forer, A. 1965. Local reduction of spindle fiber birefringence in living *Nephrotoma suturalis* (Loew) spermatocytes induced by ultraviolet microbeam irradiation. J. Cell Biol., 25:95–117.

Gall, J. G. 1954. Lampbrush chromosomes from oöcyte nuclei of the newt. J. Morph., 94:283–352.

——— 1961. Centriole replication. A study of spermatogenesis in the snail *Viviparus*. J. Biophys. Biochem. Cytol., 10:163–193.

——— 1963a. Kinetics of deoxyribonuclease action on chromosomes. Nature (London), 198:36–38.

——— 1963b. Chromosomes and cytodifferentiation. *In* Cytodifferentiation and Macromolecular Synthesis, Lock, M., ed., New York, Academic Press, Inc., pp. 119–143.

——— 1966. Microtubule fine structure.J. Cell Biol., 31:639–643.

George, P., L. J. Journey, and M. N. Goldstein. 1965. Effect of vincristine on the fine structure of HeLa cells during mitosis. J. Nat. Cancer Inst., 35:355–375.

Gibbons, I. R., and A. V. Grimstone. 1960. On flagellar structure in certain flagellates. J. Biophys. Biochem. Cytol., 7:697–716.

Haneen, W. K., and W. W. Nichols. 1966. Persistence of nucleoli in short term and long term cell cultures and in direct bone marrow preparations in mammalian materials. J. Cell Biol., 31:543–561.

Harris, P. J. 1962. Some structural and functional aspects of the mitotic apparatus in sea urchin embryos. J. Cell Biol., 14:475–487.

——— 1965. Some observations concerning metakinesis in sea urchin eggs. J. Cell Biol., (Suppl. on mitosis) 25:73–77.

——— and A. Bajer. 1965. Fine structure studies on mitosis in endosperm metaphase of *Haemanthus Katherine* Bak. Chromosoma, 16:624–636.

Hsu, T. C., and F. E. Arrighi. 1966. Chromosomal evolution in the genus *Peromyscus*. Cytogenetics, 5:355–359.

——— F. E. Arrighi, R. R. Klevecz, and B. R. Brinkley. 1965. The nucleoli in mitotic divisions of mammalian cells *in vitro*. J. Cell Biol., 26:539–553.

——— B. R. Brinkley, and F. E. Arrighi. 1967. The structure and behavior of the nucleolus organizers in mammalian cells. Chromosoma, 23:137–153.

Hughes-Schrader, S. 1942. The chromosomes of *Nautococcus schraderae* Vays and the meiotic division figure of male Llaveiella Coccids. J. Morph., 70:261–299.

——— 1947. The "prometaphase stretch" and kinetochore orientation in *Phasmids*. Chromosoma, 3:1–21.

——— 1955. The chromosomes of the giant *Aspidoproctus maximus* Louns. (Coccoidea-Margarodidae) with special reference to asynapsis and sperm formation. Chromosoma, 7:420–438.

——— and H. Ris. 1941. The diffuse spindle attachment of Coccids verified by the meiotic behavior of induced chromosome fragments. J. Exp. Zool., 87:429–456.

——— and F. Schrader. 1961. The kinetochore of the Hemiptera. Chromosoma, 12: 327–350.

Inoué, S. 1964. Organization and function of the mitotic spindle. *In* Primitive Motile Systems in Cell Biology, Allen, R. D. and Kamiya, N., eds., New York, Academic Press, Inc., pp. 549–598.

Iwata, J. 1940. Studies on chromosome structure II. Jap. J. Bot., 10:375–382.

Jarosch, R. 1964. Screw-mechanical basis of protoplasmic movement. *In* Primitive Motile Systems in Cell Biology, Allen, R. D. and Kamiya, N., eds., New York, Academic Press, Inc., pp. 599–622.

John, B., and G. M. Hewitt. 1966. Karyotype stability and DNA variability in the Acridiae. Chromosoma, 20:155–172.

Jokelainen, P. T. 1965. Fine structure aspects of kinetochores in dividing cells from fetal rat kidney. Anat. Rec., 151:367.

——— 1967. The ultrastructure and spatial organization of the metaphase kinetochore in mitotic rat cells. J. Ultrastruct. Res., 19:19–44.

Krishan, A., and R. C. Buck. 1965. Structure of the mitotic spindle in L strain fibroblasts. J. Cell Biol., 24:433–444.

Ledbetter, M. C., and K. R. Porter. 1964. Morphology of microtubules of plant cells. Science, 144:872–874.

Lima-de-Faria, A. 1950. The Feulgen test applied to centromeric chromosomes. Hereditas, 36:60–74.

——— 1956. The role of the kinetochore in chromosome organization. Hereditas, 42: 85–160.

——— 1958. Recent advances in the study of the kinetochore. *In* International Review of Cytology, Bourne, G. H., and Danielli, J. E., eds., New York, Academic Press, Inc.

Lindsley, D. L., and E. Novitski. 1958. Localization of the genetic factors responsible for the kinetic activity of the X chromosome of *Drosophila melanogaster*. Genetics, 43: 790–798.

Luykx, P. 1965a. The structure of the kinetochore in meiosis and mitosis in *Urechis* eggs. Exp. Cell Res., 39:643–657.

——— 1965b. Kinetochore-to-pole connections during prometaphase of the meiotic divisions in *Urechis* eggs. Exp. Cell Res., 39:658–668.

Markham, R., S. Frey, and G. Hills. 1963. Methods for the enhancement of image details and accentuation of structure in electron microscopy. Virology, 20:88–102.

Matsuura, H. 1941. Chromosome studies in *Trillium kamtschaticum* Pall. XII. Cytologia, 11:369–379.

Mazia, D. 1961. Mitosis and the physiology of cell division. *In* The Cell, Brachet, J., and Mirsky, A. E., eds., New York, Academic Press, Inc., Vol. 3, pp. 77–440.

Moor, H. 1967. Der Feinbau der Mikrotubuli in Hefe nach Gefrierätzung. Protoplasma, 64:89–103.

Moses, M. J. 1956. Chromosomal structures in crayfish spermatocytes. J. Biophys. Biochem. Cytol., 2:215–218.

——— 1958. The relation between the axial complex of meiotic prophase chromosomes and chromosome pairing in a salamander (*Plethodon cinereus*). J. Biophys. Biochem. Cytol., 4:633–638.

——— 1964. The nucleus and chromosomes. *In* Cytology and Cell Physiology, 3rd ed., Bourne, G. H., ed., New York, Academic Press, Inc., pp. 423–558.

Murray, R. G., A. S. Murray, and A. Pizzo. 1965. The fine structure of mitosis in rat thymic lymphocytes. J. Cell Biol., 26:601–619.

Nebel, B. R. 1939. Chromosome structure. Bot. Rev., 5:563–627.

——— and E. M. Coulon. 1962. Fine structure of chromosomes in pigeon spermatocytes. Chromosoma, 13:272–291.

Nicklas, R. B. 1967. Chromosome micromanipulation. II. Induced reorientation and the experimental control of segregation in meiosis. Chromosoma, 21:17–50.

——— and C. A. Staehly. 1967. Chromosome micromanipulation. I. The mechanics of chromosome attachment to the spindle. Chromosoma, 21:1–16.

Östergren, G. 1951. The mechanism of co-orientation in bivalents and multivalents. The theory of orientation by pulling. Hereditas, 37:85–156.

——— and A. Bajer. 1960. La mitose dans les cellules traitées par le méthanol. Étude microcinématographique. *In* L'Action Antimitotique et Caryoclasique de Substances Chimiques, Turchini, J., and Sentein, P. eds., Colloque No. 88, C.N.R.S., Paris, pp. 199–207.

Pease, D. C. 1946. Hydrostatic pressure effects upon the spindle figure and chromosome movement. II. Experiments on the meiotic division of *Tradescantia*. Pollen mother cells. Biol. Bull., 91:145–169.

Prescott, D. M., and M. A. Bender. 1963. Autoradiographic study of chromatid distribution of labelled DNA in two types of mammalian cells *in vitro*. Exp. Cell Res., 29: 430–442.

Randall, J., and C. Disbrey. 1965. Evidence for the presence of DNA at basal body sites in *Tetrahymena pyriformis*. Proc. Roy. Soc. (Biol), 162:473–491.

Reese, T. S. 1965. Olfactory cilia in the frog. J. Cell Biol., 25:209–230.

Renaud, F. L., and H. Swift. 1964. The development of basal bodies and flagella in *Allomyces arbusculus*. J. Cell Biol., 23:339–354.

Rhoades, M. M. 1952. Preferential segregation in maize. *In* Heterosis, Gowen, J. W., ed., Ames, Iowa, Iowa State University Press, pp. 66–80.

——— 1961. Meiosis. *In* The Cell, Brachet, J., and Mirsky, A. E., eds., New York, Academic Press, Inc., Vol. 2, pp. 1–75.

Ringo, D. L. 1967. Flagellar motion and fine structure of the flagellar apparatus in *Chlamydomonas*. J. Cell Biol., 33:543–571.

Robbins, E., G. Jentzsch, and A. Micali. 1968. The centriole cycle in synchronized HeLa cells. J. Cell Biol., 36:329–339.

——— and N. K. Gonatas. 1964. The ultrastructure of a mammalian cell during the mitotic cycle. J. Cell Biol., 21:429–463

Roth, L. E., and E. W. Daniels. 1962. Electron microscope studies of mitosis in amebae. I. The giant ameba *Pelomyxa carolinensis*. J. Cell Biol., 12:57–78.

Schrader, F. 1935. Notes on the mitotic behavior of long chromosomes. Cytologia, 6:422–431.

——— 1939. The structure of the kinetochore at meiosis. Chromosoma, 1:230–237.

——— 1947. Data contributing to an analysis of metaphase mechanics. Chromosoma, 3:22–47.

——— 1953. Mitosis, New York, Columbia University Press.

Sharp. L. W. 1934. Introduction to Cytology, 3rd ed., New York, McGraw-Hill Book Company.

Smith-Sonneborn, J., and W. Plaut. 1967. Evidence for the presence of DNA in the pellicle of *Paramecium*. J. Cell Sci., 2:225–234.

Sorokin, S. 1962. Centrioles and the formation of rudimentary cilia by fibroblasts and smooth muscle cells. J. Cell Biol., 15:363–377.

Sparvoli, E., H. Gay, and B. P. Kaufmann. 1964. Number and pattern of association of chromonemata in chromosomes of *Tradescantia*. Chromosoma, 16:415–435.

Staiger, H. 1950. Chromosome number variations in *P. lapillus*. Experientia, 6:140–142.

Steffensen, D. 1959. A comparative view of the chromosome. Brookhaven Symposia in Biology, Number 12, Structure and function of genetic elements, Brookhaven National Laboratory, pp. 103–124.

Stern, C. 1936. Somatic crossing-over and segregation in *Drosophila melanogaster*. Genetics, 21:625–730.

Stevens, B. J. 1964. The effects of Actinomycin D on nucleolar and nuclear fine structure in the salivary gland cell of *Chironomus thummi*. J. Ultrastruct. Res., 11:329–353.

Steward, D., and J. Shaeffer. 1967. Polysome distributions in synchronized Chinese hamster cells *in vitro*. J. Cell Biol., 35:129A.

Stubblefield, E. 1968. Centriole replication in mammalian cell. *In* The Proliferation and Spread of Neoplastic Cells, Baltimore, The Williams and Wilkins Co., pp. 185–203.

——— and B. R. Brinkley. 1967. Architecture and function of the mammalian centriole. *In* The Origin and Fate of Cell Organelles, Warren, K. B., ed., New York, Academic Press, Inc., pp. 175–218.

——— and R. R. Klevecz. 1965. Synchronization of Chinese hamster cells by reversal of Colcemid inhibition. Exp. Cell Res., 40:660–664.

Swift, H. 1965. Molecular morphology of the chromosome. *In Vitro,* 1:26–49. (Publication of the Tissue Culture Association, Inc.)

Szollosi, D. 1964. The structure and function of centrioles and their satellites in the jellyfish *Phialidium gregarium*. J. Cell Biol., 21:465–479.

Taylor, E. W. 1965. The mechanism of colchicine inhibition of mitosis. I. Kinetics of inhibition and the binding of H^3 colchicine. J. Cell Biol., 25:145–160.

Utakoji, T. 1966. Chronology of nucleic acid synthesis in meiosis of the male Chinese hamster. Exp. Cell Res., 42:585–596.

Vaarama, A. 1954. Cytological observations in *Pleurozium schreberi* with special reference to centromere evolution. Ann. Bot. Soc., 28:1–59.

Wallace, H., and M. L. Birnstiel. 1966. Ribosomal cistrons and the nucleolar organizer. Biochem. Biophys. Acta, 114:296–310.

Went, H. A. 1966. The behavior of centrioles and the structure and formation of the achromatic figure. Protoplasmatologia, 4:1–109.

Zimmerman, A. 1960. Physico-chemical analysis of the isolated mitotic apparatus. Exp. Cell Res., 20:529–547.

Lester Goldstein

Department of Molecular, Cellular and Developmental Biology, University of Colorado, Boulder, Colorado

4

On the Question of Protein Synthesis by Cell Nuclei

INTRODUCTION

That the bulk of a typical cell's protein synthesis occurs on cytoplasmic ribosomes is widely accepted, but there is considerable uncertainty as to whether, in general, protein synthesis occurs in the nucleus. To establish that protein synthesis does occur in cell nuclei would, moreover, do little to stem our curiosity about the phenomenon. If it could be established that protein synthesis by nuclei is a general phenomenon, we would immediately be interested in knowing how much and what kinds of proteins are made there.

Thus to have a satisfactory comprehension of the role of the nucleus, we need answers to the following questions. If proteins are synthesized in the nucleus, are these proteins in any way different from those synthesized in the cytoplasm? If they are the same as those synthesized in the cytoplasm, is the contribution by the nucleus in any way quantitatively significant, or is the amount made by the nucleus in proportion to its share of the cell volume? If protein synthesis does occur in the nucleus, does that synthesis involve mechanisms different from the well-characterized ribosomal mechanisms of the cytoplasm? Does protein synthesis occur in nuclei in general, or is it a feature of only a few exceptional cells?

Although in recent years increasing attention has been paid to these questions, there seems to be no satisfactory answer to any. This review is aimed at drawing together diverse data in order to focus on the current state of the problem, and hopefully, to clarify some issues needing attention. It is written with the prejudice (based on the conviction that experiments on systems isolated from broken cells are suspect until proven otherwise) that few answers are decisive unless confirmed by *in vivo* experiments.

Attention is directed to the question of the nuclear synthesis of protein, not only because the answer itself is worth knowing, but also because several interesting nuclear phenomena have been observed which perhaps may be explained by the existence of unique protein synthesis mechanisms associated with the nucleus. For example, Harris (1963) in an attempt to explain the role of rapidly turning-over, heterogeneous, large molecular weight RNA that is apparently completely confined to the nucleus, has suggested that this RNA may be involved in protein synthesis mechanisms exclusively limited to the nucleus. This protein synthesis would differ in some manner from the synthesis in the cytoplasm that presumably is not as immediately dependent on the presence of nuclear genes as the synthesis in the nucleus. This hypothesis implies, of course, that different mechanisms of polypeptide formation may be involved.

Another example of an unusual nuclear phenomenon is vividly demonstrated in cells with lampbrush chromosomes. Beermann (1967) has pointed out that a homologous pair of loops at a particular chromosome locus may have different heritable morphologies, although in the same nucleus. The striking feature of this is that, although they are in the same nucleus and presumably differ by only a point mutation, the loops maintain different morphologies *in the absence of any detectable ribosomes* associated with the loops. Thus, if we assume that the morphology of a loop is determined largely by the proteins of which it is composed, we must assume that one of two unusual mechanisms is at work. Either the proteins are made in the cytoplasm and then migrate to the loops in a highly selective manner, since they would presumably discriminate one homologous loop from another, or the proteins are made directly on the loops by a mechanism that does not involve ribosomes.

Another potentially exciting aspect of protein synthesis by nuclei is the report from a few laboratories that some amino acid incorporation is DNA- and not RNA-dependent, and apparently also independent of ribosomes. We will consider the evidence for this later.

THE SITE OF SYNTHESIS OF NUCLEAR PROTEINS

Some of the difficulty in using experimental data to reach firm conclusions about the problems under consideration is illustrated by experiments designed to determine the site of synthesis of proteins that are themselves primarily localized within the nucleus. It is sometimes assumed that if any proteins are synthesized in the nucleus, those whose primary localization is nuclear would certainly qualify. Most of the evidence bearing on this question reveals, however, that nuclear proteins are synthesized in the cytoplasm. Thus, we might conclude, according to the above logic, that the nucleus synthesized no protein. Such a conclusion would be unwise.

Histone

The class of proteins that is taken to the most characteristically nuclear is, of course, the histones—although histone-like proteins are found in the cytoplasm (see, e.g., Shepherd and Noland, 1968). In 1964, two reports on the site of histone synthesis appeared which reached opposing conclusions. Birnstiel and Flamm (1964), working with tobacco cells, concluded that histones were synthesized in the nucleoli, while Bloch and Brack (1964), working with grasshoppers, concluded that histones were synthesized in the cytoplasm of the spermatids. It is, of course, possible that the disparity is simply due to species differences, but neither experiment seems unambiguous.

Birnstiel and Flamm studied the kinetics of incorporation of ^{14}C amino acids into various protein fractions of different cell compartments, and concluded that histones are made in the nucleolus and then migrate to the nucleoplasm. They also speculated that the residual proteins follow the same path, although, if their data are subjected to the same reasoning as that employed to reach the conclusion about histone synthesis, it would lead one to speculate that the migration of residual proteins was in the opposite direction. But the interpretations are open to question on other grounds, the primary one being the assumption that newly synthesized proteins cannot migrate from cytoplasm to nucleus in less than 10 minutes from the administration of labeled precursors. This assumption is made remarkably often in the literature and, as I will try to show, is almost completely without justification. Since an average complete protein molecule is synthesized in a minute or two (cf. Dintzis, 1961) and diffusion across the cell probably would take only a fraction of a second, the assumption about rate of protein migration would obviously need experimental support; moreover, evidence will be presented that migration can proceed more rapidly than is generally assumed. (The time for the appearance of label in the nucleus following incorporation of radioactive amino acids into the cytoplasm need not, of course, be even one or two minutes. A radioactive amino acid incorporated onto the C-terminal end of a protein could appear in the nucleus in a matter of a few seconds.)

Several workers have reported the incorporation of amino acids into histones, as well as other nuclear proteins, by isolated nuclei. We will deal with evidence from studies on isolated nuclei in the section on *in vitro* evidence of protein synthesis by nuclei (see p. 193).

Block and Brack (1964), who concluded that histones are synthesized in the cytoplasm, studied some kinetics of incorporation of ^{3}H-arginine by *in situ* autoradiographic techniques. They found that, at those developmental stages of the spermatids during which arginine was incorporated into protein, there was no nuclear RNA detectable by sensitive techniques, and that incorporated ^{3}H was detectable first in the cytoplasm and later in the nucleus. Thus, they concluded that histone is made in the cytoplasm and proceeds to the nucleus from there. Alternative mechanisms are not excluded, however, and one can imagine, e.g., that arginine is incorporated into nuclear protein at a slower rate than into cytoplasmic protein, perhaps by a nuclear mechanism that does not involve direct participation of RNA. (Some experiments to be discussed later, see pp. 205-206, suggest a direct participation of DNA in protein synthesis and no involvement of RNA.)

The most convincing demonstration of the locus of histone synthesis comes from the work of Robbins and Borun (1967). They showed, by carefully controlled pulse-chase experiments with appropriate labeled amino acids, that during the S period of HeLa cells, when histone as well as DNA is being made, lysine is first incorporated into the cytoplasm and during the ensuing chase appears in the HCl-soluble (histone) fraction of the nuclei. Moreover, they showed that the initial lysine incorporation is substantially associated with small cytoplasmic polyribosomes believed to be involved in histone synthesis.

That nuclei can synthesize histones, probably has been demonstrated unambiguously for calf thymus nuclei by Reid et al. (1968), but for reasons discussed below (p. 195) it is questionable whether calf thymus nuclei can be considered representative of nuclei in general.

Nonhistone Nuclear Proteins

Where are the nonhistone nuclear proteins synthesized? The answer is unclear. (As far as certain primarily cytoplasmic proteins, such as hemoglobin and lactic dehydrogenase, that are sometimes found in the nucleus are concerned, it seems better to discuss them in another context (p. 199). Suffice to say here, that if the nuclear membrane is not a barrier to the free diffusion of nonstructural proteins, we might expect that many primarily cytoplasmic proteins will be found, to some extent, in the nucleus. It is still uncertain, however, whether "soluble" proteins can readily diffuse across the nuclear membrane.) There is some evidence that several kinds of nuclear proteins are synthesized *in vitro* by isolated nuclei, but it is possible that these isolated nuclei are atypical or that the degree of amino acid incorporation is quantitatively insignificant, and thus of small moment to the question at hand.

In *Amoeba proteus* the nucleus contains two major classes of proteins, one of which, called rapidly migrating proteins (RMP), is in relatively rapid migration between nucleus and cytoplasm, and the other, called slow turnover proteins (STP), moves from the nucleus quite slowly (Goldstein and Prescott, 1967a). RMP comprise about 40 percent of the nuclear proteins and are present in a nucleus at a concentration 25 to 50 times the cytoplasmic concentration; the remaining 60 percent of the nuclear proteins are STP, but no estimate is available on the amount in the cytoplasm—although there are indications that some are present. (Histones, which may be part of either STP, or RMP, comprise less than 5 percent of total nuclear protein in these amebas.) Byers et al. (1963) have shown that *enucleate* cells can incorporate amino acids into RMP and STP, thus providing good evidence that these nuclear proteins can be synthesized in the cytoplasm. Their experiments do not, however, exclude the possibility of some synthesis of these proteins in the nucleus.

Zetterberg (1966a) in a splendid series of experiments, involving use of such single cell techniques as cytophotometry, autoradiography, interferometry, and time-lapse cinematography, has been able to demonstrate, in cultured mouse fibroblasts, that a substantial amount of cytoplasmically synthesized protein moves into the nucleus during the S period. Subsequently he showed (Zetterberg, 1966b) that cytoplasmic synthesis of some nuclear proteins probably occurs during most of interphase and that there may very well be a continuing exchange, in both directions, of proteins between nucleus and cytoplasm, more or less akin to the movement of RMP seen in amebas.

Migration of Nuclear Proteins

The above experiments draw attention to the matter of migration of nuclear proteins, and attention on this point is needed because many conclusions about the localization and synthesis of nuclear proteins often are based on one or both of two important assumptions about experimental data: (1) if proteins are found primarily in the nucleus, they are made there; (2) proteins cannot migrate quickly (rarely expressed quantitatively) from cytoplasm to nucleus. As we have just seen, there is reason to question assumption (1) and, as we shall see, probably assumption (2). Now let us examine other evidence on these points.

Using altogether a great diversity of techniques, Richards (1960) with mouse cells, Harris (1961) with rat connective tissue cells, Martin (1961) with pollen grains, Richards and Bajer (1961) with plant endosperm, Das and Alfert (1963) with onion root tips, Beck (1962) with HeLa cells, Byers et al. (1963) and Prescott and Bender (1963) with amebas, Sims (1965) with various rat cells, and Bassleer (1966) with ascites cells, have demonstrated that proteins are liberated from cell nuclei in early prophase and migrate back in as the nuclei are reconstituted during telophase. Thus, there can be no doubt that proteins localized in interphase nuclei can migrate from cytoplasm to nucleus. This then dismisses as untenable the simple argument that proteins localized in the nucleus must ipso facto be synthesized there.

This conclusion is made even stronger, of course, by the evidence of Goldstein and Prescott (1967a) and Zetterberg (1966a, 1966b) that proteins can migrate into the nucleus, not only during telophase, but during interphase as well. Kroeger et al. (1963) also have shown that labeled proteins can migrate between cytoplasm and nucleus in explanted *Chironomus* salivary gland tissue. Recall also (p. 190) that the evidence of Robbins and Borun (1967) points to the synthesis of histone in the cytoplasm and subsequent migration to the nucleus. It is worth noting that in most, if not all, of these studies, it is probable that the proteins migrate into the nucleus against a concentration gradient, although the only objective measure of this is that of Goldstein and Prescott (1967a), who showed that the migration into the nucleus occurs despite the fact that the concentration of migrating protein in the nucleus is 25 to 50 times greater than in the cytoplasm.

Data bearing on the question of the *rate* of migration are much more sparse, but there are some that are appropriate to our discussion. I will consider here only the relatively direct approaches to the measurements of rates because I do not believe that any of the indirect measurements, which often suffer from too many unknown or unmeasurable parameters, can provide satisfactory answers. When an indirect measurement of migration rate is obtained, it may be, because of a complicated series of steps, that an estimate is being made of only the *slowest* possible rate, which in effect may reflect factors not relevant to our interests.

A number of investigators, whose work will be considered in later sections, have contended that because label is detected in both nucleus and cytoplasm within something like 10 minutes after the administration of a labeled amino acid, there must be independent synthesis in both compartments. This, of course, implies some knowledge about the rates of protein migration, but actually no useful data are ever presented to support their views. That such views may be unsound can be seen from the data of Zalokar (1960). Zalokar showed that, in *Drosophila* ovarian cells, radioactivity was detectable in cytoplasmic, but not nuclear, protein within 15 *seconds* after the administration of ^{3}H-leucine, but that nuclear protein did not become noticeably labeled until about 4 *minutes* or so later. There is no reason to exclude the possibility that the protein labeled in the cytoplasm during the first 15 seconds has migrated into the nucleus by 4 minutes and thus, although the data certainly do not prove that that mechanism is the true one, they do illustrate the need for caution in making assumptions about rates of migration.

There are some experiments that do give some crude estimate of the rate(s) of protein migration between cytoplasm and nucleus. Byers et al. (1963) determined that the Q_{10} of the migration of labeled RMP from nucleus to cytoplasm and back into the nucleus was approximately 1.3, the Q_{10} expected of a simple diffusion and not of an enzyme-facilitated transport. If the protein molecules migrated between the two compartments by diffusion without interference from the nuclear membrane, they would be able to go from cytoplasm to nucleus in a fraction of a second. The data of Byers et al. (1963) reflect, however, only the slowest step in the overall migration process and, in a sense, suggest that only that step in migration occurs by diffusion. However, in a more meaningful experiment they showed that labeled RMP can migrate from nucleus to cytoplasm and back to the nucleus in 10 minutes or less. There is no measure of the rate of migration from cytoplasm to nucleus alone, but it is conceivable that it is a very few minutes, indeed.

Feldherr and his associates (Feldherr and Harding, 1964) have injected nonphysiological particles and macromolecules into the cytoplasm of several cell types and observed the subsequent localization of the injected material. A study of rates was not their primary interest but they did, in some instances, observe the movement with time of particles into the nucleus. In some instances particles appeared to take hours to enter the nucleus from the cytoplasm and in other cases particles were detected in the nucleus in less than 10 minutes, but the meaning of these observations is hard to decipher.

In one of the most striking demonstrations of protein migration into nuclei, Coons et al. (1951) injected several kinds of proteins into mice and looked for the subsequent localization in various tissues by the fluorescent antibody technique. They found that injected proteins were detectable in the nuclei of various tissues within 10 minutes after injection (the earliest time that they looked). Clearly then, proteins can go from cytoplasm to nucleus in substantially less than 10 minutes.

We can conclude that data on the rate of migration of protein from cytoplasm to nucleus are not precise, but there seem to be rather clear limits placed on speculations about mechanisms involved. Since a complete protein molecule may be synthesized in a minute or two, it appears that any conclusion that protein synthesis occurs in the nucleus because labeled amino acids appear in the nuclear protein 10 minutes or so after administration of the precursor is unwarranted.

(For other considerations of protein migration between cytoplasm and nucleus, see Zetterberg, Chapter 5.)

IN VITRO EVIDENCE OF PROTEIN SYNTHESIS BY NUCLEI

That isolated nuclei are capable of amino acid incorporation into protein, seems to have been well established by the pioneering and extensive work of Allfrey, Mirsky, and their colleagues (Allfrey et al., 1960) on the calf thymocyte nucleus. Following them, there have been reports on amino acid incorporation by isolated nuclei from rat liver (Rendi, 1960), from pea (Birnstiel and Flamm, 1964), from HeLa cells (Bach and Johnson, 1967), as well as other reports. (It may be, however, that because some investigators seem reluctant to publish negative data, the failure to detect protein synthesis, as in the work of Magee and Burrous (1961), is more common than we know.) On the other hand, e.g., Maggio (1966) was unable to detect amino acid incorporation by isolated guinea pig nuclei but did detect such incorporation by *nucleoli* of the same material. Despite a remarkable amount of work, the picture one obtains of protein synthesis by isolated nuclei is unsatisfactory. Considerable doubt remains, at least in the mind of this reviewer, regarding the general significance of the combined observations.

Is Amino Acid Incorporation by Isolated Nuclei an Artifact?

Whenever cells are disturbed, as they are when nuclei are isolated, the question must be asked: does the resulting behavior of the isolated system reflect the true *in vivo* behavior of that system? Thus, Allfrey, Mirsky, and their colleagues, e.g., were concerned with this question and went to considerable labor to determine whether their observations of isolated nuclei reflected true physiological behavior. Their studies revealed enough interesting features, such as a dependence on sodium ions and DNA, to make it unlikely that the effects observed were anything other than what occurs in the intact cell. They report, however, a synthetic rate of 22 molecules/second of protein of average molecular weight of 50,000 per nucleus—a seemingly insubstantial rate of synthesis when compared, e.g., to an estimated rate of 30,000 molecules/second of protein of 40,000 average molecular weight calculated for HeLa cells (McConkey, 1968). (The significance of many of these features is often unclear when the reports of several investigators are compared. A case in point is the work of Uete (1967) who compared the amino acid incorporating ability of rat liver and rat thymus nuclei with respect to their responses to added nucleotides, factors influencing oxidative phosphorylation, potassium and sodium ions, ribonuclease, and deoxyribonuclease. Not only are his results occasionally in disagreement with those of others, but he also finds differences between the behavior of liver and thymus nuclei.) Regardless of how the issues are finally resolved, it is unlikely that any of these kinds of studies can be accepted unquestioningly until the product of *in vitro* activity is in fact shown to be a totally *in vitro* synthesized protein. One needs to see something like the production of an autoradiographically detectable fingerprint resembling a known protein produced following incorporation of radioactive amino acids (cf., Nathans et al., 1962), or an increase in an immunologically detectable protein, or some other kind of biological assay.

One source of artifact is the contamination of isolated nuclei by cytoplasmic ribosomes. This condition is likely to exist for many systems simply because, as seen in the electron microscope, the outer nuclear membrane resembles rough-surfaced endoplasmic reticulum. McCarthy et al. (1966) recognized this possibility and treated isolated rat liver nuclei with EDTA and ribonuclease to remove ribosomes from the outer surface of the nuclear membrane. Although they did not directly study the protein synthetic activity of such nuclei, McCarthy et al. did find that ribosomes extracted from nuclei *after* the EDTA and ribonuclease treatment were as effective as cytoplasmic ribosomes in the support of *in vitro* amino acid incorporation. (These nuclear ribosomes are insensitive to DNase, but sensitive to RNase and seem to be essentially identical to cytoplasmic ribosomes.)

It may be that too much has been made of the potential confusion resulting from the presence of cytoplasmic ribosomes on the outer nuclear membrane. That this may be an exaggerated apprehension is illustrated by the dependence on Na^+ rather than K^+. The traditional amino acid incorporating system based on cytoplasmic ribosomes requires the presence of K^+; Na^+ will not substitute effectively. Allfrey et al. (1960) found for thymus nuclei (and Rendi, 1960, confirmed in some respects for rat liver nuclei) that amino acid incorporation was strictly dependent on Na^+; K^+ would not do. This seems to be good evidence that amino acid incorporation observed with these isolated nuclei was not due to the participation of ribosomes on the exterior of the nuclear envelope.

Is Significant Protein Synthesis by Isolated Nuclei a Phenomenon of General Occurrence or Is It an Exceptional Condition of a Few Cells?

It is conceivable that intranuclear ribosomes, as found for example by McCarty et al. (1966), are merely newly synthesized ribosomes passing to the cytoplasm and are only trivially involved in amino acid incorporation while in the nucleus. Indeed, McCarty et al. (1966) report preliminary evidence that after a short pulse of ^{14}C-orotic acid, nuclear ribosomes have a higher specific activity than cytoplasmic ribosomes, suggesting that nuclear ribosomes may be precursors of those in the cytoplasm. If these above notions are true, it may be reasonable to consider that of all the systems that have been studied, only the thymus nucleus may be a worthy example of an isolated nucleus in which significant protein synthesis can occur. However, a case can be made for considering the thymus nucleus an exceptional one from which extrapolations may be unwise. These are the reasons:

1. The thymus nucleus is an unusually large one for a somatic cell; it occupies about two-thirds of the volume of the cell. For most cells the nucleus is 2 to 20 percent of the cell volume. While we do not know why the thymus nucleus is so relatively large, we can imagine that, in such a cell, many normally cytoplasmic activities are found in the nucleus. Thus, it may be that insufficient protein synthesis can be carried out by the one-third of the cell that is the cytoplasm, and the nucleus must also take on some of that activity, or that a good part of the protein synthesizing machinery never left the nucleus after it was made.
2. That the thymus nucleus has indeed taken on some normally cytoplasmic functions is suggested by the fact it is capable of a significant amount of ATP synthesis by processes involving glycolysis, the citric acid cycle, and a "type of oxidative phosphorylation" (McEwen et al., 1963). Although, in many ways, these processes resemble the same ones that function in the cytoplasm, there is some difference in susceptibilities to inhibitors. Nevertheless, it may be that, for unknown reasons, the thymus nucleus is an unusual extension of the cytoplasm in which activities that are ordinarily cytoplasmic have been somewhat modified.

How Much Protein Does an Isolated Nucleus Synthesize?

An important consideration, if nuclei do synthesize proteins, is: how much of a contribution do they make to the cell's total protein? Unfortunately, no estimate is to be found. One might expect to have some comparison of the activity of isolated nuclei with that of whole cells, but I have found no such comparison. In fact, the only quantitation I have found is that of Allfrey et al. (1960) who estimate that an isolated thymus nucleus synthesizes about 22 molecules of average 50,000 molecular weight protein per second; an estimate that, as we noted earlier, compares poorly with the rate for an intact cell. We should expect, of course, that isolated nuclei probably would be less effective than the nuclei in intact cells.

One can find other kinds of estimates in the literature, but they are not too helpful. Kuehl (1967) believes that about 10 percent of the lactic dehydrogenase of rat liver cells is made in the nucleus. We will see later why this estimate is only speculative. There is a suggestion from the work of Hammel and Bessman (1964) that most of the hemoglobin of pigeon erythrocytes is made in the nucleus. If one looks at the effect of the removal of the nucleus (see p. 197) on the rate of protein synthesis, one can estimate for *Amoeba proteus*, e.g., that there is a diminution of greater than 50 percent in the rate of amino acid incorporation and one might assume that all that lost activity would have occurred in the nucleus prior to enucleation. There may be other reasons for the effect of enucleation, of course. We should note also that if 50 percent of protein synthesis occurred in the nucleus it would mean, because the ameba nucleus occupies only 2 percent of the cell volume, that the rate of protein synthesis is about 25 times that in the cytoplasm, a conclusion for which there is no evidence whatsoever.

What Kinds of Proteins are Synthesized by an Isolated Nucleus?

Since it has been impossible thus far to measure an increase in total protein in isolated nuclei, perhaps true synthesis by nuclei can be shown by demonstrating what kinds of proteins have been made. To do this one does not need to show an absolute increase in protein mass; indirect methods may do.

As of now, qualitative descriptions of protein synthesis by isolated nuclei have been limited to the detection of labeled amino acid incorporation into such components of nuclei as residual protein, basic nonhistone proteins, proteins soluble in molar NaCl, etc.—categories too ill-defined to convince us that true protein synthesis is unquestionably established. To establish, more or less unequivccally, that true protein synthesis has occurred requires an experiment like one of those suggested on p. 194, i.e., the demonstration that there has been an increase in the amount of a specific protein or specific biological activity. One such attempt, by Maggio (1966), to demonstrate by "fingerprinting" a newly-made radioactive protein failed.

IN VIVO EVIDENCE OF PROTEIN SYNTHESIS BY NUCLEI

Because of a dearth of supporting *in vivo* experiments, I doubt that there is any decisive evidence as to whether the nucleus synthesizes protein that is qualitatively different from that produced in the cytoplasm or that is quantitatively significant for the cell. Let us examine some of the *in vivo* data.

Protein Synthesis by Enucleate Cells

Why are experiments with enucleate cells considered *in vivo,* whereas experiments with isolated nuclei are considered *in vitro?* There are several reasons. The behavior of enucleate tissue culture cells that one can see through phase-contrast microscopy is much like that of normal nucleate cells (Goldstein et al., 1960a). Enucleate *Acetabularia* behave almost identically to normal cells in almost all ways for two weeks or more, even to the extent of going through normal morphogenesis (Haemmerling, 1963). Naturally-occurring enucleate cells, such as the mammalian erythrocytes (Dintzis, 1961), do indeed show substantial synthesis of protein. Perhaps most importantly, enucleate cells, e.g., amebas (Comandon and de Fonbrune, 1939), can recover total normal behavior when a nucleus is reimplanted, whereas a nucleus that has been isolated for more than several seconds cannot recover a completely normal state when implanted into healthy cytoplasm (Jeon, 1968). Thus an enucleate cell can be considered, for a while at least, as a healthy cell that is only temporarily deprived of some functions.

If the nucleus produces protein, one might expect that the rate of protein synthesis would be reduced in enucleate cells. Indeed, the removal of the cell nucleus does lead to a reduced rate of amino acid incorporation in many cases. Thus, enucleate *Amoeba proteus* shows a reduction in rate of amino acid incorporation of more than 50 percent (Mazia and Prescott, 1955; but Goldstein and Prescott, 1967b, found only about 30 percent reduction). Enucleate *Mytilus* eggs incorporate amino acids at less than 50 percent of the rate of nucleate eggs (Abd-El-Wahab and Pantelouris, 1957). For amacronucleate *Paramecium aurelia* there is a 70 to 80 percent reduction in the capacity to incorporate amino acids (Kimball and Prescott, 1964).

It would be easy to conclude from the above evidence that the nucleus does indeed synthesize protein, but it is unlikely that any of the investigators who engaged in the work would make that claim; other interpretations seem more reasonable. For example, the most favored view is that the effect of enucleation is due to the non-replacement of rapidly decaying messenger RNAs needed for cytoplasmic protein synthesis—although the effect on amino acid incorporation is seen quite soon after enucleation in some cases. (This would argue for the existence of two classes of messenger RNAs with different stabilities, since some ability to incorporate amino acids persists for days in the absence of a nucleus.)

As mentioned earlier, if the data on the effect of enucleation are taken at face value, one might conclude that, since in *Amoeba proteus* there is reduction of about 30 to 50 percent in the rate of amino acid incorporation following enucleation and the nucleus occupies only 2 percent of the cell volume, the nucleus synthesizes protein at a rate 15 to 25 times that of the cytoplasm. There is no other support whatsoever, in *A. proteus* or any other cell, for such a marked difference between the two compartments in the protein synthetic rates.

A diminution in protein synthetic rate as a result of the removal of the nucleus is not a universal phenomenon. The most extensive studies on the relationship between enucleation and protein synthesis have been performed with *Acetabularia* (Haemmerling, 1963), and it is clear that *Acetabularia* can synthesize protein at a normal rate for up to two weeks in the absence of a nucleus. The activities of certain enzymes increase in the enucleate cell in parallel with the same activities of the whole cell, but some enzymes are not produced at the normal rate for as long as two weeks—these differences perhaps reflecting differences in the stabilities of the different mRNAs.

While on the subject of *Acetabularia,* it is interesting to recall the question raised by Harris regarding the function of the rapidly-labeled, high molecular weight, heterogenous RNA that apparently turns over completely within the nucleus. Harris (1963) proposed that this RNA is concerned with protein synthesis within the nucleus and that the RNA in the cytoplasm (which may be considered to be more stable) is involved in cytoplasmic protein synthesis. This hypothesis could be extrapolated to the idea that rapidly turning-over nuclear RNA is that used in the synthesis of inducible proteins, whereas cytoplasmic synthesis was devoted soley to constitutive proteins. This notion seems invalid, in *Acetabularia* at least, since Zetsche (1966) has shown that UDP glucose 4-epimerase could be induced during cellular "cap" morphogenesis several days after removal of the nucleus. The enzyme induction was reversibly inhibited by puromycin showing that new protein synthesis was required for the appearance of enzyme activity. (This is, thus, an example of genetic control at the level of translation, and not transcription.)

There is some other evidence that removal of the nucleus has little effect on the rate of synthesis by a cell. For enucleate human amnion cells grown in culture, there is a subjective estimate from autoradiographs that the absence of a nucleus has no *marked* effect on the rate of amino acid incorporation (Goldstein et al., 1960b). More substantial is the evidence of Ecker et al. (1968) that activated or nonactivated enucleate frog eggs incorporate amino acids at the same rate as activated or nonactivated nucleate eggs. These investigators suggest that protein synthesis during very early development is dependent upon "long-lived templates synthesized and accumulated during oogenesis." This seems like reasonable conjecture and leads me to speculate that, in those instances where enucleation results in a marked reduction in the rate of amino acid incorporation by the cell, much of the normal protein synthetic activity is dependent on a continuous supply of short-lived messages from the nucleus. The situation found in enucleate frog eggs is not a universal phenomenon, however; Clement and Tyler (1967), e.g., have found that enucleate polar lobes detached from mud snail eggs incorporate amino acids at only one-half the rate per unit volume of the whole egg.

In any event, enucleation experiments are too indirect to provide firm conclusions about the question at hand. Thus, as is often pointed out when the removal of a nucleus results in a reduced rate of protein synthesis, perhaps the absent nucleus was furnishing, not newly-synthesized protein, but rather mRNA or some simple cofactor required by the machinery in the cytoplasm.

Pulse-chase Experiments to Determine the Primary Site of Amino Acid Incorporation

A traditional method for investigating the primary localization of synthetic activity has been to administer radioactive precursors for a brief period ("pulse") followed either by immediate fixation of the cells or by a "chase" (exposure to a large excess of the unlabeled precursor) and then fixation. This is not the place to discuss the problems—such as questions of intracellular pool sizes, effectiveness of the chase, differences in synthetic rate of different compartments, etc.—that make it difficult to draw unqualified conclusions from pulse-chase experiments, but I want to reiterate what was said on p. 192: Since proteins readily migrate from cytoplasm to nucleus and vice versa, sometimes in no more than a few minutes or perhaps seconds, it is particularly difficult to draw conclusions regarding the primary site of protein synthesis from pulse-chase experiments.

Typical examples of experiments in which the localization of radioactivity following brief exposure to labeled amino acids is taken as evidence for the site of protein synthesis are those of Leblond and Amano (1962), Monesi (1964), Schultze and Maurer (1967), and Hamilton et al. (1968). Because of a significant amount of label localized in nuclei in pulses of from 5 to 30 minutes, each group concluded that protein synthesis had occurred in nuclei. As mentioned, such conclusions cannot be considered reasonable. In fact, one can cite evidence which, if analyzed by the same logic, should be taken to show that nuclear protein is synthesized in the cytoplasm. Thus, Zalokar (1960) showed that, after a 15-second exposure of *Drosophila* ovarian follicles to ^{3}H-leucine, there is substantial radioactivity in cytoplasmic protein and essentially none in nuclei. A few minutes later, however, he found appreciable label in nuclear proteins. Essentially the same observation was made for *Drosophila* spermatocytes by Hennig (1967) in more extensive experiments. These experiments led him to conclude that some chromosomal proteins, at least, are synthesized in the cytoplasm and then migrate to the nucleus. Similarly Bloch and Brack (1964), studying, as did Monesi (1964), histone synthesis during spermiogenesis, found that ^{3}H-arginine incorporation was almost exclusively cytoplasmic after one hour's exposure to the labeled precursor, but mostly nuclear several hours later. (Monesi worked with mice which may have a substantially higher metabolic rate than the grasshoppers with which Bloch and Brack worked.) By the simple reasoning often used to interpret the results of pulse-chase experiments, it can be argued that Zalokar, Hennig, and Bloch and Brack have demonstrated *cytoplasmic* synthesis of nuclear proteins, rather than the nuclear synthesis suggested by Leblond and Amano, Monesi, and Schultze and Maurer. Suffice to say, none of these experiments is likely to be the last word on this issue.

The Kinetics of Synthesis of Specific Proteins

There are two pulse-chase experiments that have a more biochemical approach and that deserve particular attention. These deal with the study of labeling kinetics of purified single proteins in nucleus and cytoplasm and, in the case of avian hemoglobin synthesis at least, come close to showing decisively that protein is indeed synthesized in the cell nucleus.

LACTIC DEHYDROGENASE (LDH) SYNTHESIS IN RAT LIVER. Kuehl (1967) injected radioactive leucine into rats and at intervals, beginning 30 seconds later, removed the livers from different groups of animals. He then extracted and purified LDH from separated nuclei and cytoplasm and determined the specific radioactivity of the enzyme. He found that the specific activity curves for LDH from both nuclei and cytoplasm rose about one minute after the injection and were essentially identical thereafter. Kuehl interprets these data as showing independent synthesis of LDH at approximately the same rate by both compartments, but, as he points out, the "interpretation depends upon the assumption that movement of the enzyme between cytoplasm and nucleus occurs slowly," an assumption for which there is little experimental support and, in fact, against which there is good evidence.

Whatever may prove to be the truth, Kuehl notes that, at best, the nucleus would be furnishing no more than 10 percent of the cell's total LDH and suggests that his data may provide evidence that LDH messenger RNA translation begins in the nucleus and continues in the cytoplasm. If true, this would be an interesting phenomenon, but would not satisfy our curiosity about whether the nucleus is making a unique contribution (but cf. the work of Nemchinskaya et al., 1968, discussed on p. 201).

Hemoglobin Synthesis in Avian Erythrocytes. Perhaps the most convincing (but not without some reservation) *in vivo* demonstration of protein synthesis by nuclei comes from a study of hemoglobin synthesis in pigeon erythrocytes by Hammel and Bessman (1964). They performed pulse-chase experiments and in essence found that the incorporation of labeled amino acids into nuclear protein (very probably hemoglobin) occurred sooner and at a more rapid rate than incorporation into cytoplasmic protein. After the chase, the activity in the nucleus fell, while that in the cytoplasm rose. It is difficult to interpret these data in any way except that hemoglobin synthesis occurs, at least in part, in the nucleus.

There are, however, a few puzzling features in the data. Following the chase with unlabeled amino acids, the specific activity of the nuclear protein falls but only by about 30 percent, and then levels off. If the hemoglobin is synthesized in the nucleus and is then transported to the cytoplasm, why doesn't the nuclear specific activity continue to fall as the cytoplasmic activity continues to rise? Another curious feature is that, in some experiments at least, following the chase, the cytoplasm acquires a good deal more activity than is lost from the nucleus. A third disturbing aspect of these data is that, in cells that were incubated in labeled precursors continuously for two to four hours, the specific activity ultimately became higher in the cytoplasm than in the nucleus. This is unusual for the usual precursor-product relationship, but perhaps may be explained by the presence in the nucleus at the outset of a higher proportion of nonturning-over and nonmoving proteins.

Even if all of these puzzling features can be satisfactorily disposed of and it is unequivocally shown that pigeon erythrocyte nuclei do synthesize protein, will this settle the issues posed in the *Introduction*? Not at all. Certainly hemoglobin synthesis by nuclei cannot be a general phenomenon of red blood cells since erythrocytes from many species synthesize hemoglobin quite well in the absence of nuclei. Rather than using these data to make some generalization about protein synthesis by nuclei, it seems more reasonable to suppose that in the avian erythrocyte, some adaptation has occurred that enables the nucleus to participate directly in hemoglobin synthesis.

One feature of hemoglobin synthesis in pigeon erythrocytes, as reported by Hammel and Bessman (1964), may prove to be an important clue. On the basis of RNA content, the nucleus of mature erythrocytes is almost twice as active in amino acid incorporation as the immature reticulocyte. Does this reflect the existence of a protein synthesizing mechanism not dependent on RNA?

Comparison of the Kinds of Proteins found in Nucleus and Cytoplasm

In general, because of the ability of proteins to migrate between nucleus and cytoplasm, if a difference is found in the kinds of proteins that are in nucleus and cytoplasm, it does not follow that that difference represents different synthetic capacities in the two compartments. (For reviews of protein, mostly enzyme, differences between nucleus and cytoplasm, see Roodyn, 1959; Siebert and Humphrey, 1965.) There are, however, two particular situations that deserve attention because they have features which suggest that there may be differences in the protein-synthesizing mechanisms in nucleus and cytoplasm.

LACTIC DEHYDROGENASE IN RAT LIVER. Recall from the discussion on p. 199 that Kuehl (1967) suggested that LDH synthesized in the nucleus may be a minor contribution resulting from the transitory translation of newly synthesized templates on their way to the cytoplasm where the bulk of the translation will be effected. If such were to prove to be generally true, we could dismiss protein synthesis by nuclei as a trivial cellular event. If, on the other hand, it could be shown that the LDH made in the nucleus is different from that made in the cytoplasm, it could hardly be pushed aside as a trivial phenomenon. Thus, the work of Nemchinskaya et al. (1968), on the isozymes of nuclear and cytoplasmic LDH from rat liver, demands attention. They found differences in the enzymes from nucleus and cytoplasm in electrophoretic patterns, denaturation by urea, heat inactivation, and optimal substrate concentration. One might conclude that these differences reflect the fact that one kind of enzyme is made in the nucleus and the other in the cytoplasm, but it is possible that the differences in the properties of the proteins result from environmental differences in the two compartments.

Unfortunately for this hypothesis, Siebert and Humphrey (1965) report that no differences of the sort described by Nemchinskaya et al. are detectable in related experiments, and Kuehl (1967) was unable to detect a difference between the nuclear and cytoplasmic enzymes by immunological techniques (although this latter procedure may not have been sensitive enough to detect differences).

THE ISOCITRATE DEHYDROGENASE ISOZYMES OF ALLOPHENIC MICE. A dramatic way in which protein synthesis by nuclei might be demonstrated would be to have two or more genetically different nuclei in a common cytoplasm and find that the nuclei contain different enzymes. One way to do this is to obtain heterokaryotic syncytial tissue from *allophenic* organisms. (Allopheny, as defined by Mintz, 1967, is a phenomenon of concurrent display of allelic phenotypes not associated with heterozygosity but rather resulting from the aggregation of eggs or embryos of different genotypes. Such aggregation leads to the development of but one organism, which may, however, display a phenotype reflecting both original genotypes.) A study of the enzymes of different nuclei in an allophenic syncytium or heterokaryon has not been done, but Mintz and Baker (1967) have carried out a study that has relevance to the questions we are asking here.

Mouse isocitrate dehydrogenase (IDH) is apparently made up of 2 subunits. Two homozygous strains of mice differing by one allele will each produce a different isozyme distinguishable by electrophoretic mobility. In a heterozygote, three isozymes are detectable in the proportion of 1:2:1, one parental form, a hybrid form, and the other parental form. If the subunits associate at random and are synthesized in the cytoplasm, one would expect that in allophenic mouse skeletal muscle, which is a syncytium formed by the fusion of mononucleate cells, all three isozymes could be present. If the isozymes are synthesized in the nuclei and the subunits do not reassociate in the cytoplasm, we could expect to find only the two parental forms and not the hybrid isozyme. Mintz and Baker found all three isozymes and thus it would appear that there is no evidence from this study for the nuclear synthesis of this enzyme—at least in skeletal muscle. However, for cardiac muscle, Mintz and Baker (1967) report preliminary evidence that *no* hybrid enzyme is present in the allophenic animal. They suggest that this supports the notion that cardiac muscle is not a syncytium, but such results might also be found if, in a heterokaryon, nuclear synthesis and dimerization of the enzyme occurred before liberation to the cytoplasm; such a mechanism could prevent the formation of hybrid enzyme. Obviously more work must be done to establish whether such is the case.

EVIDENCE REGARDING PROTEIN SYNTHESIS MECHANISMS IN CELL NUCLEI

Do Nuclei Contain Ribosomes?

The question of the existence of nuclear ribosomes is still a matter of debate. Even in those cases where it seems probable that nuclear ribosomes do occur, the significance of that occurrence is not clear—except, of course, in the case of calf thymus nuclei, which are thought to be atypical. No attempt will be made here to review all the relevant literature, primarily because much of the evidence has been presented in uncritical fashion.

The evidence in favor of the existence of nuclear ribosomes is of two general kinds. One kind is based on extraction from nuclei of ribonucleoprotein particles that in one way or another are thought to resemble ribosomes. Such evidence has been offered by Rendi (1960), Wang (1961), Frenster et al. (1960), Birnstiel et al. (1963), Winckelmans et al. (1964), and Traub et al. (1964), to name just a sampling of the available contributors. Hnilica (1967) provides further discussion of this literature. The other kind of evidence comes from electron microscope observations in which "ribosome-like" granules are seen in sections of various kinds of cells. The conclusion that these particles are ribosomes, is based almost entirely on size measurements and hence cannot be considered more than suggestive. Various features of nuclear granules of various sizes are discussed by Swift (1963), Smetana et al. (1963), and Davies and Tooze (1966).

The evidence that ribosome-like particles can be extracted from nuclei is subject to two major kinds of criticism: (1) the ribosomes are cytoplasmic contaminants and/or (2) the particles have not, in fact, been shown to be true ribosomes.

Electron microscopists have known for a long time that particles similar to those attached to the rough-surfaced endoplasmic reticulum, are attached to the cytoplasmic side of the nuclear envelope, and more recently there have been reports (cf., Weston, 1968) of ribosome-like particles *within* the nuclear envelope. Such reports led McCarty et al. (1966) and Sadowski and Howden (1968) to be concerned with such "contamination" in their preparations of isolated nuclei. They have urged that special measures must be taken to remove these outer nuclear envelope ribosomes before experiments on nuclei can lead to meaningful conclusions about the existence of nuclear ribosomes. Thus, McCarty et al. (1966) used ribonuclease and EDTA to remove the membrane ribosomes from isolated nuclei, and Sadowski and Howden (1968) used Triton X-100. On the other hand, a confirmation of sorts came from Lawford et al. (1967) who insured recovery of polysome-like particles from rat liver nuclei by using a ribonuclease inhibitor during their preparative procedure. Similar polysome-like material was extracted from isolated nuclei, apparently from the nuclear membrane, by Bach and Johnson (1967).

With regard to the criticism that there is a dearth of evidence to support the conclusion that the isolated ribonucleoprotein particles are ribosomes, we can say that, aside from the material extracted from the calf thymus nuclei, it is likely that only McCarty et al. (1966) have clearly shown that true ribosomes may be found within nuclei. McCarty et al., having taken the precaution to remove the outer nuclear membrane ribosomes from isolated rat liver nuclei, extracted ribosome-like particles that were able to support *in vitro* amino acid incorporation much like cytoplasmic ribosomes. These workers, nevertheless, raise the question as to whether the intranuclear ribosomes really support protein synthesis in the intact cell or are simply newly-manufactured ribosomes on the way to the cytoplasm. Indeed, in support of the latter notion, they report preliminary experiments in which after a short exposure of cells to ^{14}C-orotic acid, the extracted nuclear ribosomes are found to have a higher specific activity than cytoplasmic ribosomes from the same cells. This suggests that one may be the precursor of the other. In a similar study, also with rat liver material, Sadowski and Howden (1968) provide indirect evidence that intranuclear ribosomes may be less effective in amino acid incorporation *in situ* than are cytoplasmic ribosomes, and this, too, may be interpreted as evidence that nuclear ribosomes are immature precursors of those in the cytoplasm.

One more point about the above kind of experiment. The role of intranuclear proteins would undoubtedly be much better understood if we were able to obtain some data comparing the concentration of ribosomes in nucleus and cytoplasm.

There are other experiments that are taken to show that the nucleus contains ribosomes, but none are truly adequate to prove the point. One other kind of experiment, however, is worth mentioning because it may provide useful information on this question. Mundell (1967) compared the acrylamide gel disc electropherograms of starfish oocyte nucleolar proteins with cytoplasmic ribosomal proteins and concluded that many of the proteins of both organelles are identical. From this (and because the nucleolus is the site of ribosomal RNA synthesis) he suggested that ribosome assembly occurs in the nucleolus—an idea shared by many. Examination of the figures given in the paper shows, however, that, when consideration is given to both the *quantitative* and qualitative features of the electropherograms, there seems to be true acrylamide gel identity between perhaps 25 percent of the proteins of either fraction, and this implies that there is likely to be no simple relationship between nucleolar proteins and ribosomal proteins. Indeed, Izawa and Kawashima (1968) preliminarily report similar data for mouse ascites tumor cells but caution that such identity between nucleolar and ribosomal proteins does not mean that nucleolar material is ribosomal, particularly since they find other significant differences between the two fractions.

An increasingly popular notion is that certain nuclear ribonucleoprotein particles are some form of precursor of cytoplasmic ribosomes. This is the view expressed by Izawa and Kawashima (1968) based on a study of the centrifugal sedimentation patterns of various fractions found in the isolated nucleoli of mouse ascites tumor cells. They found that newly labeled RNA-containing particles became progressively more like ribosomal particles. A similar conclusion could be drawn from the work of Davies and Tooze (1966), since their morphological study of nuclear granules could be interpreted to show ribosomal maturation processes—although they do not make that point. Likewise Sadowski and Howden (1968), who studied ribosomes and polysomes from rat liver nuclei with and without the outer membrane particles removed, provide evidence for a precursor-product relationship between nuclear and cytoplasmic ribonucleoprotein particles. They found that the nuclear material incorporated RNA precursors more rapidly than the cytoplasmic material. Perhaps most interesting is their finding that intranuclear ribosomes and polysomes probably support protein synthesis less well than their cytoplasmic counterparts, an observation that can be interpreted as evidence for the immaturity of the nuclear particles. In the same vein, is the evidence of Rogers (1968) that amphibian oocyte nuclei, which probably contain no 78S ribosomes, do contain a 50 to 55S particle that is a precursor of the 60S ribosomal subunit. Should it be firmly established that the nucleus contains essentially only ribosomal precursor particles, we will probably conclude that protein synthesis by nuclei is a trivial phenomenon or else occurs by mechanisms different from those in the cytoplasm.

Evidence against the existence of intranuclear ribosomes seems impressive but probably can be questioned on grounds of technique and/or uniqueness of the material used. Penman (1966) prepared detergent-cleaned isolated HeLa cell nuclei in which he could find no 16S RNA and concluded, therefore, that no ribosomes were associated with these nuclei. A similar result was obtained by Casola and Agranoff (1968) with nuclei from goldfish brain. Fisher, as reported by Harris (1963), looked for ribosomes in isolated HeLa cell nuclei treated with DNase and sodium deoxycholate and found no ribosomal particles. It may be argued that, in these experiments, the various treatments to which isolated nuclei were exposed would result in the selective loss of the 16S RNA of the ribosomes or perhaps the loss of mature ribosomes.

It is unlikely that the work of Rogers (1968) on the contents of oocyte nuclei could be criticized on the grounds that the isolation treatments could produce an artifact; the manual isolation procedure she used is probably the gentlest procedure possible. Not only did she not find any 78S ribosomes in the amphibian oocyte nuclei she examined, but, like Penman and Casola and Agranoff, she was unable to find the smaller ribosomal RNA fraction.

Finally, Craig (1967) transplanted ^{3}H-protein labeled nuclei into unlabeled amebas and looked for the appearance of radioactive ribosomes in the host cytoplasm. No labeled ribosomes were found under these conditions, but in a parallel experiment using transplanted ^{3}H-RNA nuclei, the host cytoplasm did acquire radioactive ribosomes. This demonstrated that no detectable *complete* ribosomes go from nucleus to cytoplasm but does not exclude the possibility that nonmigrating ribosomes may be present in the ameba nucleus.

Is the Remainder of the Usual Protein Synthesizing Machinery Present in the Nucleus?

There have been few reports of the presence in nuclei of other components of the traditional protein synthesizing system. The most complete study is probably that of Allfrey et al. (1960) on the isolated thymus nucleus. Their studies show that nuclei probably contain all the components normally thought to be associated with cytoplasmic ribosomal systems including susceptibility to the same inhibitors (but cf. Wang, 1961). One could argue, of course, that this supports our previous hypothesis that the thymus nucleus is atypical in that it has assumed many normally cytoplasmic functions.

Is There a Unique System for the Synthesis of Protein in the Nucleus?

No consequence of the study of protein synthesis by nuclei is likely to be more fascinating than the discovery that such synthesis is effected by mechanisms fundamentally different from the mechanisms responsible for protein synthesis in the cytoplasm. There have been a few reports that suggest that there may indeed be a unique system(s) for protein synthesis in the nucleus.

From the early work on calf thymus cells, it was thought that, in contrast to cytoplasmic protein synthesis, protein synthesis in the nucleus was dependent on sodium ions and DNA. Subsequently it was shown (Allfrey et al., 1960) that sodium ions were necessary for amino acid transport into the nucleus and DNA was needed for the generation of ATP within the nucleus; neither process reflecting anything particularly unique about the basic machinery of protein synthesis.

There have been occasional reports of unusual activities related to amino acid incorporation by nuclei, but generally they are too nebulous or preliminary, or lacking in some other way to be treated seriously here. There are, however, two reports that are highly suggestive and deserve a good deal of attention. One report is that of Patel and Wang (1965) on a system that incorporates ^{14}C-tryptophan into thymus nuclear protein *after* the removal of ribosomal material. This system incorporates labeled tryptophan into nonhistone protein without the lag period (attributable to the need for the production of new messenger RNA) usually observed with isolated thymus nuclei. Most striking is the finding that this incorporation is *not* sensitive to ribonuclease but is sensitive to deoxyribonuclease. Finally, they report that the incorporation of tryptophan is stimulated by the action of trypsin, which presumably digests histone under the experimental conditions they employed. The evidence can be interpreted to mean that some amino acid incorporating system in the nucleus is immediately dependent on DNA and is independent of RNA or ribosomes.

Another system, perhaps related to the one above, was found by Sekeris et al. (1966) in rat liver nuclei. The authors state that this system actually is the same as the "aggregate enzyme" of Weiss (1960) that possesses DNA-directed RNA polymerase activity. Sekeris et al. find that their system, which is approximately quantitatively equivalent to a ribosomal system in the ability to incorporate amino acids, is insensitive to ribonuclease or added template RNA but is sensitive to deoxyribonuclease and is, in that way, similar to the preparation of Patel and

Wang (1965). There is a suggestion that this system, as well as that of Patel and Wang, may have some relation to DNA replication.

It may be that the work of Hammel and Bessman (1964) on hemoglobin synthesis by avian nuclei has some bearing here. Recall (p. 200) that the amount of hemoglobin synthesized was not proportional to the amount of RNA present in the nucleus. Does that mean some non-RNA mechanism is at work?

CONCLUSION

We can conclude that at least some nuclei probably do synthesize protein. For the present, however, true protein synthesis probably has been shown only for nuclei of cells like those of the thymus, which may be atypical in having assumed what are normally cytoplasmic functions in other cells, or for nuclei of the pigeon erythrocyte, which synthesizes a protein (hemoglobin) that is also synthesized by the cytoplasm (obviously exclusively in the cytoplasm for nonnucleated reticulocyte). It may be that many other cell nuclei perform some kind of protein synthesis but, for the moment, it remains to be shown that such synthesis has any significance for the overall pattern of cell activity. We must bear in mind that no interpretation of nuclear activity based on the localization of proteins can be taken seriously unless consideration is given to the possibility that proteins can migrate readily from one part of the cell to another.

If we were to grant for the moment that protein synthesis is a phenomenon common to all or most nuclei, we ought then to ask what the value of this assumption is. One is hard put to find any insight that we have gained as a consequence of assuming that what Mirsky, Allfrey, and their co-workers have found for calf thymus nuclei is a general phenomenon. It certainly has not helped us understand such special nuclear phenomena, as mentioned in the *Introduction,* as the acquisition of specific chromosome loop morphologies or the role of the rapidly-labeled, heterogeneous nuclear RNA.

If the protein synthesis by nuclei is a significant general circumstance, we ought to ask: Is there a special mechanism for effecting this synthesis? Is there a special mechanism for retaining particular proteins within the nucleus? Is there a special mechanism for retaining particular mRNAs within the nucleus? Why are particular proteins synthesized in the nucleus?

The main conclusion to be drawn from this essay is that the decisive experiment is yet to be done. I am not sure what the decisive experiment would be, but a possible one is the following. Assume, as discussed on p. 201, that the LDH in rat liver nuclei is different from that in the cytoplasm, and further assume that one can obtain mutants of the nuclear LDH and that the enzyme is made of subunits. By some version of the cell fusion technique of Harris and Watkins (1965), produce a heterokaryon by fusion of cell types differing by the one gene in question for which each cell is homozygous. If synthesis of the nuclear enzyme is cytoplasmic, I would expect to find that at least some of the enzyme extracted from the nuclei is hybrid, i.e., composed of both mutant types. If synthesis of the nuclear enzyme is nuclear, no hybrid enzyme ought to be found; unless, of course, the nuclear enzyme subunits are free to migrate to the cytoplasm and back into the nucleus. This experiment could settle the question only if the nuclei contained no hybrid enzyme.

Note added in proof

The notion that protein synthesis within nuclei may be more or less directly dependent on DNA is based on the sensitivity to deoxyribonuclease of the amino acid-incorporating ability of isolated nuclei, and it therefore requires careful reconsideration. Gallwitz and Mueller (1969. Europ. J. Biochem., 9:431) report that when isolated HeLa cell nuclei are mixed with microsomes and then treated with deoxyribonuclease, the microsomes, as well as the nuclei, lose much of their capacity to support amino acid incorporation—although in the absence of nuclei the microsomes are insensitive to deoxyribonuclease. It is likely that the effect is due to deoxyribonuclease causing the liberation (and activation) of a ribonuclease that is bound to chromatin. This possibility is supported by the work of Stevens et al. (1969. Biochim. Biophys. Acta, 186:124), in which it was shown that DNase acting on *Amoeba proteus* nuclei does indeed cause the stimulation of ribonuclease activity.

REFERENCES

Abd-El-Wahab, A., and E. M. Pantalouris. 1957. Synthetic processes in nucleated and non-nucleated parts of *Mytilus* eggs. Exp. Cell Res., 13:78.

Allfrey, V. G., J. W. Hopkins, J. H. Frenster, and A. E. Mirsky. 1960. Reactions governing incorporation of amino acids into the proteins of the isolated cell nucleus. Ann. N. Y. Acad. Sci., 88:722.

Bach, M. K., and H. G. Johnson. 1967. The isolation and characterization of a novel microsome fraction from washed and detergent-treated nuclei of HeLa cells by the use of solutions containing deoxyribonucleic acid. Biochemistry, (ACS) 6:1916.

Bassleer, R. 1966. Cytologie—Étude cytologique et cytochimique de la tumeur d'ascite d'Ehrlich. Comportement des protéines nucléaires totales et des acides désoxyribonucléiques au cours de la preparation à la mitose et de la division cellulaire. C. R. Acad. Sci. (Paris), Ser. D, 262:1117.

Beck, J. S. 1962. The behaviour of certain nuclear antigens in mitosis. Exp. Cell Res., 28:406.

Beerman, W. 1967. Gene action at the level of the chromosome. *In* Heritage from Mendel, Brink, R. A., ed., Madison, University of Wisconsin Press, p. 779.

Birnstiel, M. L., M. I. H. Chipchase, and R. J. Hayes. 1962. Incorporation of L-(^{14}C) leucine by isolated nuclei. Biochim. Biophys. Acta, 55:728.

——— M. I. H. Chipchase, and B. B. Hyde. 1963. The nucleolus, a source of ribosomes. Biochim. Biophys. Acta, 76:454.

——— and W. G. Flamm. 1964. Intranuclear site of histone synthesis. Science, 145:1435.

Bloch, D. P., and S. D. Brack. 1964. Evidence for the cytoplasmic synthesis of nuclear histone during spermiogenesis in the grasshopper *Chortophaga viridifasciata* (De Geer). J. Cell Biol., 22:327.

Byers, T. J., D. B. Platt, and L. Goldstein. 1963. The cytonucleoproteins of amoeba. II. Some aspects of cytonucleoprotein behavior and synthesis. J. Cell Biol., 19:467.

Casola, L., and B. W. Agranoff. 1968. Release of RNA from goldfish brain nuclei by sodium dodecyl sulfate. Biochem. Biophys. Res. Commun., 30:262.

Clement, A. C., and A. Tyler. 1967. Protein-synthesizing activity of the anucleate polar lobe of the mud snail *Ilyanassa obsoleta*. Science, 158:1457.

Comandon, J., and P. de Fonbrune. 1939. Ablation du noyau chez une amibe, réactions cinétiques à la piqûre de l'amibe normale ou denucléé. C. R. Soc. Biol., 130:740.

Coons, A. H., E. H. Leduc, and M. H. Kaplan. 1951. Localization of antigens in tissue cells. VI. The fate of injected foreign proteins in the mouse. J. Exp. Med., 93:173.

Craig, N. C. 1967. The origin of ribosomal RNA and protein in *Amoeba proteus*. J. Cell Biol., 35:28A.

Das, N. K., and M. Alfert. 1963. Silver staining of a nucleolar fraction, its origin and fate during mitotic cycle. Ann. Histochim., 8:109.

Davies, H. G., and J. Tooze. 1966. Electron and light microscope observations on the spleen of the newt *Triturus cristatus*: The surface topography of the mitotic chromosomes. J. Cell Sci., 1:331.

Dintzis, H. M. 1961. Assembly of the peptide chains of hemoglobin. Proc. Nat. Acad. Sci. U.S.A., 47:247.

Ecker, R. E., L. D. Smith, and S. Subtelny. 1968. Kinetics of protein synthesis in enucleate frog oocytes. Science, 160:1115.

Feldherr, C. M., and C. V. Harding. 1964. The permeability characteristics of the nuclear envelope at interphase. Protoplasmatologia Bd. V., No. 2:35.

Frenster, J. H., V. G. Allfrey, and A. E. Mirsky. 1960. Metabolism and morphology of ribonucleoprotein particles from the cell nucleus of lymphocytes. Proc. Nat. Acad. Sci. U.S.A., 46:432.

Goldstein, L., R. Cailleau, and T. T. Crocker. 1960a. Nuclear-cytoplasmic relationships in human cells in tissue culture. II. The microscopic behavior of enucleate human cell fragments. Exp. Cell Res., 19:332.

——— J. Micou, and T. T. Crocker. 1960b. Nuclear-cytoplasmic relationships in human cells in tissue culture. IV. A study of some aspects of nucleic acid and protein metabolism in enucleate cells. Biochim. Biophys. Acta, 45:82.

——— and D. M. Prescott. 1967a. Proteins in nucleocytoplasmic interactions. I. The fundamental characteristics of the rapidly migrating proteins and the slow turnover proteins of the *Amoeba proteus* nucleus. J. Cell Biol., 33:637.

——— and D. M. Prescott. 1967b. Protein interactions between nucleus and cytoplasm. *In* The Control of Nuclear Activity, Goldstein, L., ed., Englewood Cliffs, N. J., Prentice-Hall, Inc., p. 273.

Haemmerling, J. 1963. Nucleo-cytoplasmic interactions in *Acetabularia* and other cells. Ann. Rev. Plant Physiol., 14:65.

Hamilton, T. H., C.-S. Teng, and A. R. Means. 1968. Early estrogen action: Nuclear synthesis and accumulation of protein correlated with enhancement of two DNA-dependent RNA polymerase activities. Proc. Nat. Acad. Sci. U.S.A., 59:1265.

Hammel, C. L., and S. P. Bessman. 1964. Hemoglobin synthesis in avian erythrocytes. J. Biol. Chem., 239:2228.

Harris, H. 1961. The nucleolus. Nature (London), 190:1077.

——— 1963. Nuclear ribonucleic acid. Progr. Nucl. Acid Res., 2:19.

——— and J. F. Watkins. 1965. Hybrid cells derived from mouse and man: Artificial heterokaryons of mammalian cells from different species. Nature, 205:640.

Hennig, W. 1967. Untersuchungen zur Struktur und Funktion des Lampenbürsten-Y-Chromosoms in der Spermatogenese von *Drosophila*. Chromosoma, 22:294.

Hnilica, L. S. 1967. Proteins of the cell nucleus. Progr. Nucl. Acid Res., 7:25.

Izawa, M., and K. Kawashima. 1968. RNA synthesis in the nucleoli of mouse ascites tumor cells in relation to nucleolar components. Biochim. Biophys. Acta, 155:51.

Jeon, K. W. 1968. Nuclear control of cell movement in amoebae: Nuclear transplantation study. Exp. Cell Res., 50:467.

Kimball, R. F., and D. M. Prescott. 1964. RNA and protein synthesis in amacronucleate *Paramecium aurelia*. J. Cell Biol., 21:496.

Kroeger, H., J. Jacob, and J. R. Sirlin. 1963. The movement of nuclear protein from the cytoplasm to the nucleus of salivary cells. Exp. Cell Res., 31:416.

Kuehl, L. 1967. Evidence for nuclear synthesis of lactic dehydrogenase in rat liver. J. Biol. Chem., 242:2199.

Lawford, G. R., P. Sadowski, and H. Schachter. 1967. Use of a ribonuclease inhibitor from rat liver supernatant fraction in the preparation of polyribosome-like particles from isolated rat liver nuclei. J. Molec. Biol., 23:81.

Leblond, C. P., and M. Amano. 1962. Symposium: Synthetic processes in the cell nucleus. IV. Synthetic activity in the nucleolus as compared to that in the rest of the cell. J. Histochem. Cytochem., 10:162.

Magee, W. E., and M. J. Burrous. 1961. The preparation and properties of intact nuclei from cells grown in tissue culture. Biochim. Biophys. Acta, 49:393.

Maggio, R. 1966. Progress report on the characterization of nucleoli from guinea pig liver. U.S. Nat. Cancer Inst. Monogr., 23:213.

Martin, P. G. 1961. Evidence for the continuity of nucleolar material in mitosis. Nature, (London), 190:1078.

Mazia, D. and D. M. Prescott. 1955. The role of the nucleus in protein synthesis in amoeba. Biochim. Biophys. Acta, 17:23.

McCarty, K. S., J. T. Parsons, W. A. Carter, and J. Laszlo. 1966. Protein-synthetic capacities of liver nuclear subfractions. J. Biol. Chem., 241:5489.

McConkey, E. H. 1968. Personal communication.

McEwen, B. S., V. G. Allfrey, and A. E. Mirsky. 1963. Studies on energy-yielding reactions in thymus nuclei. II. Pathways of aerobic carbohydrate catabolism. J. Biol. Chem., 238:2571.

Mintz, B. 1967. Gene control of mammalian-pigmentary differentiation. I. Clonal origin of melanocytes. Proc. Nat. Acad. Sci. U.S.A., 58:344.

——— and W. W. Baker. 1967. Normal mammalian muscle differentiation and gene control of isocitrate dehydrogenase synthesis. Proc. Nat. Acad. Sci. U.S.A., 58:592.

Monesi, V. 1964. Autoradiographic evidence of a nuclear histone synthesis during mouse spermiogenesis in the absence of detectable quantities of nuclear ribonucleic acid. Exp. Cell Res., 36:683.

Mundell, R. D. 1967. The occurrence of ribosomal proteins in nucleoli of starfish oocytes. Biochem. Biophys. Res. Commun., 28:117.

Nathans, D., G. Notani, J. H. Schwartz, and N. D. Zinder. 1962. Biosynthesis of the coat protein of coliphage f2 by *E. coli* extracts. Proc. Nat. Acad. Sci. U.S.A., 48:1424.

Nemchinskaya, V. L., L. Sh. Ganelina, and A. D. Braun. 1968. Differences in isozyme patterns of nucleic and cytoplasmic lactate dehydrogenase of rat liver cells. Nature, (London), 217:251.

Patel, G., and T. Y. Wang. 1965. Protein synthesis in nuclear residual protein. Biochim. Biophys. Acta, 95:314.

Penman, S. 1966. RNA metabolism in the HeLa cell nucleus. J. Molec. Biol., 17:117.

Prescott, D. M., and M. A. Bender. 1963. Synthesis and behavior of nuclear proteins during the cell life cycle. J. Cell. Physiol., 62 (Suppl.), 1:175.

Reid, B. R., R. H. Stellwagen, and R. D. Cole. 1968. Further studies on the biosynthesis of very lysine-rich histones in isolated nuclei. Biochim. Biophys. Acta, 155:593.

Rendi, R. 1960. *In vitro* incorporation of labeled amino acids into nuclei isolated from rat liver. Exp. Cell Res., 19:489.

Richards, B. M. 1960. Redistribution of nuclear proteins during mitosis. *In* The Cell Nucleus, Mitchell, J. S., ed., London, Butterworth, p. 138.

——— and A. Bajer. 1961. Mitosis in endosperm. Exp. Cell Res., 22:503.

Robbins, E., and T. W. Borun. 1967. The cytoplasmic synthesis of histones in HeLa cells and its temporal relationship to DNA replication. Proc. Nat. Acad. Sci. U.S.A., 57:409.

Rogers, M. E. 1968. Ribonucleoprotein particles in the amphibian oocyte nucleus. J. Cell Biol., 36:421.

Roodyn, D. B. 1959. A survey of metabolic studies on isolated mammalian nuclei. Int. Rev. Cytol., 8:279.

Sadowski, P. D., and J. A. Howden. 1968. Isolation of two distinct classes of polysomes from a nuclear fraction of rat liver. J. Cell Biol., 37:163.

Schultze, B., and W. Maurer. 1967. Nuclear and cytoplasmic protein synthesis in various cell types from rats and mice. *In* The Control of Nuclear Activity, Goldstein, L., ed., Englewood Cliffs, N. J., Prentice-Hall, Inc., p. 319.

Sekeris, C. E., W. Schmid, D. Gallwitz, and I. Lukacs. 1966. Protein synthesis in the cell nucleus. I. Amino acid incorporation into protein by the aggregate enzyme of Weiss. Life Sci., 5:969.

Shepherd, G. R., and B. J. Noland. 1968. The intracellular distribution of basic proteins in the Chinese hamster ovary cell. Exp. Cell Res., 49:238.

Siebert, G., and G. B. Humphrey. 1965. Enzymology of the nucleus. Adv. Enzym., 27:239.

Sims, R. T. 1965. The synthesis and migration of nuclear proteins during mitosis and differentiation of cells in rats. Quart. J. Micr. Sci., 106:229.

Smetana, K., W. J. Steele, and H. Busch. 1963. A nuclear ribonucleoprotein network. Exp. Cell Res., 31:198.

Swift, H. 1963. Cytochemical studies on nuclear fine structure. Exp. Cell Res. (Suppl)., 9:54.

Traub, A., E. Kaufmann, and Y. Ginzburg-Tietz. 1964. Studies on nuclear ribosomes. I. Association of DPN-pyrophosphorylase with nuclear ribosomes in normal and neoplastic tissues. Exp. Cell Res., 34:371.

Uete, T. 1967. Comparison of amino acid incorporation into nuclear protein of rat liver and thymus in cell-free systems. J. Cytochem., 61:251.

Wang, T. Y. 1961. Incorporation of (^{14}C) amino acids into ribonucleoprotein fraction of isolated thymus cell nuclei. Biochim. Biophys. Acta, 49:108.

Weiss, S. B. 1960. Enzymatic incorporation of ribonucleoside triphosphates into the interpolynucleotide linkages of ribonucleic acid. Proc. Nat. Acad. Sci. U.S.A., 46:1020.

Weston, J. C. 1968. Ribosome-like granules within areas of the perinuclear space in cells of 13-14 somite chick embryos. Z. Zellforsch., 87:199.

Winckelmans, D., M. Hill, and M. Errera. 1964. Ribosomes from HeLa cells. Biochim. Biophys. Acta, 80:52.

Zalokar, M. 1960. Sites of ribonucleic acid and protein synthesis in *Drosophila*. Exp. Cell Res., 19:184.

Zetsche, K. 1966. Regulation der UDP-Glucose 4-Epimerase Synthese in kernhaltigen und kernlosen *Acetabularien*. Biochim. Biophys. Acta, 124:332.

Zetterberg, A. 1966a. Synthesis and accumulation of nuclear and cytoplasmic proteins during interphase in mouse fibroblasts *in vitro*. Exp. Cell Res., 42:500.

——— 1966b. Protein migration between cytoplasm and cell nucleus during interphase in mouse fibroblasts *in vitro*. Exp. Cell Res., 43:526.

Anders Zetterberg

Institute for Medical Cell Research and Genetics, Medical Nobel Institute, Karolinska Institutet, Stockholm, Sweden

5

Nuclear and Cytoplasmic Growth During Interphase in Mammalian Cells

INTRODUCTION

Growth in tissues and organs is the result of two physiologically different processes; one involves cell growth, i.e., an increase in the fundamental substances of the cell, the other a division of the enlarged cell into two daughter cells. Cell growth occurs during a relatively long period, the interphase, while the process of cell division, mitosis, is relatively short. Interphase and mitosis form together what is generally termed the cell life cycle.

Although a certain growth of the cell appears to be necessary for cell division, it is well known that the cell growth as such is not directly linked to the control of cell division. This control is considered to be based on a series of metabolic events in interphase linked in an ordered sequence that terminates with signals for mitosis. Interphase metabolic activities can in principle thus be divided into those involved in the direct control of mitosis, such as DNA replication, and into those responsible for the cell growth. However, under normal conditions an approximate doubling of the cell mass occurs during interphase, i.e., a balance between cell growth and division is maintained, indicating also that cell growth as such is somehow related to the events concerned with control of cell division.

During the last fifteen years our knowledge about cell growth and cell division processes has increased considerably and in a number of articles different aspects of the cell life cycle have been discussed (Andersson, 1956a, 1956b; Baserga, 1965; Bassleer, 1968; Burns, 1961; Campbell, 1957; Hughes, 1952; Mazia, 1958;

Mitchison, 1963; Prescott, 1961b, 1964b; Stern, 1956; Swann, 1957, 1958; Zetterberg, 1966d; Zeuthen, 1958). In the present review the discussion has been restricted to certain aspects of nuclear and cytoplasmic growth phenomena during interphase. Special attention has been devoted to the initiation of DNA synthesis, the overall rate control of RNA and protein synthesis, and the interplay between protein accumulation in cell nucleus and cytoplasm.

SYNTHESIS OF DNA

G1, S, and G2 Periods

Quite early in the investigations of interphase it was shown, both with microspectrophotometry (Swift, 1950; Walker and Yates, 1952) and with autoradiography (Howard and Pelc, 1953), that DNA synthesis occurs in interphase and not during prophase and metaphase, as was originally believed. The period of DNA synthesis is generally limited to a part of interphase. Using DNA synthesis as a marker, interphase can be subdivided into the G1, S, and G2 periods (Howard and Pelc, 1953). S designates the period of DNA synthesis, while G1 and G2 denote the intervals of interphase during which no nuclear DNA synthesis takes place; G1 precedes, and G2 follows, the S period. Exceptions to this general pattern of DNA synthesis in interphase have been described, e.g., in bacteria (Schaechter et al., 1959) and in grasshopper neuroblasts (Gaulden, 1956) the S period seems to fill all of interphase.

The duration and the position of S in interphase varies considerably. In vertebrate cells, the S period, which generally lasts for 6 to 8 hours, is usually preceded by a relatively long G1 period and followed by a relatively short G2 period (Bender and Prescott, 1962; Cameron, 1964; Edwards et al., 1960; Harris, 1959; Killander and Zetterberg, 1965a; Mendelsohn et al., 1960; Painter and Drew, 1959; Painter et al., 1960; Richards et al., 1956; Sisken and Kinosita, 1961; Stanners and Till, 1960; J. H. Taylor, 1960b; Walker and Yates, 1952; R. W. Young, 1962). However, in some cells, such as ascites tumor cells (Baserga, 1963) and one line of Chinese hamster cells (Robbins and Scharff, 1967), DNA synthesis starts immediately after telophase. In mouse fibroblasts *in vitro* (L-cells), both the duration and the position of the S period can vary from one culture occasion to another (Killander and Zetterberg, 1965b). Although the length of S is generally about 6 to 8 hours in mammalian cells, considerably longer durations have been reported (Baserga, 1964; Bresciani, 1964; Sherman et al., 1961).

In plant root cells the condition is similar to that of vertebrate cells (Deeley et al., 1957; Howard and Pelc, 1951; Sisken, 1959; Wimber, 1960). A different situation, however, prevails in the generative nucleus of pollen grain (J. H. Taylor, 1958; J. H. Taylor and McMaster, 1954; Woodard, 1958), in the slime mold (Nygaard et al., 1960), and in the micronucleus of *Tetrahymena* (McDonald, 1958; Prescott, 1960b), in which the period of DNA synthesis is restricted to the early part of interphase, i.e., G1 is very short or nonexistent. In contrast macronuclear DNA synthesis in *Paramecium* occurs during the last half of interphase (Kimball and Barka, 1959; Kimball and Perdue, 1962).

Initiation of DNA Synthesis

It has been found that the length of the S and G2 periods is relatively independent of the total interphase time, and large variations in interphase time are mainly ascribable to the G1 period (Baserga, 1965; Defendi and Manson, 1963; Ford and Young, 1963; Pilgrim and Maurer, 1965; Sisken, 1963, 1965; Sisken and Kinosita, 1961; Swann, 1957; Terasima and Tollmach, 1963; Watanabe and Okada, 1967). G1 is also the phase in which cells with very long or indefinite cycle times generally rest, e.g., contact inhibited cells (Nilausen and Green, 1965; Todaro et al., 1965). This means that once a cell has initiated its DNA synthesis, and the cell has made the decision to divide, it takes a relatively constant time to reach mitosis. The initiation of DNA synthesis therefore appears to be a key event in the control of cell division. Experiments with various inhibitors have shown that both RNA synthesis (Caspersson et al., 1963; Fujioka et al., 1963; Lieberman et al., 1963) and protein synthesis (Harris, 1959; Littlefield and Jacobs, 1965; Maaløe and Kjeldgaard, 1966; Mueller et al., 1962; Powell, 1962) are necessary prerequisites for the successful initiation of DNA synthesis. This means that initiation of DNA synthesis is preceded by a de novo synthesis of protein that is involved in the control of DNA synthesis, and consequently is concerned with the progression of cells through the life cycle. Besides the necessary enzymes mediating the biosynthesis of DNA, the de novo protein synthesis perhaps also provides the postulated "initiator protein" (Jacob et al., 1963). Little is known, however, about stimuli that are involved either in the induced synthesis of "DNA enzymes" or "initiator proteins." A factor of importance in this respect could be the intracellular concentration of deoxyribosidic compounds. In anthers of *Lilium longiflorum*, for example, a twenty-fivefold increase in the concentration of deoxyribosides has been found to precede the de novo synthesis of thymidylic kinase in the relatively short interval just prior to DNA synthesis (Foster and Stern, 1959; Hotta and Stern, 1963).

Protein synthesis during G1 may also affect the initiation of DNA synthesis in a way different from the above described induction of specific "DNA enzymes" or "initiator proteins." In L-cells it appears that a certain total amount of protein must be synthesized during the G1 period in order for DNA synthesis to begin. Cells entering the interphase with a relatively small mass synthesize more protein during the G1 period, and also spend a longer time in G1, than cells entering the interphase with a larger mass (Killander and Zetterberg, 1965b). Consistent with these findings are other observations showing that if protein synthesis is inhibited for a certain time during the G1 period, the initiation of DNA synthesis is delayed for an equally long period, irrespective of the stage in G1 during which the protein synthesis was inhibited (Terasima and Yasukawa, 1966). Since the major quantity of the proteins synthesized in the cell during G1 accumulates in the cytoplasm (Zetterberg, 1966a), it may be that DNA synthesis is not initiated until the cell nucleus is surrounded by a certain amount of cytoplasmic protein. Growth of the cytoplasm can thus be considered as a factor advancing the cell towards mitosis, and the integration between growth and division processes might thus, in some cases, be based on the link between cytoplasmic growth and initiation of DNA synthesis.

There are further observations which indicate that the cytoplasm plays an important role in the onset of DNA synthesis. In binucleated cells, in which the two nuclei are well separated from each other by cytoplasm, DNA synthesis starts synchronously in both nuclei (Kimball and Prescott, 1962). The cytoplasmic role in the onset, as well as in the maintenance of DNA synthesis, has been demonstrated in transplantation experiments in *Amoeba proteus* by the fact that DNA synthesis starts in a G2 nucleus introduced into an S cytoplasm and declines in an S nucleus grafted into a G2 cytoplasm (Goldstein and Prescott, 1967).

In addition to cytoplasmic factors, the physico-chemical properties of DNA, or the deoxyribonucleoprotein complex, are also of considerable importance for the initiation of DNA synthesis. For bacteria, it has been proposed that replication of the chromosome is initiated at a specific site termed the "replicator" (Jacob, 1966; Jacob et al., 1963), which appears to be attached to the bacterial cell membrane (Ganesan and Lederberg, 1965; K. G. Lark, 1966; C. Lark and K. G. Lark, 1964; Ryter and Jacob, 1963). In agreement with this idea are data on mammalian cells, based on electron microscopic autoradiography, indicating that at the beginning of the S period DNA synthesis starts in the portion of the chromatin located at the nuclear membrane (Comings and Kakefuda, 1968). Autoradiographic investigations of mammalian chromosomes have revealed that DNA synthesis is initiated asynchronously at the different foci in a nonrandom and definite pattern (German, 1964; Gilbert et al., 1962; Hsu, 1964; Hsu et al., 1964; Lima-de-Faria, 1959; Mueller et al., 1962; Schmid, 1962; J. H. Taylor, 1960b). The asynchrony is particularly evident in the heterochromatic X chromosome which displays a characteristic delay in the onset of DNA replication (German, 1962; Lima-de-Faria, 1959; J. H. Taylor, 1960b). The complete asynchrony in DNA synthesis between macronucleus and micronucleus, demonstrated in *Tetrahymena* and *Euplotes* (Prescott, 1964b), gives further support to the idea that the physical properties of the DNA, or deoxyribonucleoprotein complex, is of importance for the initiation of DNA synthesis.

Protein synthesis is also necessary in maintaining DNA synthesis at a normal rate after its initiation in mammalian cells (Littlefield and Jacobs, 1965; Mueller et al., 1962; Shah, 1963; E. W. Taylor, 1965; Young, 1966). Since inhibition of protein synthesis during the S period does not lead to a reduction in the activity of some of the "DNA enzymes" (Baserga et al., 1965; Powell, 1962; E. W. Taylor, 1965), it has been suggested that the necessary proteins play an important role in converting specific sites of the chromosomes to a "primer" form competent for DNA synthesis (Billen, 1962).

SYNTHESIS AND ACCUMULATION OF RNA

Three functionally different types of RNA have been found to operate in the synthesis of protein: messenger RNA, transfer RNA, and ribosomal RNA (Chantrenne, 1961; Wiseman, 1965). In bacterial cells all three types have been shown to be synthesized with a DNA template (Yankofsky and Spiegelman, 1962a, 1962b), and the cell nucleus is generally believed to be the main site of RNA synthesis in most organisms (Feinendegen and Bond, 1963; Feinendegen et al., 1960; Goldstein and Micou, 1959a, 1959b; Goldstein and Plaut, 1955; Graham and Rake, 1963; Perry et al., 1961a, 1961b; Prescott, 1959, 1960a, 1961a, 1964a; Seed, 1965).

Biochemical studies on various animal cell types have revealed great differences in physicochemical and biochemical properties between nuclear and cytoplasmic RNA. The RNA fractions of the cell nucleus show a great heterogeneity in size, chemical composition, rate of synthesis, and turnover (Hiatt, 1962; Houssais and Attardi, 1966; Perry, 1962; Scherrer et al., 1963). Cytoplasmic RNA, which is mainly ribosomal in type (Georgiev and Mantieva, 1961), has a more homogeneous composition and a considerably longer average half-life than the nuclear RNA (Penman et al., 1963). Pulse labeling with nucleotides during interphase thus represents the synthesis of a number of different RNA fractions, while continuous labeling, on the other hand, primarily reflects synthesis of stable RNA, mainly ribosomal RNA (rRNA).

Nuclear transplantation experiments in *Amoeba* (Goldstein and Plaut, 1955) and time-course studies of incorporation of labeled precursors into RNA have given considerable evidence that RNA is transferred from the cell nucleus to the cytoplasm throughout interphase (Amano and Leblond, 1960; Feinendegen et al., 1960; Goldstein and Micou, 1959a, 1959b; Perry, 1960; Perry et al., 1961a, 1961b; Prescott, 1961a; Woods, 1959; Zalokar, 1960). Cytophotometric measurements on L-cells have added quantitative evidence to these observations by showing that the major portion of the cellular RNA, mainly rRNA, accumulates continuously in the cytoplasm of the growing cell during interphase (Zetterberg, 1966b). Since most, or all, of this RNA has its origin in the cell nucleus, this accumulation in the cytoplasm also reflects a continuous transfer of rRNA from the site of synthesis, in the nucleus, to the cytoplasm over the entire period of interphase. This transfer of rRNA across the nuclear membrane takes place despite a considerably higher rRNA concentration in the cytoplasm (Zetterberg, 1966b), suggesting that the cytoplasmic rRNA does not, to any great extent, migrate back to the nucleus to equilibrate with the rRNA there. More direct evidence of this finding has been reported from experiments with transplanted nuclei in *Amoeba* (Goldstein and Plaut, 1955). The explanation for this may be that the cytoplasmic ribosomes are either attached to structures in the cytoplasm, e.g., the membranes of the endoplasmic reticulum, or are free in the cytoplasm but unable to pass back through the nuclear membrane.

The nuclear RNA content, which constitutes only a small part of the total cellular RNA content, has been shown to remain constant, or nearly so, throughout interphase in mammalian fibroblast *in vitro* (Sundelin, 1968; Zetterberg, 1966b). This lack of RNA accumulation in the cell nucleus during interphase is probably the result of the rapid turnover of nuclear RNA, as well as the fact that the relatively stable ribosomal RNA fraction is transferred from nucleus to cytoplasm in proportion to synthesis.

Most of the RNA is released from the nucleus during prophase, at a time when the nuclear membrane dissolves or becomes fragmented (Jacobson and Webb,

1952; Prescott, 1964a; Prescott and Bender, 1962, 1963; Rabinovitch and Plaut, 1956; Srinivasan et al., 1963). During chromosome condensation in mitosis, all RNA synthesis in the nucleus ceases (Das, 1963; Feinendegen et al., 1960; Prescott and Bender, 1962; J. H. Taylor, 1960a). However, the RNA content characteristic of the interphase nucleus is found in this compartment very early in interphase (Zetterberg, 1966b). To some extent at least, this rapid RNA accumulation probably represents de novo synthesis of RNA in the telophase and early interphase nucleus, and not return of RNA to the reconstituting nucleus from the cytoplasm. This is presumably true for the nucleolar and chromatin-associated RNA, which has been shown to be synthesized rapidly (Pelling, 1959; Perry et al., 1961a; Rho and Bonner, 1961; Sirlin et al., 1961). It is not certain, however, whether *all* RNA to be found in the early interphase nucleus is the result of de novo synthesis. Some portion of the RNA in the "nuclear sap" (Georgiev and Samarina, 1961; Georgiev et al., 1961) may have migrated from the cytoplasm to the cell nucleus during mitosis, before the nuclear membrane was reformed. This idea is, to some extent, supported by the finding that the cytoplasmic RNA content increases more than twofold during interphase, i.e., the preprophase cytoplasm contained *more* RNA than the two posttelophase cytoplasms of the daughter cells taken together (Zetterberg, 1966b).

The kinetics of RNA synthesis during interphase have been studied by different techniques in various types of cells. Cytophotometric analyses of *Paramecium aurelia* (Kimball et al., 1959; Woodard et al., 1961a) and *Vicia faba* (Woodard et al., 1961b), and autoradiographic studies on *Schizosaccharomyces pombe* (Mitchison, 1963) have demonstrated an increase in the rate of RNA synthesis during interphase. Furthermore, nucleotide incorporation data on *Tetrahymena* (Prescott, 1960b), on synchronously multiplying HeLa cells (Terasima and Tolmach, 1963), and on Chinese hamster fibroblast (Stubblefield et al., 1967), as well as cytophotometric data on L-cells (Zetterberg and Killander, 1965a), have demonstrated a temporal relationship in interphase between the duplication of DNA and the increased rate of RNA synthesis. Some evidence for a proportionality between the amount of DNA and the rate of RNA synthesis has also been reported for *Paramecium* (Kimball and Perdue, 1962). Although no definite conclusions can be drawn from this temporal relationship between DNA duplication and increased rate of RNA synthesis, it seems plausible that the duplication of those DNA regions (ribosomal RNA cistrons) which are responsible for ribosomal RNA synthesis (Yankofsky and Spiegelman, 1962a, 1962b) and which may be located on the nucleolus-associated chromatin (Caspersson et al., 1963; McConkey and Hopkins, 1964; Wallace and Birnstiel, 1966; Ritosa and Spiegelman, 1965), may lead to a twofold increase in the rate of synthesis and subsequent transfer of the ribosomal RNA to the cytoplasm.

It has been postulated that DNA engaged in replication is unable to serve as a template for RNA synthesis. This assumption has been supported by incorporation data obtained from autoradiographic examinations of the macronucleus in *Euplotes* (Prescott and Kimball, 1961). While RNA synthesis was found to take place throughout the nucleus, it was undetectable in the reorganization bands, i.e., in the regions of DNA synthesis. Although only a small fraction of the DNA molecules in an individual cell nucleus is undergoing replication at any given moment, a depression in the rate of nucleotide incorporation into RNA has been detected in various cell types by means of autoradiographic techniques (Moses and J. H. Taylor, 1955; Nygaard et al., 1960; Sisken, 1959; J. H. Taylor, 1958; J. H. Taylor and McMaster, 1954). Alternatively, depletion of the intracellular nucleotide pools as a result of DNA synthesis may significantly depress the rate of incorporation of labeled nucleotides into RNA. Contrary to these incorporation data, no decrease in the rate of RNA accumulation was observed in L-cells during the S period (Zetterberg and Killander, 1965a), suggesting that the rate of ribosomal RNA synthesis was not much depressed by the DNA synthesis in the cell nucleus. It thus appears that the DNA regions responsible for ribosomal RNA synthesis are able to serve as templates during the major part of the S period, i.e., the replication of these DNA regions is probably limited to a very short interval of S.

SYNTHESIS AND ACCUMULATION OF PROTEIN

Kinetics of Protein Synthesis

Cellular growth during interphase is quantitatively dominated by synthesis and accumulation of protein. The increase in protein content of the growing cell (net synthesis) represents the sum of de novo synthesis and breakdown. In exponentially multiplying mammalian cell populations, about 25 percent of the cellular protein content is broken down in one generation (Eagle et al., 1959; Sköld and Zetterberg, 1969). The net synthesis of protein during interphase has been studied by the use of cytophotometric methods for dry mass and protein determinations of individual cells at different stages of interphase. The cellular content of protein increases throughout the whole of interphase in contrast to the behavior of DNA. Quantitative differences in the time course of net protein synthesis during interphase have been demonstrated in various types of cells. In *Amoeba,* the rate of net protein synthesis is highest at the beginning of interphase and levels off towards the end of interphase (Prescott, 1955). Dry mass measurements of HeLa cells in tissue culture suggest that protein accumulates in these cells at a roughly constant rate in interphase (Sandritter et al., 1960). In *Paramecium aurelia,* mass measurements by interference microscopy and x-ray absorption methods (Kimball et al., 1959) and protein measurements by microspectrophotometric methods (Woodard et al., 1961a) demonstrate an accelerating rate of net protein synthesis over the interphase period. The situation is similar in mouse fibroblasts in tissue culture, as demonstrated by dry mass measurements (Killander and Zetterberg, 1965a; Zetterberg and Killander, 1965a).

Under certain conditions, relatively detailed information about the kinetics of net protein synthesis during interphase can be obtained from the frequency distribution of cellular mass in the population (Zetterberg and Killander, 1965a). The mathematical and statistical relationship between frequency distribution and time course of synthesis during interphase has been discussed by several authors (Richards et al., 1956; Scherbaum and Rasch, 1957; Walker, 1954; Walker and Mitchison, 1957; Zetterberg and Killander, 1965a). During the last three years a large number of L-cells have been analyzed in the microinterferometer in our laboratory. By pooling 45 randomly analyzed L-cell populations, grown on different occasions, a frequency distribution of 3,781 cells was constructed. In Figure 1 is illustrated the "rate curve" derived from this large sample according to our previously employed method (Zetterberg and Killander, 1965a). The curve clearly demonstrates a *continuously* increasing rate of net protein synthesis during interphase in mouse fibroblasts *in vitro.*

Autoradiographic techniques for the quantitation of radioactive amino acid incorporation into the growing cell after a pulse incubation, have been used for kinetic analyses of de novo protein synthesis during interphase in various types of cells. In *Tetrahymena,* the rate of protein synthesis, measured by incorporation of ^{14}C-methionine, was found to be constant during interphase (Prescott, 1960b). In *Schizosaccharomyces pombe,* on the other hand, pulse labeling experiments with ^{14}C-leucine, ^{14}C-glycine, and ^{35}S-methionine showed that protein was synthesized at increasing rate throughout the interphase period (Mitchison and Wilbur, 1962). In mouse fibroblast *in vitro,* incorporation of ^{14}C-leucine and ^{35}S-methionine revealed

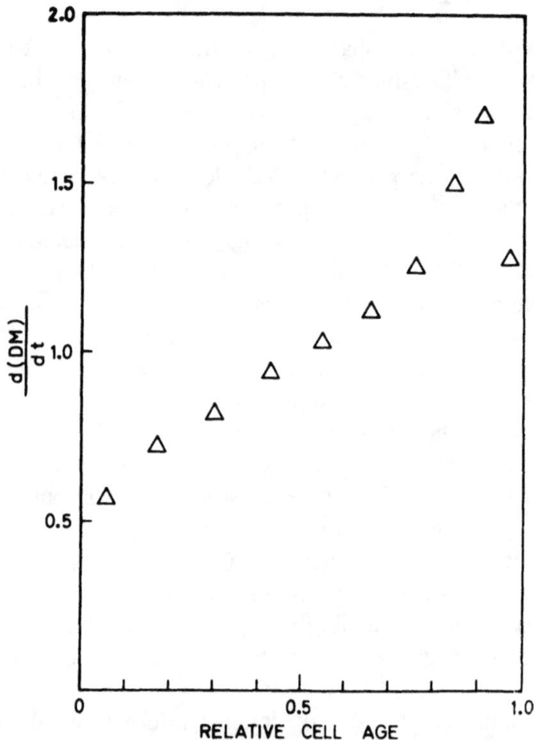

FIG. 1. Rate of mass accumulation (d(DM)/dt) plotted against interphase time (relative cell age). The plots are derived from a histogram based on 3,781 mass measured cells (cf. text).

that the rate of protein synthesis increased throughout the entire period of interphase by a total factor of approximately two (Zetterberg and Killander, 1965b). Similar results were obtained when ^{3}H-leucine incorporation into the mouse fibroblast cytoplasm was investigated (Zetterberg, 1966a, 1966b).

The factors to be considered in understanding the cellular control of the overall rate of protein synthesis are (1) the rate of peptide polymerisation at each synthesizing unit, i.e., the ribosome primed with the genetic message, and (2) the total number of such synthesizing units in the cell. In bacterial cell populations growing under steady-state conditions, but with different growth rates, a close correlation has been demonstrated between the ribosomal RNA content and the rate of protein synthesis, indicating that the average rate of synthesis per ribosome is constant and thus independent of growth rate (Maaløe and Kjeldgaard, 1966; Neidhardt and Magasanik, 1960; Schaechter et al., 1958). In "shift up" experiments, in which a bacterial cell population is transferred from a poor to a rich growth medium, it has been demonstrated that the rate of protein synthesis increases directly with the number of ribosomes. This suggests that the constant average rate of protein synthesis per ribosome is close to the maximum rate of synthesis, since flooding the cells with amino acids did not increase the rate of synthesis per ribosome (Maaløe and Kjeldgaard, 1966). If this is the case in each cell in a population irrespective of its interphase stage, one would expect to find that, during interphase, the overall rate of protein synthesis in the cell would increase directly with the cellular content of ribosomal RNA. By using a combination of quantitative cytophotometric and autoradiographic methods, it was possible to demonstrate that this was in fact the case in log phase L-cell populations; at all stages of the interphase period a strong correlation existed between the overall rate of protein synthesis per cell and its total amount of ribosomal RNA (Zetterberg and Killander, 1965b). Since the major portion of the cellular ribosomal RNA is located in the cytoplasm, this result primarily reflects the situation in the cytoplasm. Experimental evidence in favor of this idea was also furnished by separate analyses on the cytoplasmic compartment of the L-cell (Zetterberg, 1966b).

It thus appears that the average rate of protein synthesis per ribosome remains the same in L-cells at different stages of interphase. This suggests, in turn, that the microenvironment of the ribosome in L-cells in log phase is saturated with respect to messenger RNA, transfer RNA, various enzymes, and amino acids; and that the overall capacity for protein synthesis in this case is limited by the number of ribosomes in the cell.

Protein Accumulation in Cell Nucleus and Cytoplasm

Microinterferometric analyses on the cell nucleus and cytoplasm have yielded information about the growth, in terms of protein accumulation, of these two major cellular compartments during the cell life cycle. In various types of mammalian cell lines *in vitro,* the protein content of the cell nucleus remains constant, or nearly so, during the G1 period, and the predominant increase in nuclear protein content occurs during the period of DNA synthesis (Seed, 1962; Zetterberg, 1966a). In the cytoplasm the situation is the reverse. The major increase in the amount of cytoplasmic protein occurs during the G1 period. During the S period, on the other hand, the protein content of the cytoplasm shows only a very small increase (Zetterberg, 1966a). This reduced cytoplasmic growth is not due to a reduced rate of protein synthesis in the cytoplasm (see p. 221). For L-cells, G1 can thus be characterized as the period of cytoplasmic growth, and S as the period of nuclear growth. In the reorganization band of *Euplotes,* a doubling of the mass content was shown to be associated with the synthesis of DNA (Ringertz and Hoskins, 1965). Measurements performed on isolated nuclei of *Paramecium aurelia* have also shown that the content of mass and DNA remain constant during the first part of interphase, whereafter both parameters increase simultaneously (Kimball et al., 1960). In contrast to this result for the isolated nucleus, the mass of the whole cell, predominantly cytoplasm, increased continuously throughout the entire interphase period (Kimball et al., 1959) indicating differences in the nuclear and cytoplasmic growth similar to the findings on L-cells.

An increased protein content in cell nucleus or cytoplasm need not necessarily be the result of protein synthesis within that compartment. The relationship between synthesis and accumulation of protein in the cell nucleus and in the cytoplasm has been investigated in L-cells by means of a combination of microinterferometric and autoradiographic methods (Zetterberg, 1966a, 1966d). In Figure 2 some of

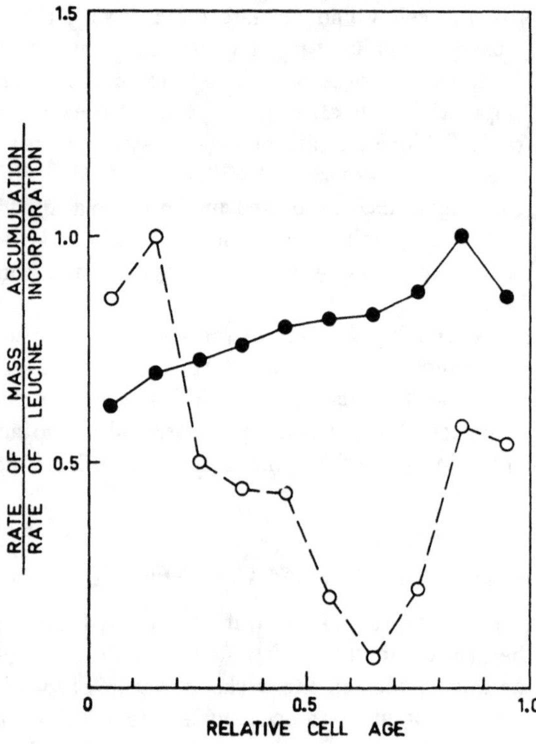

FIG. 2. Rate of mass accumulation (d(DM)/dt) divided by rate of leucine incorporation, plotted against interphase time (relative cell age). Whole cell ●; cytoplasm ○.

these data are collected. The ratio between rate of mass accumulation and rate of protein synthesis (^{14}C-leucine or ^{3}H-leucine incorporation) at different stages of interphase was calculated for the whole cell and compared with the corresponding ratios for the cytoplasm. It is clear from Figure 2 that there are striking differences in the relation between synthesis and accumulation of protein when comparing the cytoplasmic compartment with the whole cell. As illustrated by the solid line in Figure 2, the amount of protein accumulating in the growing L-cell per unit time is, at all stages of interphase, related to the rate of protein synthesis in the cell. This is not, however, the case in the cytoplasmic compartment, as shown by the broken line in Figure 2. Very little protein accumulates in the cytoplasm in relation to synthesis in the beginning of the second half of interphase, i.e., during the S period. This cannot be ascribed to an extensive breakdown of protein in the cytoplasm or to protein losses from the cell, e.g., secretion, since the rate of protein accumulation in relation to synthesis is not reduced when considering the whole cell during this period of interphase. The result therefore appears to demonstrate a redistribution of protein *within* the cell, i.e., protein synthesized in the cytoplasm is transferred to the nucleus. This interpretation is further supported by the finding that large quantities of protein are accumulated by the cell nucleus during the S period. It has been estimated that 70 percent, or more, of the nuclear proteins in L-cells is synthesized in the cytoplasm and transported into the nucleus during the second half of interphase. (Zetterberg, 1966a, 1966d). The above described correlation between rate of protein synthesis and amount of ribosomal RNA in the cell, together with the fact that the major portion of the ribosomal RNA is located in the cytoplasmic compartment, gives further support to the idea that the cytoplasm is the major site of protein formation in L-cells.

Autoradiographic studies on the kinetics of protein migration from cytoplasm to cell nucleus at different stages of interphase have revealed that a substantial amount of protein enters the cell nucleus not only during the period of nuclear growth, the S period, but also during G1 and G2 (Zetterberg, 1966c). Since little or no protein accumulates in the L-cell nucleus during the G1 period, as pointed out above, the migration of protein from the cytoplasm to the cell nucleus must be balanced by an equally pronounced migration of protein in the opposite direction during this part of interphase. This idea is, in addition, supported by observation on *Euplotes,* showing that labeled amino acids incorporated in protein accumulate in the macronucleus during G1 (Prescott, 1966) and thus in the absence of macronuclear growth (Ringertz and Hoskins, 1965). The relation between nuclear growth and protein migration across the nuclear membrane is illustrated in Figure 3. During G1 this migration is a reversible process; equal amounts of protein enter and leave the nucleus simultaneously. During the S period this balance is upset; more protein is now transferred to the nucleus than from it, which results in nuclear growth.

The finding of a reversible protein migration between cytoplasm and cell nucleus in L-cells is consistent with similar results from *Amoeba proteus* (Byers et al., 1963a, 1963b; Goldstein, 1963; Prescott, 1963; Prescott and Bender, 1963), in which a protein fraction termed "rapidly migrating proteins," was found to shuttle back and forth between nucleus and cytoplasm. The functional significance of a reversible protein migration between nucleus and cytoplasm is as yet obscure. One suggestion is that it is part of a signal system in the cell (Goldstein, 1963). Many of the

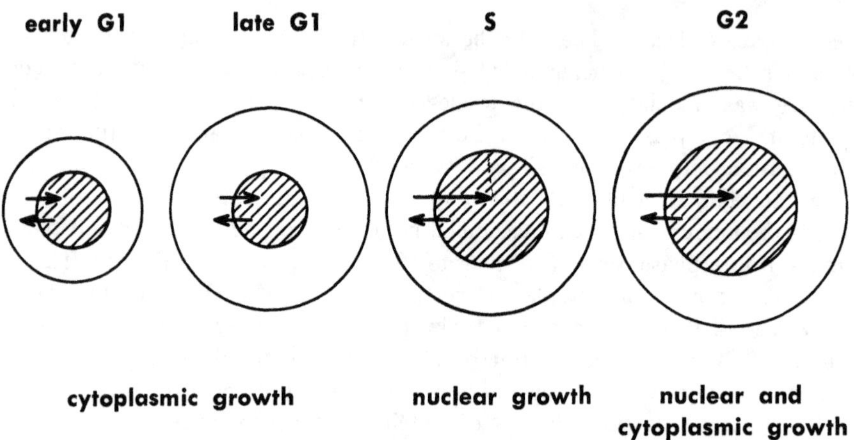

FIG. 3. Direction and quantity of protein flow across the nuclear membrane in relation to nuclear and cytoplasmic growth during interphase.

cellular functions are carried out in the cytoplasm under the direct or indirect control of the nucleus. Accordingly, one might suppose that a proper control of the cytoplasmic function requires feedback of signals from cytoplasm to nucleus. These signals could very well be mediated by the protein molecules involved in the reversible migration between nucleus and cytoplasm. It may also be that some of these proteins are involved in the transfer of RNA from nucleus to cytoplasm, i.e., after synthesis in the cytoplasm these protein molecules enter the nucleus, combine with RNA molecules, and leave the nucleus together with the RNA.

As described above, the balance between the amount of protein entering and leaving the nucleus was found to be changed during the period of DNA synthesis, which resulted in nuclear growth. The role of DNA synthesis for nuclear growth has been studied in L-cells exposed to 5-fluorodeoxyuridine (FUDR), i.e., in a situation where DNA synthesis was inhibited but protein accumulation in the cell was allowed to continue (Auer et al., 1969; Auer and Zetterberg, 1969). When the DNA synthesis was inhibited, the protein accumulation in the cell nucleus was inhibited also. In the cytoplasm, however, the protein content continued to increase relatively independently of the inhibited DNA synthesis, resulting in cells having cytoplasmic protein quantities far exceeding those characteristic of the mitotic cells. When DNA synthesis was again restimulated, the nuclear content of DNA and protein increased in close synchrony (Auer and Zetterberg, 1969). The protein accumulation in the L-cell nucleus thus appears to be strongly dependent on DNA synthesis or increase in DNA quantity, contrary to the protein accumulation in the cytoplasm. There are at least two principally different ways in which the nuclear DNA synthesis could affect the protein accumulation in the cell nucleus. One is that the increasing nuclear DNA content leads to an increasing protein binding capacity of the cell nucleus. This affects the balance between protein quantities entering and leaving the nucleus, characteristic of the G1 cell, by reducing the protein migration from the nucleus and thereby inducing nuclear growth. This need not necessarily preclude all protein migration from the cell nucleus. The other possible explanation is that the de novo synthesis of protein with nuclear affinity is dependent on continued DNA synthesis. Such proteins with nuclear affinity, described as nucleus-specific, have been demonstrated in *Amoeba proteus* (Goldstein, 1958; Prescott and Bender, 1963) and in salivary cells (Kroeger et al., 1963). The reversible migration of protein between cell nucleus and cytoplasm, characteristic of the G1 cell, may very well continue during the S period, parallel to the transfer of nucleus-specific proteins to the nucleus.

At present, no distinction can be made between these two possible explanations. Moreover, the situation may prove to be much more complex than outlined above, and the protein accumulation in the cell nucleus during S may very well also be a combined effect of de novo synthesis of nucleus-specific proteins and an increased binding capacity of the cell nucleus for these, and other proteins.

Although the major growth of the cell nucleus occurs in connection with DNA synthesis, nuclear growth in the absence of DNA synthesis has been reported. In Ehrlich ascites tumor cells, accumulation of protein takes place in nuclei blocked in G2 by a dose of x-radiation, which prevents mitosis (Killander et al., 1962). In HeLa cells, protein accumulation has been observed in the cell nucleus during G1 (Seed, 1962). Whether this accumulation includes nucleus-specific proteins, or simply reflects a net increase of the protein fraction involved in reversible migration between nucleus and cytoplasm during G1 remains to be established.

Synthesis of Histones

In multiplying mammalian cells, more than 50 percent of the nuclear dry mass consists of nonchromosomal proteins, defined as proteins which are not components of the metaphase chromosomes (Richards, 1960). The results discussed above therefore reflect the behavior of at least the major portion of these nonchromosomal proteins. The chromosomal proteins, and in particular the nuclear histones, which can be assumed to make up less than 10 percent of the nuclear mass, have been studied by cytochemical staining techniques, autoradiographic techniques, and biochemical techniques. Early cytochemical data have shown a correlation between the amount of nuclear histones, as demonstrated by the alkaline fast green reaction, and the amount of DNA, as demonstrated by the Feulgen reaction (Bloch and Godman, 1955). The use of these staining reactions on the macronucleus of *Euplotes,* have shown that the histone content doubles together with DNA in the reorganization band (Gall, 1959). In cells arrested in G2 for some time by x-radiation, the amount of cytochemically determined histone did not increase beyond the normal G2 level, even though other proteins continued to accumulate in the nucleus (Ringertz, 1963). These results suggest that histone accumulates in the nucleus in close synchrony with DNA replication during the cell cycle.

Autoradiographic studies on the incorporation of labeled amino acids into the macronuclear histone of *Euplotes* suggest that the synthesis of histone is also well synchronized with DNA replication (Gall, 1959; Prescott, 1966). Further support for this view is furnished by Robbins and Borun (1967) who have studied the incorporation of labeled amino acids into the HCl-soluble nuclear fraction in synchronized HeLa cells. The synthesis of histones, which was found to take place in the cytoplasm, was initiated simultaneously with DNA synthesis (Robbins and Borun, 1967). Other biochemical experiments have also demonstrated that FUDR blocks the increase in nuclear histone accumulation at the same time that DNA synthesis is blocked. Some incorporation of labeled amino acids into the histone fraction can, however, also be observed in the absence of any increase in histone amount (Littlefield and Jacobs, 1965). Although both the synthesis and the subsequent accumulation of histone in the cell nucleus are well synchronized with DNA replication, autoradiographic results on labeled chromosomes in human leucocytes indicate that the DNA replication is not intimately associated with the histone accumulation at the chromosomal level (Cave, 1967).

It has not yet been clearly established whether histones are synthesized in the cytoplasm or in the nucleus. Autoradiographic studies on spermiogenesis suggest that some basic proteins ("protamines") are synthesized in the cytoplasm and migrate into the nucleus (Bloch and Brack, 1964). Also, in HeLa cells, evidence has been presented in favor of the idea that histones are synthesized in the cytoplasm (Robbins and Borun, 1967). However, both the cell nucleus (Reid and Cole, 1964) and the nucleolus (Birnstiel and Flamm, 1964) have been described as important loci of nuclear basic protein synthesis. (See Chapter 4 for further discussion of nuclear protein synthesis.)

SUMMARY

Synthesis and accumulation of DNA, RNA, and protein in cell nucleus and cytoplasm during interphase of growing mammalian cells show the following general characteristics:

1. Nuclear DNA synthesis usually occurs in a limited interval of the interphase, the S period. This is in most cases preceded by a relatively long interval, the G1 period, and followed by a relatively short interval, the G2 period. Large variations in interphase time are mainly ascribable to variations in the length of the G1 period.
2. The cytoplasmic mass quantity of the cell appears to be of importance for the initiation of DNA synthesis; a cell starting in the initial stage of interphase with a small cytoplasmic mass has to make more cytoplasm during the G1 period than a cell starting in interphase with a larger cytoplasmic mass. It may perhaps be that a minimal "critical" amount of cytoplasm must surround the cell nucleus before DNA synthesis can start.
3. RNA and protein are synthesized and accumulated in the cell throughout all of interphase.
4. The overall rate of ribosomal RNA synthesis appears to double in close synchrony with nuclear DNA content during interphase.
5. Ribosomal RNA is continuously transferred to the cytoplasm, where it accumulates throughout the entire interphase.
6. The overall rate of protein synthesis in the growing cell seems to be proportional to (and perhaps limited by) the amount of ribosomal RNA in the cell at all stages of interphase.

7. The major amount of the cellular proteins—also the major portion of the nuclear proteins—is synthesized in the cytoplasm, which contains most of the ribosomal RNA of the cell.
8. The time course of cytoplasmic growth during interphase differs from that of the cell nucleus. During the G1 period, protein is almost exclusively accumulated by the cytoplasm. During the S period, however, a large portion of the proteins synthesized in the cytoplasm accumulates in the cell nucleus, which leads to nuclear growth. This protein accumulation in the cell nucleus seems to be controlled by nuclear DNA synthesis and/or increase in nuclear DNA quantity.
9. Protein is continuously exchanged between cytoplasm and cell nucleus during interphase. In G1 equal amounts of protein enter and leave the cell nucleus. In S more protein migrates into the cell nucleus than from it, leading to protein accumulation in the nucleus.

REFERENCES

Amano, M., and C. P. Leblond. 1960. Comparison of the specific activity time curves of ribonucleic acid in chromatin, nucleolus and cytoplasm. Exp. Cell Res., 20:250.

Anderson, N. G. 1956a. Cell division. I. A theoretical approach to the primeval mechanism, the initiation of cell division, and chromosomal condensation. Quart. Rev. Biol., 31:169.

——— 1956b. Cell division. II. A theoretical approach to chromosomal movements and the division of the cell. Quart. Rev. Biol., 31:243.

Auer, G., G. E. Folcy, and A. Zetterberg. 1969. The effect of FUDR on protein accumulation in cell nucleus and cytoplasm. (To be published.)

——— and A. Zetterberg. 1969. The relationship between nuclear DNA content and protein accumulation in the cell nucleus during interphase. (To be published.)

Baserga, R. 1963. Mitotic cycle of ascites tumor cells. Arch. Path., 75:156.

——— 1964. Uptake of radioactive thymidine and cytidine by Ehrlich ascites tumor cells in different stages of growth. Z. Zellforsch., 64:1.

——— 1965. The relationship of the cell cycle to tumor growth and control of cell division: A review. Cancer Res., 25:581.

——— R. O. Estensen, R. O. Petersen, and J. P. Layde. 1965. Inhibition of DNA synthesis in Ehrlich ascites cells by actinomycin D. II. Delayed inhibition by low doses. Proc. Nat. Acad. Sci., 54:745.

Bassleer, R. 1968. Recherches sur les protéin nucléaires totales et les acides déoxyribonucléiques dans des fibroblastes cultivés *in vitro* et dans des cellules tumorales d'Ehrlich. Arch. Biol. (Liège). 79:181.

Bender, M. A., and D. M. Prescott. 1962. DNA synthesis and mitosis in cultures of human peripheral leukocytes. Exp. Cell Res., 27:221.

Billen, D. 1962. Alteration in deoxyribonucleic acid synthesizing capacity in bacteria: An *in vivo—in vitro* study. Biochim. Biophys. Acta 55:960.

Birnstiel, M. L., and W. G. Flamm. 1964. Intranuclear site of histone synthesis. Science, 145:1435.

Bloch, D. P., and S. Brack. 1964. Evidence for the cytoplasmic synthesis of nuclear histone during spermio-genesis in grasshopper *Chortophaga viridifasciata* (De Geer). J. Cell Biol., 22:327.

——— and G. C. Godman. 1955. A microphotometric study of the syntheses of desoxyribonucleic acid and nuclear histone. J. Biophys. Biochem. Cytol. 1:17.

Bresciani, F. 1964. DNA synthesis in alveolar cells of the mammary gland: Acceleration by ovarian hormones. Science, 146:653.

Burns, V. W. 1961. Relations among DNA and RNA synthesis and synchronized cell division in *L. acidophilus*. Exp. Cell Res., 23:582.

Byers, T. J., D. B. Platt, and L. Goldstein. 1963a. The cytonucleoproteins of *Amebae*. I. Some chemical properties and intracellular distribution. J. Cell Biol., 19:453.

——— D. B. Platt, and L. Goldstein. 1963b. The cytonucleoproteins of *Amebae*. II. Some aspects of cytonucleoprotein behavior and synthesis. J. Cell Biol., 19:467.

Cameron, I. L. 1964. Is the duration of DNA synthesis in somatic cells of mammals and birds a constant? J. Cell Biol., 20:185.

Campbell, A. 1957. Synchronization of cell division. Bact. Rev., 21:263.

Caspersson, T., S. Farber, G. E. Foley, and D. Killander. 1963. Cytochemical observations on the nucleolus-ribosome system. Exp. Cell Res., 32:529.

Cave, M. D. 1967. Chromosomal ^{3}H-lysine incorporation and patterns of deoxyribonucleic acid synthesis in human cells. Exp. Cell Res., 45:631.

Chantrenne, H. 1961. The Biosynthesis of Proteins. Oxford, Pergamon Press, Inc.

Comings, D. E., and T. Kakefuda. 1968. Initiation of deoxyribonucleic acid replication at the nuclear membrane in human cells. J. Molec. Biol., 33:225.

Das, N. K. 1963. Chromosomal and nucleolar RNA synthesis in root tips during mitosis. Science, 140:1231.

Deeley, E. M., H. G. Davies, and J. Chayen. 1957. The DNA content of cells in the root of *Vicia faba*. Exp. Cell Res., 12:582.

Defendi, V., and L. A. Manson. 1963. Analysis of the life-cycle in mammalian cells. Nature (London), 198:359.

Eagle, H., K. A. Piez, R. Fleischman, and V. I. Oyama. 1959. Protein turnover in mammalian cell cultures. J. Biol. Chem., 234:592.

Edwards, J. L., A. L. Koch, P. Youcis, H. L. Freese, M. B. Laite, and J. T. Donalson. 1960. Some characteristics of DNA synthesis and the mitotic cycle in Ehrlich ascites tumor cells. J. Biophys. Biochem. Cytol., 7:273.

Feinendegen, L. E., and V. P. Bond. 1963. Observations on nuclear RNA during mitosis in human cancer cells in culture (HeLa-S_3) studied with tritiated cytidine. Exp. Cell. Res., 30:393.

——— V. P. Bond, W. W. Shreeve, and R. B. Painter. 1960. RNA and DNA metabolism in human tissue culture cells studied with tritiated cytidine. Exp. Cell Res., 19:443.

Ford, J. K., and R. W. Young. 1963. Cell proliferation and displacement in the adrenal cortex of young rats injected with tritiated thymidine. Anat. Rec., 146:125.

Foster, T. S., and H. Stern. 1959. The accumulation of soluble deoxyribosidic compounds in relation to nuclear division in anthers of *Lilium longiflorum*. J. Biophys. Biochem. Cytol., 5:187.

Fujioka, M., M. Koga, and I. Lieberman. 1963. Metabolism of ribonucleic acid after partial hepatectomy. J. Biol. Chem., 238:3401.

Gall, J. G. 1959. Macronuclear duplication in the ciliated protozoan *Euplotes*. J. Biophys. Biochem. Cytol., 5:295.

Ganesan, A. T., and J. Lederberg. 1965. A cell membrane bound fraction of bacterial DNA. Biochem. Biophys. Res. Comm., 18:824.

Gaulden, M. E. 1956. DNA synthesis and X-ray effects at different mitotic stages in grasshopper neuroblasts. Genetics, 41:645.

Georgiev, G. P., and V. L. Mantieva. 1961. On the cytochemistry of ribonucleic acid synthesis in Ehrlich ascites carcinoma. Biokhimia, 26:165.

——— and O. P. Samarina. 1961. Metabolic activity of nuclear juice activity. Biokhimia, 26:454.

——— O. P. Samarina, V. L. Mantieva, and I. B. Zbarskii. 1961. Nuclear ribonucleoproteins. Biochim. Biophys. Acta 46:339.

German, J. L. 1962. DNA synthesis in human chromosomes. Trans N. Y. Acad. Sci., 24:395.

——— 1964. The pattern of DNA synthesis in the chromosomes of human blood cells. J. Cell Biol., 20:37.

Gilbert, C. W., S. Mulldal, L. G. Lajtha, and J. Rowley. 1962. Time-sequence of human chromosome duplication. Nature (London), 195:869.

Goldstein, L. 1958. Localization of nucleus-specific protein as shown by transplantation experiments in *Amoeba proteus*. Exp. Cell Res., 15:635.

——— 1963. RNA and protein in nucleocytoplasmic interactions *In* Cell Growth and Cell Division, Harris, R. J. C., ed., New York, Academic Press, Inc.

——— and J. Micou. 1959a. Nuclear-cytoplasmic relationships in human cells in tissue culture. III. Autoradiographic study of interrelation of nuclear and cytoplasmic ribonucleic acid. J. Biophys. Biochem. Cytol., 6:1.

——— and J. Micou. 1959b. On the primary site of nuclear RNA synthesis. J. Biophys. Biochem. Cytol., 6:301.

——— and W. Plaut. 1955. Direct evidence for nuclear synthesis of cytoplasmic ribose nucleic acid. Proc. Nat. Acad. Sci. U.S.A., 41:874.

——— and D. M. Prescott. 1967. Nucleocytoplasmic interactions in the control of nuclear reproduction and other cell cycle stages. *In* The Control of Nuclear Activity, Goldstein, L., ed., Englewood Cliffs, Prentice-Hall, Inc.

Graham, A. F., and A. V. Rake. 1963. RNA synthesis and turn-over in mammalian cells propagated *in vitro*. Ann. Rev. Microbiol., 17:139.

Harris, H. 1959. The initiation of deoxyribonucleic acid synthesis in the connective-tissue cell, with some observations on the function of the nucleolus. Biochem. J., 72:54.

Hiatt, H. H. 1962. A rapidly labeled RNA in rat liver nuclei. J. Molec. Biol., 5:217.

Hotta, Y., and H. Stern. 1963. Molecular facets of mitotic regulation. I. Synthesis of thymidine kinase. Proc. Nat. Acad. Sci. U.S.A., 49:648.

Houssais, J. F., and G. Attardi. 1966. High molecular weight nonribosomaltype nuclear RNA and cytoplasmic messenger RNA in HeLa cells. Proc. Nat. Acad. Sci. U.S.A., 56:616.

Howard, A., and S. R. Pelc. 1951. Nuclear incorporation of P^{32} as demonstrated by autoradiographs. Exp. Cell Res., 2:178.

——— and S. R. Pelc. 1953. Synthesis of deoxyribonucleic acid in normal and irradiated cells and its relation to chromosome breakage. *In* Symposium on Chromosome Breakage, Springfield, Charles C Thomas, Suppl. to Heredity, Vol. VI, p. 261.

Hsu, T. C. 1964. Mammalian chromosomes *in vitro*. XVIII. DNA replication sequence in the Chinese hamster. J. Cell Biol., 23:53.

——— W. Schmid, and E. Stubblefield. 1964. DNA replication sequences in higher animals. *In* The Role of Chromosomes in Development, Locke, H., ed., New York, Academic Press, Inc., p. 91.

Hughes, A. F. W. 1952. The Mitotic Cycle. London, Butterworths.

Jacob, F. 1966. Genetics of the bacterial cell. Science, 152:1470.

——— S. Brenner, and F. Cuzin. 1963. On the regulation of DNA replication in bacteria. Cold Spring Harbor Symp. Quant. Biol., 28:329.

Jacobson, W., and M. Webb. 1952. The two types of nucleoproteins during mitosis. Exp. Cell Res., 3:163.

Kimball, R. F., and T. Barka. 1959. Quantitative cytochemical studies on *Paramecium aurelia*. II. Feulgen microspectrophotometry of the macronucleus during exponential growth. Exp. Cell Res., 17:173.

——— T. O. Caspersson, G. Svensson, and L. Carlson. 1959. Quantitative cytochemical studies on *Paramecium aurelia*. I. Growth in total dry weight measured by the scanning interference microscope and X-ray absorption methods. Exp. Cell Res., 17:160.

——— and S. W. Perdue. 1962. Quantitative cytochemical studies on *Paramecium aurelia*. V. Autoradiographic studies on nucleic acid syntheses. Exp. Cell Res. 27:405.

——— and D. M. Prescott. 1962. DNA synthesis and distribution during growth and amitosis of the macronucleus of *Euplotes*. J. Protozool., 9:88.

——— L. Vogt-Köhne, and T. O. Caspersson. 1960. Quantitative cytochemical studies on *Paramecium aurelia.* III. Dry weight and ultraviolet absorption of isolated macronuclei during various stages of the interdivision interval. Exp. Cell Res., 20:368.

Killander, D., K. Ribbing, N. R. Ringertz, and B. M. Richards. 1962. The effect of X-radiation on nuclear synthesis of protein and DNA. Exp. Cell Res., 27:63.

——— and A. Zetterberg. 1965a. Quantitative cytochemical studies on interphase growth. I. Determination of DNA, RNA and mass content of age determined mouse fibroblasts *in vitro* and of intercellular variation in generation time. Exp. Cell Res., 38:272.

——— and A. Zetterberg. 1965b. A quantitative cytochemical investigation of the relationship between cell mass and initiation of DNA synthesis in mouse fibroblasts *in vitro.* Exp. Cell Res., 40:12.

Kroeger, H., J. Jacob, and J. L. Sirlin. 1963. The movement of nuclear protein from the cytoplasm to the nucleus of salivary cells. Exp. Cell Res., 31:416.

Lark, C., and K. G. Lark. 1964. Evidence for two distinct aspects of the mechanism regulating chromosome replication in *Escherichia coli.* J. Molec. Biol., 10:120.

Lark, K. G. 1966. Regulation of chromosome replication and segregation in bacteria. Bact. Rev., 30:3.

Lieberman, I., R. Abrams, N. Hunt, and P. Ove. 1963. Levels of enzyme activity and deoxyribonucleic acid synthesis in mammalian cells cultured from animal. J. Biol. Chem., 238:3955.

Lima-de-Faria, A. 1959. Differential uptake of tritiated thymidine in hetero and euchromatin in *Melanoplus* and *Secale.* J. Biophys. Biochem. Cytol., 6:457.

Littlefield, J. W., and P. S. Jacobs. 1965. The relation between DNA and protein synthesis in mouse fibroblasts. Biochim. Biophys. Acta, 108:652.

Maaløe, O., and N. O. Kjeldgaard. 1966. *In* Control of Macromolecular Synthesis. New York and Amsterdam, W. A. Benjamin, Inc.

Mazia, D. 1958. Cell division. *In* The Harvey Lectures. New York, Academic Press, Inc., p. 130.

McConkey, E. H., and J. W. Hopkins. 1964. The relationship of the nucleolus to the synthesis of ribosomal RNA in HeLa cells. Proc. Nat. Acad. Sci. U.S.A., 51:1197.

McDonald, B. B. 1958. Quantitative aspects of deoxyribose nucleic acid (DNA) metabolism in an amicronucleate strain of *tetrahymena.* Biol. Bull., 114:71.

Mendelsohn, M. L., F. C. Dohan, Jr., and H. A. Moore, Jr. 1960. Autoradiographic analysis of cell proliferation in spontaneous breast cancer of C3H mouse. I. Typical cell cycle and timing of DNA synthesis. J. Nat. Cancer Inst., 25:477.

Mitchison, J. M. 1963. Patterns of synthesis of RNA and other cell components during the cell cycle of *Schizosaccharomyces pombe.* J. Cell. Physiol. (Suppl.), 62:1.1.

——— and K. M. Wilbur. 1962. The incorporation of protein and carbohydrate precursors during the cell cycle of a fission yeast. Exp. Cell Res., 26:144.

Moses, M. J., and J. H. Taylor. 1955. Desoxypentose nucleic acid synthesis during microsporgenesis in *Tradescantia.* Exp. Cell Res., 9:474.

Mueller, G. C., K. Kajiwara, E. Stubblefield, and R. R. Rueckert. 1962. Molecular events in the reproduction of animal cells. I. Cancer Res., 22:1084.

Neidhardt, F. C., and B. Magasanik. 1960. Studies on the role of ribonucleic acid in the growth of bacteria. Biochim. Biophys. Acta, 42:99.

Nilausen, K., and H. Green. 1965. Reversible arrest of growth in G1 of an established fibroblast line (3T3). Exp. Cell Res., 40:166.

Nygaard, O., S. Güttes, and H. P. Rusch. 1960. Nucleic acid metabolism in a slime mold with synchronous mitosis. Biochim. Biophys. Acta, 38:298.

Painter, R. B., and R. M. Drew. 1959. Studies on deoxyribonucleic acid metabolism in human cancer cell cultures (HeLa). I. The temporal relationships of deoxyribonucleic acid synthesis to mitosis and turnover time. Lab. Invest., 8:278.

——— R. M. Drew, and B. G. Giaque. 1960. Further studies on deoxyribonucleic acid metabolism in mammalian cell cultures. Exp. Cell Res., 21:98.

Pelling, G. 1959. Chromosomal synthesis of ribonucleic acid as shown by incorporation of uridine labelled with tritium. Nature (London), 184 (Suppl. 9):655.

Penman, S., K. Scherrer, Y. Becker, and J. Darnell. 1963. Polyribosomes in normal and poliovirus-infected HeLa cells and their relationship to messenger RNA. Proc. Nat. Acad. Sci. U.S.A., 49:654.

Perry, R. P. 1960. On the nucleolar and nuclear dependence of cytoplasmic RNA synthesis in HeLa cells. Exp. Cell Res., 20:216.

——— 1962. The cellular sites of synthesis of ribosomal and 4S RNA. Proc. Nat. Acad. Sci. U.S.A., 48:2179.

——— M. Errera, A. Hell, and H. Dürwald. 1961a. Kinetics of nucleoside incorporation into nuclear and cytoplasmic RNA. J. Biophys. Biochem. Cytol., 11:1.

——— A. Hell, and M. Errera. 1961b. The role of the nucleolus in ribonucleic acid and protein synthesis. I. Incorporation of cytidine into normal and nucleolar inactivated HeLa cells. Biochim. Biophys. Acta, 49:47.

Pilgrim, C. H., and W. Maurer. 1965. Autoradiographische Untersuchung über die Konstanz der DNA-Verdopplungsdauer bei Zellarten von Maus und Ratte durch Doppelmarkierung mit ^{3}H- und ^{14}C- Thymidine. Exp. Cell Res., 37:183.

Powell, W. F. 1962. The effects of ultraviolet irradiation and inhibitors of protein synthesis on the initiation of deoxyribonucleic acid synthesis in mammalian cells in culture. I. The overall process of deoxyribonucleic acid synthesis. Biochim. Biophys. Acta, 55:969.

Prescott, D. M. 1955. Relation between cell growth and cell division. I. Reduced weight, cell volume, protein content, and nuclear volume of *Amoeba proteus* from division to division. Exp. Cell Res., 9:328.

——— 1959. Nuclear synthesis of cytoplasmic ribonucleic acid in *Amoeba proteus*. J. Biophys. Biochem. Cytol., 6:203.

——— 1960a. Nuclear dependence of RNA synthesis in *Acanthamoeba* sp. Exp. Cell Res., 19:29.

——— 1960b. Relation between cell growth and cell division. IV. The synthesis of DNA, RNA and protein from division to division in *Tetrahymena*. Exp. Cell Res., 19:228.

——— 1961a. RNA synthesis in the nucleus and RNA transfer to the cytoplasm in *Tetrahymena pyriformis*. *In* Biological Structure and Function, Goodwin, T. W., ed., New York, Academic Press, Inc., p. 527.

——— 1961b. The growth-duplication cycle of the cell. *In* International Review of Cytology, XI, Bourne, G. H., and Danielli, J. F., eds., New York, Academic Press, Inc., p. 255.

——— 1963. RNA and protein replacement in the nucleus during growth and division and the conservation of components in the chromosome. *In* Cell Growth and Cell Division, Harris, R. J. C., ed., New York, Academic Press, Inc.

——— 1964a. Cellular sites of RNA synthesis. *In* Progress in Nucleic Acid Research and Molecular Biology, Davidson, J. N., and Cohn, W. E., eds., New York, Academic Press, Inc., Vol. 3, p. 33.

——— 1964b. The normal cell cycle. *In* Synchrony in Cell Division and Growth, Zeuthen, E., ed., New York, Interscience Publishers, Inc., p. 71.

——— 1966. Synthesis of total macronuclear protein, histone and DNA during cell cycle in *Euplotes eurystomus*. J. Cell Biol., 31:1.

——— and M. A. Bender. 1962. Synthesis of RNA and protein during mitosis in mammalian tissue culture cells. Exp. Cell Res., 26:260.

——— and M. A. Bender. 1963. Synthesis and behavior of nuclear proteins during the cell life cycle. J. Cell. Physiol., 62: (Suppl. 1) 175.

——— and R. F. Kimball. 1961. Relation between RNA, DNA and protein synthesis in the replicating nucleus of *Euplotes*. Proc. Nat. Acad. Sci. U.S.A., 47:686.

Rabinovitch, M., and W. Plaut. 1956. Cytochemical and autoradiographic observations in nuclear ribonucleic acid in *Amoeba proteus*. Exp. Cell Res., 10:120.

Reid, B. R., and R. D. Cole. 1964. Biosynthesis of a lysine-rich histone in isolated calf thymus nuclei. Proc. Nat. Acad. Sci. U.S.A., 51:1044.

Rho, J. H., and J. Bonner. 1961. The site of ribonucleic acid synthesis in the isolated nucleus. Proc. Nat. Acad. Sci. U.S.A., 47:1611.

Richards, B. M. 1960. Redistribution of nuclear proteins during mitosis. *In* The Cell Nucleus, London, Butterworths, p. 138–141.

——— P. M. B. Walker, and E. M. Deeley. 1956. Changes in nuclear DNA in normal and ascites tumor cells. Ann. N. Y. Acad. Sci., 63:831.

Ringertz, N. R. 1963. The effect of X-radiation on nuclear histone content. Exp. Cell Res., 32:401.

——— and G. C. Hoskins. 1965. Cytochemistry of macronuclear reorganization. Exp. Cell Res., 38:160.

Ritossa, F. M., and S. Spiegelman. 1965. Localization of DNA complementary to ribosomal RNA in the nucleolus organizer region of *Drosophila melanogaster*. Proc. Nat. Acad. Sci. U.S.A., 53:737–745.

Robbins, E., and T. Borun. 1967. The cytoplasmic synthesis of histone in HeLa cells and its temporal relationship to DNA replication. Proc. Nat. Acad. Sci. U.S.A., 57:409.

——— and M. D. Scharff. 1967. The absence of a detectable G1 phase in a cultured strain of Chinese hamster lung cell. J. Cell Biol., 34:684.

Ryter, A., and F. Jacob. 1963. Étude au microscope electronique des relations entre mêsosomes et noyaux chez *Bacillus subtilis*. C. R. Acad. Sci. (Paris), 257:3060.

Sandritter, W., H. G. Schiemer, H. Kraus, and U. Dörrien. 1960. Interferenzmikroskopische Unterschungen über das Wachstum von Einzelzellen (HeLa-Zellen) in der Gewebekultur. Frankfurt. Z. Pathol., 70:271.

Schaechter, M., M. W. Bentzon, and O. Maaløe. 1959. Synthesis of deoxyribonucleic acid during the division cycle of bacteria. Nature (London), 183:1207.

——— O. Maaløe, and N. O. Kjeldgaard. 1958. Dependency on medium and temperature of cell size and chemical composition during balanced growth of *Salmonella typhimurium*. J. Gen. Microbiol., 19:592.

Scherbaum, O., and G. Rasch. 1957. Cell size distribution and single cell growth in *Tetrahymena pyriformis* GL. Acta Pathol. Microbiol. Scand., 41:161.

Scherrer, K., H. Latham, and J. E. Darnell. 1963. Demonstration of an unstable RNA and of a precursor to ribosomal RNA in HeLa cells. Proc. Nat. Acad. Sci. U.S.A., 49:240.

Schmid, W. 1962. DNA replication of patterns of the heterochromosomes in *Gallus domesticus*. Cytogenetics, 1:344.

Seed, J. 1962. The synthesis of deoxyribonucleic acid and nuclear protein in normal and tumour strain cells. Proc. Roy. Soc., B156:41.

——— 1965. Deoxyribonucleic acid and ribonucleic acid synthesis in cell cultures. *In* Cells in Tissues and Culture, Willmer, E. N., ed., New York, Academic Press, Inc.

Shah, V. C. 1963. Autoradiographic studies of the effects of antibiotics, amino acid analogs and nucleases on the synthesis of DNA in cultured mammalian cells. Cancer Res., 23:1137.

Sherman, F. G., H. Quastler, and D. R. Wimber. 1961. Cell population kinetics in the ear epidermis of mice. Exp. Cell Res., 25:114.

Sirlin, J. L., K. Kato, and K. W. Jones. 1961. Synthesis of ribonucleic acid in the nucleolus. Biochim. Biophys. Acta, 48:421.

Sisken, J. E. 1959. The synthesis of nucleic acids and proteins in the nuclei of *Tradescantia* root tips. Exp. Cell Res., 16:602.

——— 1963. Analyses of variation in intermitotic time. *In* Cinemicrography in Cell Biology. Rose, G. C., ed., New York, Academic Press, Inc.

——— 1965. Exposure time as a parameter in the effects of temperature on the duration of metaphase. Exp. Cell Res., 40:436.

——— and R. Kinosita. 1961. Timing of DNA synthesis in the mitotic cycle *in vitro*. J. Biophys. Biochem. Cytol., 9:509.

Sköld, O., and A. Zetterberg. 1969. Studies on the amino acid regulation of RNA-synthesis in mammalian cells in tissue culture. Exp. Cell Res., 55:289.

Srinivasan, P. R., A. Miller-Faures, M. Brunfaut, and M. Errera. 1963. Kinetics or pulse-labeling of ribonucleic acid in HeLa cells. Biochim. Biophys. Acta, 72:209.

Stanners, C. P., and J. E. Till. 1960. DNA synthesis in individual L-strain mouse cells. Biochim. Biophys. Acta, 37:406.

Stern, H. 1956. The physiology of cell division. Ann. Rev. Plant Physiol., 7:91.

Stubblefield, E., R. Klevecz, and L. Deaven. 1967. Synchronized mammalian cell cultures. I. Cell replication cycle and macromolecular synthesis following brief colcemid arrest of mitosis. J. Cell. Physiol., 69:345.

Sundelin, P. 1968. DNA and RNA at successive postmitotic intervals in chick embryo fibroblasts infected with Rous sarcoma virus. Exp. Cell Res., 50:233.

Swann, M. M. 1957. The control of cell division: A review. I. General mechanisms. Cancer Res., 17:727.

——— 1958. The control of cell division: A review. II. Special mechanisms. Cancer Res., 18:1118.

Swift, H. H. 1950. The desoxyribose nucleic acid content of animal nuclei. Physiol. Zool., 23:169.

Taylor, E. W. 1965. Control of DNA synthesis in mammalian cell in culture. Exp. Cell Res., 40:316.

Taylor, J. H. 1958. Incorporation of phosphorus-32 into nucleic acids and proteins during microgametogenesis of *Tulbaghia*. Amer. J. Botany, 45:123.

——— 1960a. Nucleic acid synthesis in relation to the cell division cycle. Ann. N. Y. Acad. Sci., 90:409.

——— 1960b. Asynchronous duplication of chromosomes in cultured cells of Chinese hamster. J. Biophys. Biochem. Cytol., 7:455.

——— and R. D. McMaster. 1954. Autoradiographic and microphotometric studies of desoxyribose nucleic acid during microgametogenesis in *Lilium longiflorum*. Chromosoma, 6:489.

Terasima, T., and L. J. Tolmach. 1963. Growth and nucleic acid synthesis in synchronously dividing population of HeLa cells. Exp. Cell Res., 30:344.

——— and M. Yasukawa. 1966. Synthesis of G1 protein preceding DNA synthesis in cultured mammalian cells. Exp. Cell Res., 44:669.

Todaro, G. J., G. K. Lazar, and H. Green. 1965. The initiation of cell division in a contact-inhibited mammalian cell line. J. Cell. Physiol., 66:325.

Walker, P. M. B. 1954. The mitotic index and interphase processes. J. Exp. Biol., 31:8.

——— and J. M. Mitchison. 1957. DNA synthesis in two ciliates. Exp. Cell Res., 13:167.

——— and H. B. Yates. 1952. Nuclear components of dividing cells. Proc. Roy. Soc. Edingb., 140:274.

Wallace, H., and M. L. Birnstiel. 1966. Ribosomal cistrons and the nucleolar organizer. Biochim. Biophys. Acta, 114:296–310.

Watanabe, I., and S. Okada. 1967. Effects of temperature on growth rate of cultured mammalian cells (L5178Y). J. Cell Biol., 32:309.

Wimber, D. E. 1960. Duration of the nuclear cycle in *Tradescantia paludosa* root tips as measured with ^{3}H-thymidine. Amer. J. Bot., 47:828.

Wiseman, A. 1965. Organization for Protein Biosynthesis. Oxford, Blackwell Scientific Publications.

Woodard, J. W. 1958. Intracellular amounts of nucleic acids and protein during pollen grain growth in *Tradescantia*. J. Biophys. Biochem. Cytol., 4:383.

——— B. Gelber, and H. Swift. 1961a. Nucleoprotein changes during the mitotic cycle in *Paramecium aurelia*. Exp. Cell Res., 23:258.

——— E. Rasch, and H. Swift. 1961b. Nucleic acid and protein metabolism during the mitotic cycle in *Vicia Faba*. J. Biophys. Biochem. Cytol., 9:445.

Woods, P. S. 1959. RNA in nuclear-cytoplasmic interaction. Brookhaven Sympos. Biol., No. 12:153.

Yankofsky, S. A., and S. Spiegelman. 1962a. The identification of the ribosomal RNA cistron by sequence complementarity. I. Specificity of complex formation. Proc. Nat. Acad. Sci. U.S.A., 48:1069.

——— and S. Spiegelman. 1962b. Saturation of and competitive interaction at the RNA cistron. Proc. Nat. Acad. Sci., U.S.A., 48:1466.

Young, C. W. 1966. Inhibitory effects of acetoxycycloheximide, puromycin and pactamycin upon synthesis of protein and DNA in asynchronous populations of HeLa cells. Molec. Pharmacol., 2:50.

Young, R. W. 1962. Cell proliferation and specialization during endochondral osteogenesis in young rats. J. Cell Biol., 14:357.

Zalokar, M. 1960. Sites of protein and ribonucleic acid synthesis in the cell. Exp. Cell Res., 19:559.

Zetterberg, A. 1966a. Synthesis and accumulation of nuclear and cytoplasmic proteins during interphase in mouse fibroblasts *in vitro*. Exp. Cell Res., 42:500.

——— 1966b. Nuclear and cytoplasmic nucleic acid content and cytoplasmic protein synthesis during interphase in mouse fibroblasts *in vitro*. Exp. Cell Res., 43:517.

——— 1966c. Protein migration between cytoplasm and cell nucleus during interphase in mouse fibroblasts *in vitro*. Exp. Cell Res., 43:526.

——— 1966d. Nuclear and cytoplasmic growth during interphase. Almqvist & Wiksells Boktryckeri AB, Uppsala. Thesis.

——— and D. Killander. 1965a. Quantitative cytochemical studies on interphase growth. II. Derivation of synthesis curves from the distribution of DNA, RNA and mass values of individual mouse fibroblasts *in vitro*. Exp. Cell Res., 39:22.

——— and D. Killander. 1965b. Quantitative cytophotometric and autoradiographic studies on the rate of protein synthesis during interphase in mouse fibroblasts *in vitro*. Exp. Cell Res., 40:1.

Zeuthen, E. 1958. Artificial and induced periodicity in living cells. *In* Advances in Biological and Medical Physics, Tobias, C. A. and Lawrence, J. H., eds., New York, Academic Press, Inc., Vol. VI, p. 37.

Richard J. Goss

Division of Biological and Medical Sciences, Brown University, Providence, Rhode Island

6

Turnover in Cells and Tissues

INTRODUCTION

With the possible exception of the DNA molecule, and some of the proteins trapped in the lens of the eye, there is hardly any component of the body which is not replaced from time to time. This is not to say that all tissues or cells turn over. They do not. But many of them do, and even those that do not are able to renew their constituent molecules, if not their organelles, as a matter of course.

This state of affairs raises some formidable problems for an organism, not the least of which is the need to govern the rates of input and output in the various systems undergoing renewal. This control must operate on the mechanisms of synthesis and destruction by which new things are added and subtracted from the population. The resultant homeostasis finds expression in the remarkable equilibrium normally observed in living systems, an equilibrium all the more extraordinary for its adaptation to the ever-changing physiological demands constantly impinging on it.

Offhand, it would appear that a population of molecules, organelles, or cells should be held constant if the birth and death rates were equal. Things are not that simple, however. The size of a population is also affected by the lifespan of its constituent units. These units may be programmed to self-destruct after some predetermined length of time, or they may, like a group of hostages, be randomly dispensed with, young and old alike, by some external agency. In either case, the net result will still be a population maintained in a steady state by virtue of controlled births, deaths, and longevities.

Over and above these considerations, the fact remains that population size may actually fluctuate widely. Under certain conditions enzyme concentrations in a cell may suddenly increase; the number of ribosomes or mitochondria may proliferate; the population of red cells in the blood may expand. Hence, a new plateau may be reached and maintained, indicating that one set of mechanisms exists for the purpose of holding the population *constant* and another to determine the *level* at which this balance shall be established. It is like that paradigm of feedback control, the thermostat. The heating and refrigeration units are turned on and off by a device which reacts to temperature variations, but someone still has to set the thermostat to the desired temperature level in the first place.

How do these controls work? They may be either intrinsic or adaptive. In an intrinsically controlled system the population of units is genetically predetermined. Here the production of new units goes on at a fixed rate regardless of what physiological conditions might prevail. Obviously such a rigid setup has certain disadvantages in the living system. Its inflexibility cannot cope with the vicissitudes of the environment, both internal and external. Yet it is not all bad, for it has the security of money in the bank. It insures that a basic rate of interest shall be paid no matter what the purchasing value of the currency may be.

Like the wise investor she is, however, nature has evolved control systems which are closely linked to the physiological market. They can adapt to inflations and recessions by adjusting the rates of production and consumption according to the dictates of supply and demand. The logical way to do this is to arrange that physiological stimuli play dual roles, one role being to increase functional activity, the other to enhance the synthesis of those specific products upon which function depends. In this way the number of functioning units, whether they are cells or molecules, is automatically geared to the physiological demands for their services.

Although rates of turnover are undeniably subject to functional demand, it is still far from certain whether or not this is their exclusive means of regulation. Many systems undergo atrophy when experimentally prevented from functioning. Denervated muscles, cryptorchid testes, and vitamin A deficient retinas are cases in point. Sometimes these, and other structures, degenerate altogether when chronically unemployed. More often, however, they persist indefinitely in a state of depreciation, suggesting the existence of some form of intrinsic maintenance which operates despite long-term physiological inactivity. In other words, the qualitative existence of biological structures is inherent in any living system, while quantitative dimensions depend upon the work load. By playing it both ways, unused structures can be held in reserve against the day when they may again be called upon to perform. For the most part, however, growth and turnover are morphological expressions of physiological adaptation.

If physiological factors influence rates of turnover, there must be some pathway by which the demands for increased function are communicated to the sites where functional units are produced. At least two steps must be involved. First, the physiological deficiency (or sufficiency) must be monitored by some sensory mechanism. Second, the production facilities must adjust their output according to the information received. The most direct arrangement is for the reacting component to act as its own detecting device. Living systems, however, are seldom that simple. More commonly there are intermediate steps in the feedback loop, steps involving neural or humoral agents by which the message is refined and passed on. These extra measures are important when the function of one part of an organism affects other parts, as is almost always the case. In any organization, it makes good sense for all interested parties to influence the decisions that affect them, even if this complicates the chain of command. It is the very complexity of biological systems, however, which accounts for the marvelous precision with which they adapt, both physiologically and morphologically, to variable circumstances.

SUBCELLULAR TURNOVER

For purposes of comparison it might be instructive to examine some instances of turnover at the molecular and organelle levels before turning our attention to the more complex phenomenon of renewal at the cellular level of organization. It goes without saying that when cells divide, their constituent organelles are duplicated, and that this can be achieved only by the synthesis of molecules. Yet even the substance of nondividing cells in both static and expanding tissues is constantly being produced and degraded.

This state of ceaseless upheaval raises some interesting questions concerning the level of organization at which turnover is really going on. Tracer studies too numerous to mention leave no room for doubt that practically every kind of molecule in the cell will sooner or later be destroyed and replaced. Many of these molecules, however, are known to be organized into cytological structures: either granules, fibers, or membranes. What we do not know is the extent to which these formed elements turn over. Are they immortal so long as their host cell survives, or do some of them degenerate now and then in the natural course of events?

If organelles turn over, then their constituent molecules do likewise. But if a population of molecules undergoes renewal, it does not necessarily follow that the organelles containing them are transient. Unfortunately, we lack adequate techniques to find out, in many instances, if the formed elements within cells are subject to destruction and reconstruction. Only when there are marked fluctuations in their populations, as established by microscopic observation, can we be sure that a given organelle has a finite lifespan. Hence, anything that disappears during part of the mitotic cycle, for example, or in the course of successive rounds of functional activity, may be presumed to be turned over from time to time. Such entities as mitotic spindles, the nuclear envelope, the Golgi apparatus, lysosomes, and secretory granules all come and go according to the conditions that prevail in their host cells. Even the membranes and ribosomes of the endoplasmic reticulum go through periods of increase and decrease.

There are other cytological structures, however, which are more persistent. Once they have differentiated inside a cell they are not known to disappear except under unusual conditions. Certain organelles, such as mitochondria, chromosomes, and the cell membrane itself, are so ubiquitous that any cell lacking them could hardly be called a true cell. Others, however, are not formed in every cell, but are organelles designed for the performance of specialized functions. They are the structures by which certain cell types are diagnosed and upon which their physiological activities depend. In most cases they are the structural proteins peculiar to a given species of cell. To cite a few examples, there are keratin fibers in epidermal cells, pigment granules in chromatophores, neurofibrils in nerve cells, crystallin proteins in lens fibers, and hemoglobin in erythrocytes. Among the most interesting specializations, however, are cilia and flagella, the outer segments of retinal rods, and the myofilaments of muscle fibers. These three systems are worth exploring in some detail since the dynamics of their existence are, in some degree, known.

Regeneration of Cilia and Flagella

Of all the organelles which adorn the surfaces of cells, few inspire as much wonder as do cilia and flagella. One can hardly admire the exquisite arrangement and anatomy of these delicate appendages, however, without becoming curious about how they can develop in such orderly patterns and precise orientations, and how they grow and regenerate to such constant lengths. Despite their structural complexities and incessant activity, recent evidence indicates that cilia and flagella not only undergo molecular turnover within, but that, in some cases (Fig. 1), they are

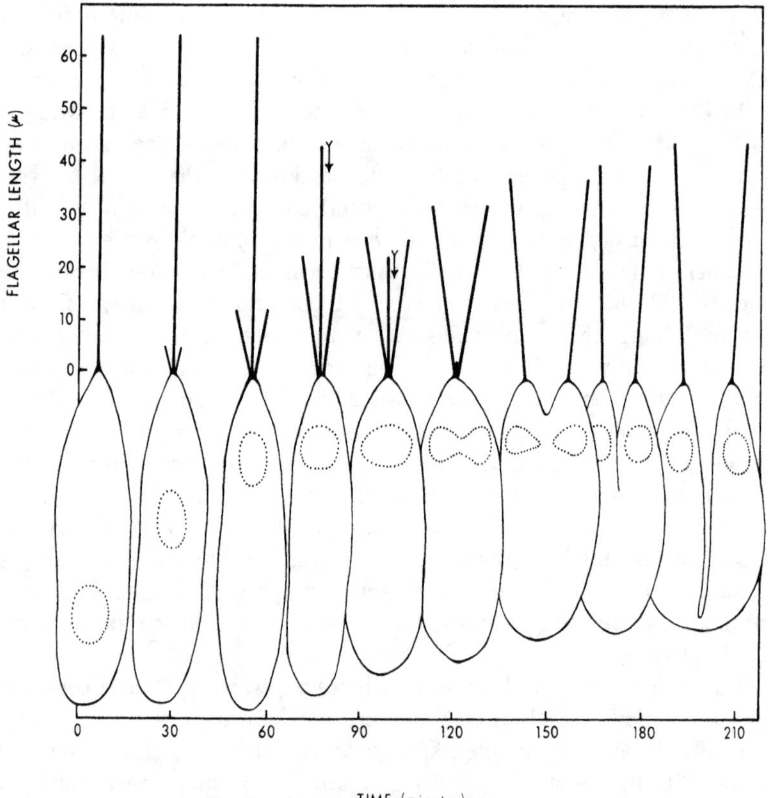

FIG. 1. Flagellar regression and regeneration during division of *Peranema trichophorum*. New flagella begin to elongate before the old one shortens. (Diagram adapted from Tamm, 1967.)

periodically resorbed and replaced by new outgrowths from the cell surface (Tamm, 1967). Indeed, they are even able to regenerate following amputation. Clearly, their capacities for renewal at both the molecular and the organelle levels can only mean that they are under the control of regulatory mechanisms by which the amounts of protein synthesized are balanced against the rates of destruction.

The control of ciliary and flagellar protein synthesis can be analyzed either in growing or in nongrowing structures. In nongrowing ones it is possible to assess the rates of turnover in a morphologically static system for comparison with the same kind of data during regeneration. Rosenbaum and Child (1967) have carried out experiments along these lines in the flagellates, *Euglena gracilis* and *Ochromonas danica*. When these organisms are cultured in a medium containing ^{3}H-L-leucine, their flagella incorporate the labeled amino acid in substantial quantities throughout several hours' exposure time. To be sure, they do not synthesize as much protein as do those which are regenerating flagella, but after 5 to 6 hours even the intact flagella take up from 25 to 60 percent as much leucine as do the growing ones. In still other experiments, *Euglena* may be allowed to regenerate flagella in the presence of labeled leucine after which they are washed and transferred to culture medium containing unlabeled leucine. After 18 hours their new flagella lose 36 to 37 percent of the labeled amino acid originally taken up during the course of regeneration. These are surprisingly high rates of turnover for an organelle which is only maintaining its morphological status quo. Such important results as these bear witness to the interesting fact that living systems are far more dynamic than their apparent anatomical stability would lead one to believe. Their growth and regeneration are only special cases of the incessant turnover which is going on at all times in all parts of all living organisms.

Although nongrowing cilia and flagella renew themselves at far greater rates than could ever have been suspected before their turnover was isotopically measured, the most rewarding approach to the problem has been to study the growing cilium or flagellum as it pushes out from the cell surface. This phenomenon takes place during the normal course of differentiation of ciliate or flagellate cells. It also occurs periodically in some ciliated epithelia in response to such cyclical changes as occur in the oviduct at different phases of the menstrual cycle. Finally, it can be induced by amputation of the cilia and flagella, which is a stimulus to regeneration. No matter what the circumstances may be that initiate growth, there is little reason to believe that the mechanisms of ciliary or flagellary elongation are not identical in all cases.

It is well established that cilia and flagella are derived from basal bodies, themselves descended from centrioles, and that the profusion of cilia which sprout from the cell membrane is made possible by the self-replication of their basal bodies (Sleigh, 1962). It is a curious coincidence that centrioles should take part in such diverse cytological events as mitosis and ciliogenesis. Yet these two processes may turn out to have more in common than just centrioles and basal bodies. Already it has been discovered by Auclair and Siegel (1966) that colchicine, which arrests mitosis, also inhibits the regeneration of cilia in sea urchin embryos. The implication, of course, is that the proteins present in cilia may be similar to those in mitotic spindles. Although ontogenetically, the basal bodies of cilia and flagella are derived from centrioles, it is not certain which came first during phylogeny. Indeed, it is quite possible that cilia and flagella might have evolved before mitosis did (Sagan, 1967; Margulis, 1968).

Once they have been formed, cilia usually persist indefinitely, even surviving mitotic episodes intact. Sometimes, however, they disappear and regrow at regular intervals, as in the case of the cilia in certain parts of the oviducts of rhesus monkeys. In this system, Brenner (1968) has chronicled the ciliary changes which occur during the menstrual cycle. Briefly, he has shown that cilia grow out from the epithelial cells early in the cycle, and are then lost after ovulation. This recurrent phenomenon is profoundly affected by hormones, for ovariectomized monkeys lose their cilia but grow them back following injections of estradiol benzoate. Here, then, is a case in which ciliogenesis is triggered by exogenous hormones. In most ciliate cells, however, their development is more exclusively under the control of intracellular agencies.

This is the case in sea urchin embryos which can be deciliated by exposure for a few minutes to hypertonic sea water (Auclair and Siegel, 1966) or sodium citrate solution (Iwaikawa, 1967). In either case, after only a 10-minute delay, new cilia begin to regenerate and in about 2 hours they reach the original lengths of almost 25 μ. The rate of elongation approximates 12 μ per hour at 20°C. According to Iwaikawa (1967), the growing cilia begin to beat long before they attain their full lengths.

Although regenerating cilia readily incorporate ^{14}C-L-leucine and ^{14}C-L-glutamic acid into their proteins, as reported by Auclair and Siegel (1966), their regeneration is inhibited neither by actinomycin D nor puromycin. Since these antimetabolites interfere with protein synthesis elsewhere in the cell, it must be assumed that ciliary growth takes place at the expense of a backlog of prefabricated proteins. This pool of proteins in the cell must be remarkably redundant, for Auclair and Siegel (1966) have shown that cilia can be induced to regenerate four times in a row over a 10-hour period even in the presence of puromycin, a compound capable of preventing the synthesis of all new proteins. It seems that the margin for safety in cilium regeneration is wide enough to permit extensive development simply by the assembly of preexisting proteins. Yet regenerating cilia normally incorporate newly synthesized proteins too, as their uptake of labeled amino acids amply testifies. Evidently the pool of ciliary proteins undergoes continual turnover as molecules are drawn from it to regenerate missing cilia or just to maintain intact ones.

Flagella have much in common with cilia. Although fewer in number and often longer in length, their growth kinetics resemble ciliary regeneration, to the extent of what is known in that area. In order to study flagellar regeneration, Rosenbaum and Child (1967) amputated the flagella from several species of protozoans by agitation in a homogenizer, or by reducing the pH of their culture medium from 6.8 to 4.7 for a couple of minutes. Following these treatments new flagella can be seen growing out after about half an hour. They are mobile from the very outset and their elongation, rapid at first, gradually decelerates as the original length is approached in about 5 hours.

Studies of the relation of protein synthesis to flagellum regeneration by Rosenbaum and Child (1967) have revealed that cycloheximide not only prevents the incorporation of amino acids into proteins, but also inhibits flagellar regeneration. The inhibition is total when the deflagellated protozoans are exposed to the antimetabolite during the lag phase, but is delayed in organisms treated during the elongation phase of regeneration. These results could be taken to indicate that on-the-spot protein synthesis is necessary for flagellar regeneration, but such an interpretation is not entirely consistent with comparable investigations using other antimetabolites in other systems.

It will be recalled that puromycin inhibits protein synthesis in deciliated sea urchin embryos while permitting the normal regeneration of cilia. Similar results have been obtained in studies of regenerating flagellates. These findings argue in favor of an extensive pool of precursor proteins from which cilia and flagella can regenerate. Although both puromycin and cycloheximide inhibit protein synthesis at the translational level, the above inconsistencies are apparently to be resolved in terms of differences in the efficiency with which these compounds are absorbed by various kinds of cells or organisms. In any case, it is apparent that a precursor pool does exist, but the extent to which it must be supplemented by *de novo* synthesis varies with the system under investigation.

Aside from the question of whether or not protein synthesis is essential for flagellar regeneration, the fact remains that labeled amino acids are built into growing flagella as they elongate. Because of this fortunate circumstance, Rosenbaum and Child (1967) were able to attack the problem of whether the flagellum grows by adding new material at its base or at its tip. Accordingly, they allowed flagella to begin to regenerate in the absence of labeled amino acid, and then exposed them to ^{3}H-leucine as regeneration went on to completion. Autoradiographs of such regenerates revealed that more label was incorporated distally in the flagella than proximally. Although lesser amounts of label nevertheless appeared in the already formed portions of the regenerating flagella, indicating that turnover commences in the new parts as soon as they develop, it is clear from these data that flagella elongate by the terminal addition of proteins. Hence, a flagellum regenerates by means of an apical growth zone, but maintains its structural integrity by turnover along its entire length. These important results may take on added significance in view of the patterns of turnover and regeneration, to be described below, in the highly specialized cilia that make up the outer rod segments in the vertebrate retina.

Renewal in Retinal Photoreceptors

Photoreceptors in the retina are highly specialized cells both biochemically and morphologically. The outer segments in the rods are actually modified cilia. Their most conspicuous features are the many closely-packed membranes folded upon each other in transverse orientation along the length of the segment. This is where the visual pigment, rhodopsin, is located. It was originally discovered by Droz (1963) that if labeled amino acids were injected into rats, the subsequently synthesized retinal proteins gradually migrated distally along the lengths of the outer rod segments. More recent autoradiographic studies by Young (1967), on a variety of

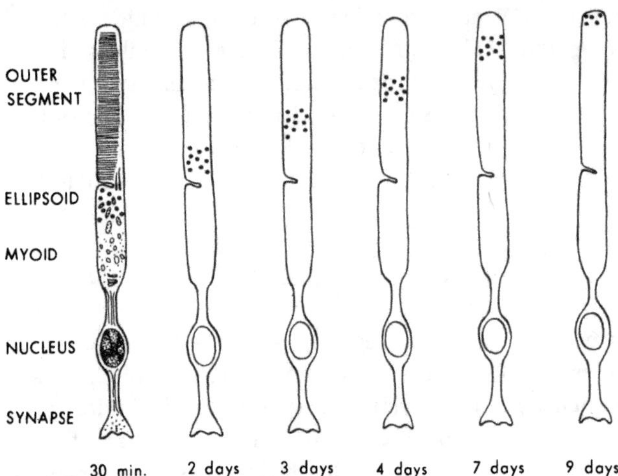

FIG. 2. Distal displacement of protein in a retinal rod. ^{3}H-methionine is taken up by the proximal segment, and can be traced autoradiographically in a reaction band (black dots) as it migrates to the ends of the outer rod segments. (Adapted from Young, 1967.)

vertebrate eyes, have confirmed and extended these original observations (Fig. 2).

When rats and mice were treated with ^{3}H-methionine and sacrificed at intervals afterward, their retinal rods were found by Young to incorporate the labeled amino acid into the inner segments first. Later on, however, some of these proteins move into the outer rod segments, as revealed by the position of radioactive bands seen in autoradiographs made a day after injection. At this time the label is located in the proximal ends of the outer segments. By the fourth day it has moved about half way out, and after nine or ten days the reaction band reaches the distal tips of the outer segments, which are in contact with the pigment epithelium. Here the label disappears.

Clearly some of the proteins into which the labeled amino acid is incorporated migrate from their sites of synthesis through the outer rod segments, eventually to be escalated off the far ends. As might be expected the rate of movement is temperature-dependent. In the retinas of frogs no migration can be detected at 4°C, but at 13, 25, and 34°C, progressively greater distances are traversed per unit time. Furthermore, it has been shown (Young, 1967) that in both rats and frogs the rate of migration is slightly accelerated by constant illumination as compared with total darkness.

Exciting as these results are, they leave certain questions unanswered. We do not know whether the proteins visualized in these experiments are broken down when they disappear or are secreted terminally from the outer segments. Neither do we know if the movement of the label signifies simple molecular diffusion or the translocation of the entire system of membranes which make up the outer segment. We do not even know what proteins are being labeled in the first place. Yet the spatial displacement of cellular components during the course of their lifespan is a particularly impressive demonstration of turnover at the subcellular level of organization, an example equalled only by the centrifugal flow of axoplasm in the neuron (Droz and Leblond, 1962).

It is tempting to speculate that the photoreceptor proteins that can be followed autoradiographically might include rhodopsin. Visual physiology depends upon the recurrent breakdown of rhodopsin into retinene and opsin, and their subsequent recombination. If opsin, the proteinaceous component of this combination, is indeed subject to turnover, it follows that the cell's entire inventory of this protein, upon which photoreception depends, should become depleted if its formation were blocked. There is now evidence that this may be what happens under chronic vitamin A deficiency. Without this vitamin, retinene cannot be produced. As a consequence, it is conceivable that opsin synthesis might also be arrested. In any event, rats maintained without vitamin A for prolonged periods not only suffer from night blindness but also exhibit degeneration of their retinal rods. Experiments by Dowling (1966) have shown that during the second month of the deficient state the rhodopsin levels in the retinas of rats declined, as did their sensitivity to light. This was followed by the morphological degeneration of the outer rod segments. Although such rodless photoreceptors can persist for many months in their undifferentiated condition, after ten months or so most of the cells disappear altogether. These striking results are especially significant because they suggest that the structural integrity of a cell cannot be maintained when deprived of the wherewithal to carry on its intended functions. They further suggest that under conditions of chronic unemployment the very survival of a cell is seriously jeopardized.

Up to a point, these degenerative changes are reversible. If vitamin A is restored to deficient rats before it is too late, both morphological and physiological recovery is possible. Hence, rod cells which have lost their outer segments, but which have not yet died, can be saved by replacement therapy with vitamin A. In such cases, the outer rod segments regenerate and vision returns, a phenomenon that would probably not be possible were it not for the normal everyday turnover of their components.

It is no coincidence that intracellular turnover and regeneration are prominent features of neurons and derivative cell types, such as retinal photoreceptors. These cells are incapable of dividing in the fully differentiated state and therefore lack provisions for renewal at the cellular level. In contrast, tissues which do renew themselves by cellular proliferation tend not to turn over their intracellular components to such an extent. Hence, the keratin in epidermal cells and the hemoglobin in red blood cells are both static components once they have been formed in the cytoplasm. It would be superfluous for these to be degraded and resynthesized intracellularly since the cells themselves are due to be replaced anyway.

Molecular Turnover in Muscle Fibers

Muscle fibers, like neurons and photoreceptors, are mitotically static. Unable to renew themselves by the periodic loss and replacement of cells, they must resort to turnover at the molecular level of organization in order to maintain their structure. They can also regenerate parts of their cytoplasmic mass which may have been lost or destroyed, a capacity also shared in common with neurons and retinal rod cells.

Unlike the latter, however, the material being replenished in muscle fibers does not change its position within the cell, which means that the rate of turnover cannot be as easily measured by the examination of autoradiographs.

Rates of turnover in muscle can be determined, however, by measuring specific activities of labeled proteins at successive intervals after injection of ^{14}C-glycine, as Dreyfus et al. (1960) have done (Fig. 3). Their studies have revealed that myosin has a lifespan of about 30 days in the skeletal muscles of the rat and about 20 days in the mouse, a divergence which suggests a relationship between turnover rates and metabolic rates. Other molecules have different lifespans. Aldolase (Schapira et al., 1960) and myoglobin (Akeson et al., 1960) have lifespans of about 20 days in rat skeletal muscle. Other things being equal, proteins in muscle are constantly being broken down as fast as they are synthesized, but each one seems to exist for a specific length of time. Structural proteins, such as myosin, appear to turn over as fast as the rest. Whether entire myosin filaments are bodily dismantled in the process, or whether their constituent molecules slip in and out without disturbing the form or function of the filament, is not clear. Although there is no evidence from electron micrographs indicating that myofilaments are ordinarily destroyed and rebuilt in a normal muscle, there is ample reason to believe that their synthesis, if not their demise, can occur under appropriate circumstances.

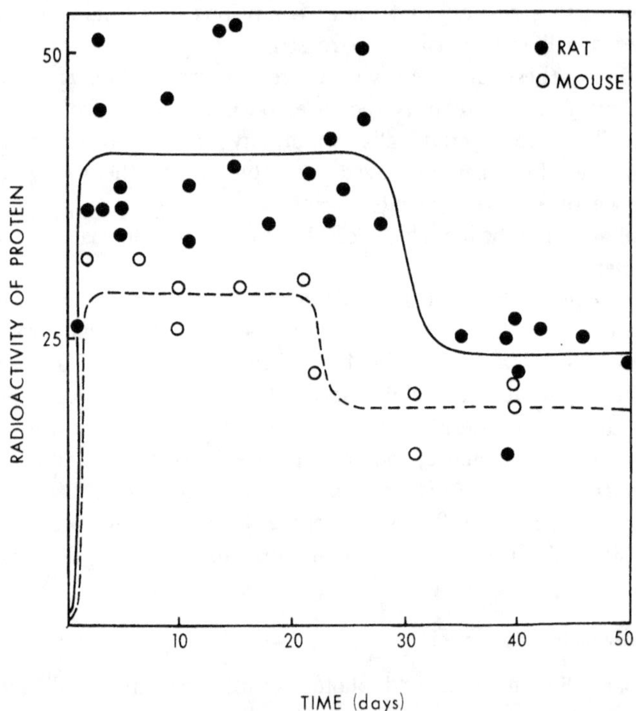

FIG. 3. Radioactivity of myosin after injection of ^{14}C-glycine into rats (●) and mice (○). Molecules of myosin are degraded about 30 days after being synthesized in the rat, and 20 days in the mouse. (Adapted from Dreyfus et al., 1960; and Kruh et al., 1960.)

Myofilaments, and the fibrils into which they are organized, are formed de novo in the course of ontogeny. Inasmuch as the size ultimately attained by a muscle fiber is in part a function of its physiological activity, it might be expected that hypertrophy of muscle fibers should be achieved either by making more fibrils or by enlarging those already present. Since myofibrils are more or less the same diameter in all striated muscles, it stands to reason that the muscle fiber enlarges by multiplying its armamentarium of fibrils. This can be studied under experimental as well as pathological conditions.

MUSCULAR HYPERTROPHY. Compulsory exercise is the most direct, but not the only, way to stimulate muscle growth. Morpurgo (1897) showed long ago that in the sartorius of the dog, exercise causes hypertrophy of the muscle fibers. Many other investigators have since confirmed these findings. More recently, Helander (1961) has shown that the overworked calf muscles in exercised guinea pigs have greater amounts of myofilament protein than do those of animals whose activity is normal or restricted. Drahota and Gutmann (1962) promoted muscular hypertrophy in a more subtle manner. By denervating one leg of a rat the muscles in the contralateral limb can be overburdened. Consequently, the extensor, digitorum longus, for example, will become hypertrophic within two weeks and its total nitrogen will increase correspondingly. The experiments of Hamosh et al. (1967) confirm and extend these results. Induced to become hypertrophic only four days after tenotomy of the synergistic muscles, the rat soleus will increase 30 percent in wet weight while its RNA concentration rises as much as 50 percent. *In vitro* assays disclose that the RNA-rich microsomal fraction from the hypertrophic muscles can synthesize protein more rapidly than equivalent preparations from normal control muscles.

Not all muscles respond directly to overwork. Some, like the levator ani of rats and mice, are sensitive to testosterone levels. This muscle, located in the perineal region, can be made to hypertrophy or atrophy by hormone treatments or castration, respectively (Fig. 4). Venable (1966) recognized in this system a unique opportunity to study the fine structure of myofibril formation and destruction. Under the stimulus of testosterone, especially if administered to previously castrated animals, the muscle fibers of the levator ani can be made to increase almost 50 percent in diameter after only a week. This remarkable growth is achieved largely by augmenting the preexisting population of myofibrils. Exactly how new fibrils come into existence, however, is still a matter for conjecture. In the normally differentiating muscle fiber they are known to arise de novo by the aggregation of filaments in the sarcoplasm. In mature muscles, however, they may develop by the subdivision of old ones. Even with the electron microscope, Venable (1966) was unable to find loci of de novo fibril formation in hypertrophying levator ani muscles. He tentatively concluded, therefore, that myofilaments can be added to the surface of already formed fibrils and that the latter tend to be split lengthwise by the sarcoplasmic reticulum when their diameters exceed some critical dimension by which the average fibril size is defined.

Something similar may occur in cardiac hypertrophy, a condition brought on by heightened demands for functional activity. Any condition which creates a greater

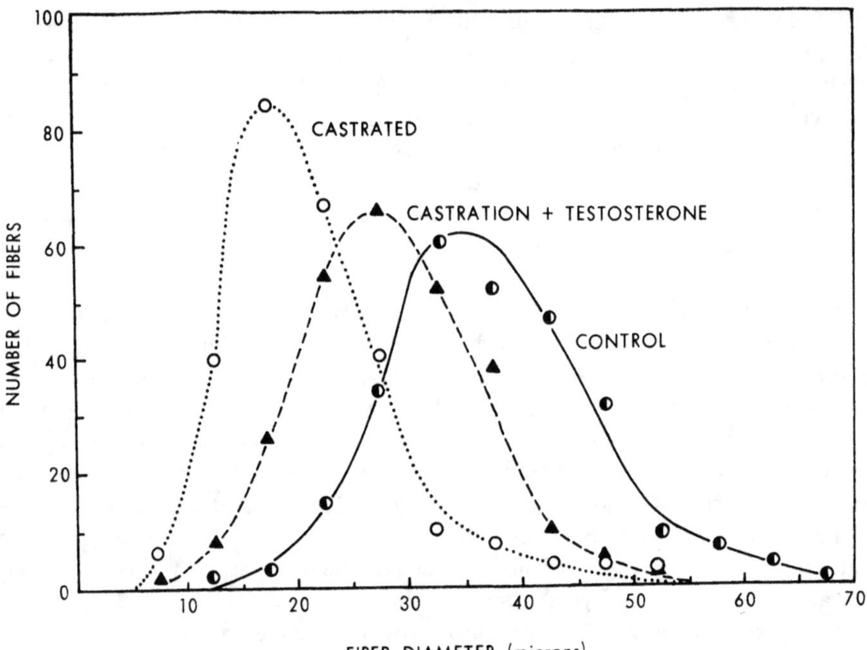

FIG. 4. Frequency distributions of muscle fibers in the levator ani of the mouse showing atrophy 45 days after castration, and reenlargement 8 days following testosterone treatment. (Adapted from Venable, 1966.)

need for cardiac output than the heart can deliver will tend to cause overwork and overgrowth of the heart. Pathologically, this is what happens in cases of hypertension. Experimentally, it can be duplicated by constricting the aorta, a method which induces hypertrophy of the left ventricle. Other procedures sometimes used to promote enlargement of the heart include forced exercise (Whitehorn and Grimmenga, 1956; Drasnin et al., 1958), hypoxia (Hultgren and Miller, 1965; Sulkin and Sulkin, 1965), chronic exposure to carbon monoxide (Campbell, 1932), anemia (Paplanus et al., 1958; Norman and McBroom, 1958; Sumner and McIntosh, 1963), the infliction of lesions on the aortic valve (Tomita, 1966), treatment with 3,3′,5-triiodo-L-thyronine (Gemmill, 1958), cholesterol-rich diets (Spraragen et al., 1962), and even audiogenic stress (Geber and Anderson, 1967).

Whatever the method used to enlarge the heart, the cytological changes seem to be very much the same as those encountered in hypertrophic skeletal muscle. According to studies of hypertrophic human hearts by Richter and Kellner (1963) and of cardiac enlargement in hypertensive rats by Carney and Brown (1964), the muscle fibers of the left ventricles enlarge but the myofibrils do not. Compared with normal controls, the number of myofibrils per cell increases over 50 percent. The myofilaments are likewise multiplied, as judged by their unchanged dimensions in hypertrophic fibers. Accordingly, Carney and Brown (1964) concluded that new myofilaments are probably added to myofibrils at their peripheries.

In view of the microscopic evidence that muscle fibers become hypertrophic by augmenting the numbers of fibrils and filaments, it comes as no surprise that this growth is accompanied by stepped-up protein synthesis. Not only is the total protein content of the ventricles increased (Hoyle et al., 1963; Tomita, 1966), but its rate of synthesis is accelerated. In rabbit hearts rendered hypertrophic by aortic stenosis, Meerson and Zayats (1960) noted that the incorporation of ^{35}S-methionine was doubled during the first post-operative week, and Gudbjarnason et al. (1963) measured a 350 percent increase in the specific activity of intravenously injected glycine-2-^{14}C in such hearts. In keeping with these changes, the total RNA content of hypertrophic hearts is also elevated (Hoyle et al., 1963; Sumner and McIntosh, 1963; Kleitke and Sydow, 1965). Since DNA does not increase in total amount, its concentration in hypertrophic heart muscle declines (Grimm et al., 1966) and the RNA/DNA ratio rises (Matsumoto et al., 1965).

In myocardial hypertrophy the mitochondria increase in number (Tobian et al., 1960), size (Wollenberger and Schulze, 1961; McCallister and Brown, 1965), and enzyme activity (Rossi and Dianzani Mor, 1958). According to McCallister and Brown (1965), the mitochondria in the hearts of hypertensive rats developed swollen cristae mitochondriales and thickened membranes. Meerson et al. (1964) postulated that in the overloaded heart it is essential that mitochondrial activity be promptly augmented in order to meet the demands for the extra energy needed, among other things, for protein synthesis. They further proposed that the increased rate of mitochondrial destruction in hypertrophying hearts yields breakdown products which might specifically stimulate the synthesis of extra mitochondrial proteins. There is no proof, however, that such a direct feedback control actually operates to govern the rate at which new mitochondria are produced.

Mitochondria and myofibrils respond in similar fashion to physiological overload in the heart. Athough they both increase in numbers, their concentrations in heart muscle remain unchanged. Grimm et al. (1966) reported that the concentration of proteins in general, and of actomyosin in particular, in hypertrophic rat hearts remains constant over a fourfold rise in ventricular weights. In order to achieve this, the rates of protein synthesis must be increased during enlargement. But despite this, there is no change in the rate of turnover (Gudbjarnason et al., 1964). What this means, of course, is that protein is being broken down faster at the same time its synthesis is accelerated, but which is cause and which is effect is not clear from the evidence as we now know it.

Muscular Atrophy. The other side of the coin is atrophy, a phenomenon which has revealed some very significant features of turnover in skeletal muscle fibers. Atrophy can be induced in many ways. Fasting alone will cause skeletal muscles to become smaller (Hines and Knowlton, 1934). So will immobilization (Summers and Hines, 1951), provided the muscles are fixed in the shortened position. Stretched muscles prevented from moving often gain weight instead of losing it (Thomsen and Luco, 1944; Imig et al., 1953; Feng et al., 1963). When a tendon is severed its muscle also atrophies (Lipschütz and Audova, 1921; Hines and Knowlton, 1937; Beránek et al., 1963), perhaps because this operation tends to immobilize the muscle in a shortened position. Some muscles are affected by hormones. The perineal musculature, for example, is sensitive to testosterone levels. Castration in rodents brings about marked atrophy of the levator ani muscle (Wainman and Shipounoff, 1941; Saunders and Leonard, 1955), a degeneration which goes on for 45 days before the weight of the muscle becomes stabilized at the reduced level. The cremaster muscles, from which the testes are suspended, are also subject to atrophy following castration (Stewart, 1966), although this may be more directly the result of reduced tension than reduced hormones. These and other methods all cause atrophy in normally innervated muscles, an atrophy which can also be caused by cutting the nerve supply.

Denervation is by far the most common way to induce muscle atrophy. Without nerves, the muscles in adult animals lose weight (Suftin et al., 1954; Kohn, 1964; Guth et al., 1966) due to a gradual decline in the diameter of their fibers (Gutmann and Zelená, 1962). This process can be delayed, but not prevented altogether, by daily electric stimulation (Fischer and Ramsey, 1946). Yet denervated muscles are not totally incapable of growth, as studies on immature animals have shown. Myoblasts regularly differentiate into muscle fibers even in tissue culture (Konigsberg, 1963). *In vivo,* however, the development of muscles, while not arrested completely, is considerably retarded in the absence of nerves (Zelená, 1962). Thus the treatment of chick embryos with Botulinum toxin, which blocks acetylcholine release at the myoneural junctions, results in severe atrophy of their skeletal musculature (Drachman, 1964). In infant rabbits, denervation of the gastrocnemius causes only a relative atrophy (Schapira et al., 1950). As the rabbit grows up, its denervated gastrocnemius becomes heavier, but not to the extent that the intact one grows. Like walking down an escalator that is moving up, the denervated muscles of young animals can share in the overall growth of the body at the same time their relative weights are declining.

In addition to the foregoing methods by which muscles may be made to atrophy, there are some interesting and instructive pathological conditions which result when the rate of turnover of muscle components is out of balance. In muscular dystrophy, for example, the skeletal muscles undergo a relentless degeneration for reasons that remain unexplained. Although a similar condition is mimicked by the effects of vitamin E deficiency, in which muscle fibers undergo histolysis and eventual necrosis (West, 1963), attempts to cure muscular dystrophy by vitamin E treatments have not proved successful. Still another so-called "natural atrophy" of skeletal muscles occurs in advanced age. In senile rats, for example, Berg (1956) has described the degeneration of muscles in the hind legs. As postulated by Drahota and Gutmann (1962), this condition may result from the decreased innervation typical of muscles in older animals.

At the ultrastructural level, muscular atrophy has been most thoroughly examined in the levator ani muscles of castrated rats and mice by Venable (1966) and Gori et al. (1967). They found that the decrease in fiber diameter occurs at the expense of the myofibrils, some of which disappear altogether. The degeneration of fibrils is preceded by their reduction in diameter owing to the loss of peripheral myofilaments. In many respects, this is the reverse of hypertrophy. Clearly, myofilaments and fibrils can be added to or subtracted from the population within a muscle fiber under a variety of experimental and pathological circumstances. Whether these organelles come and go in normal adult muscle which is neither hypertrophying nor atrophying cannot easily be determined with existing methods of investigation.

As was mentioned earlier, there is considerable turnover of the molecular constituents of nongrowing muscle. When induced to atrophy there is, of necessity, a net loss of substance. In the denervated diaphragm, for example, Stewart (1955) reported that about half the protein is lost over a period of 45 days. Theoretically, this depletion can be accounted for either by a decrease in the rate of molecular synthesis or by an acceleration of degradation. In the case of proteins, however, it turns out that synthesis may actually be increased during the course of atrophy. In the hind limbs of rats which are both denervated and immobilized, Slack (1954), noted that protein synthesis does not decline despite a 60 percent reduction in weight over a period of 16 weeks. More recently, Pater and Kohn (1967) confirmed this in the denervated gastrocnemius and tibialis anterior muscles of the rat leg. Their studies of ^{14}C-glycine uptake into structural proteins of such muscles showed that the rate of synthesis is normal or even greater than that in controls. In dystrophic mice, Simon et al. (1962) found that the rate of protein synthesis is about twice that in normal animals, but that the rate of proteolysis is tripled. Dreyfus (1963) also measured enhanced rates of protein synthesis in the muscles of dystrophic mice, findings which were coupled, however, with even higher rates of breakdown. Hence, it seems that atrophy is the result of proteins being destroyed faster than they are made.

It follows from this that the rate of protein turnover in atrophic muscles should be faster than normal, and that the lifespan of myosin molecules should be reduced, conclusions which have been verified by Kruh et al. (1960), Simon et al. (1962), and Dreyfus (1963) in their studies on dystrophic mouse muscle. Hence, the lesion responsible for muscular dystrophy appears to lie in whatever mechanisms at the molecular level serve to eliminate structural proteins in muscle. If so, then the accelerated synthesis and turnover rates of protein are secondary responses to the rapid and uncontrolled depletion of these molecules. This could mean, of course, that even in normal muscles the rates of protein synthesis might be related indirectly to the rates of destruction and, perhaps more directly, to the size of the protein pool. This, in turn, may feed back to the appropriate synthetic mechanisms via physiological pathways known to cause atrophy or hypertrophy, namely, exercise, innervation, hormones, tonus, and such other factors as may determine the workload imposed upon a muscle.

In atrophy, as in hypertrophy, the heightened rates of protein synthesis are made possible by relatively more abundant production facilities. Although the total content of RNA in an atrophying muscle may not change, its concentration in the diminishing mass of muscle increases (Gutmann and Zelená, 1962; Lissak et al. 1963). Atrophy, therefore, is not all bad since the muscle's synthetic mechanisms remain intact and operating at top speed. The problem is that even this is not sufficient to offset the abnormally high rate of molecular degradation. Atrophy and hypertrophy thus have much in common. In both cases the muscle fibers attempt to compensate for a deficiency by increasing the output of proteins. Hypertrophying muscles do so in response to physiological overload. Atrophying muscles may react to overload also, but the increased physiological demands in this case are probably brought on by the functional inadequacy of the muscle fibers which cannot stop destroying themselves.

Atrophy and hypertrophy are relative conditions which differ from the norm only in degree. If they are abnormal, it is because the cells involved are responding to stimuli of extreme intensities. The fact that muscle fibers alter their dimensions according to varying circumstances is living testimony to the competence of their homeostatic mechanisms to adapt to changing conditions. Such responses become pathological only when the demands for change exceed the adaptive capabilities of the reacting system.

TURNOVER AT THE CELLULAR LEVEL

Thus far we have explored the ways in which cells restore themselves when they cannot divide—which is to turn over their molecules, if not their subcellular organelles, rapidly and constantly. The vast majority of cells in the body, however, can divide mitotically when necessary and thus do not have to rely solely on intracellular renewal.

Static, Expanding, and Renewing Tissues

It is to his everlasting credit that Bizzozero (1894) first pointed out the distinction between various tissues according to their mitotic capabilities. Some tissues, such as neurons and muscle fibers, lose the capacity to divide as they become fully differentiated early in life. Since neuroblasts and myoblasts seldom persist after late embryonic stages, there remains no provision for replacing or augmenting neurons or muscle fibers beyond infancy in most vertebrates. Because they totally lack the potential for turnover at the cellular level of organization, these tissues have come to be referred to as mitotically static (even though they are far from static at the molecular level).

Of the remaining kinds of cells in the body, Bizzozero recognized two categories: those which proliferate only to expand their populations within a growing tissue or organ, and those which must multiply continuously to offset the ceaseless depletion of their ranks by programmed cell death. These two types of tissues are generally referred to as expanding and renewing, respectively.

Expanding tissues are characteristic of many glands in the body, both exocrine and endocrine. These organs are made up of cells which can survive indefinitely but which also retain their mitotic competence throughout life. They only divide to enable the tissue or organ to keep pace with the overall growth of the body as a whole, or when stimulated to proliferate in compensation for tissue loss or increased functional demands. Such tissues are of particular interest because their cells can still divide after becoming fully differentiated. Consequently, there is no need to establish a growth zone to separate the generative compartment from the differentiated one. Furthermore, there is no cell turnover to speak of in expanding tissues, since cells are never lost except by accident. When cell division does take place in expanding tissues, however, it can occur anywhere throughout the organ. Although we have learned a great deal about how the cells of expanding organs are stimulated to divide by physiological overload (Goss, 1965), we know almost nothing about what factors decide which cell, in a seemingly homogeneous population, shall divide, and when.

Unlike static and expanding organs, the cells of most renewing tissues are born to die. They may be broken down *in situ* and phagocytized, as in the case of blood cells. They may be sloughed from the body surface, like epidermis or epithelial cells of the intestinal mucosa. They may even be shed, as viable eggs and sperm, through ducts to the outside. And in at least one instance, namely, the lens of the eye, they are literally buried alive by future generations of cells that are laid down around them. Whatever their fates, the cells of renewing tissues are effectively removed from the scene by death, exile, or imprisonment.

The elimination of such cells in some kinds of renewing tissues is insured when they opt out of the turnover game. No cell can continue as an active member of a living organism unless it is part consumer and part producer. Those destined to die owe their suicidal tendencies to their failure to synthesize proteins once they have matured. In the case of the epidermal cell, keratin once deposited is not degraded, but the cell continues to make new keratin as long as it can. Whether such cells die because they choke on their own end products or stop synthesizing keratin because of their impending demise we have no way of knowing.

Red blood cells, on the other hand, cease making hemoglobin long before they die. Once they are released from the hemopoietic centers as reticulocytes they soon lose all capacity to make more proteins. Yet they also lose the capacity to break down what proteins they have. The half-lives of hemoglobin molecules in man (Shemin and Rittenberg, 1945) and rats (Akeson et al., 1960) do not differ significantly from the average lifespans of their respective erythrocytes. Set adrift in the blood stream with no means of self-repair, these fragile corpuscles eventually succumb to the depreciations of old age.

Lens fibers suffer a different fate. Like epidermal cells and erythrocytes, they manufacture quantities of specific proteins, in this case the so-called lens crystallins. When fully differentiated they cease production and dismantle their manufacturing equipment. Ribosomes as well as nuclei disappear, and the cells are left in a highly differentiated state with no provisions for making more proteins or getting rid of those already present. Sequestered within the lens without benefit of blood supply, the lens fibers are dependent upon such nutrients as may diffuse through to them from the surrounding ocular fluids. They may survive indefinitely this way by living in a state of metabolic limbo, if that can be called living.

Factors Affecting Mitosis

The key to understanding cellular turnover lies in the problem of mitogenesis. It is self-evident that this is affected by the rate at which cells are dispensed from renewing tissues, but the death rate is not directly responsible for controlling the birth rate. Mitotic activity is profoundly affected by functional demands, which are in turn related to the functional output of the tissue concerned. Since this output depends upon the state of differentiation in the cells, it is hardly surprising that mitotic competence should be influenced by a cell's degree of specialization.

It has been proposed elsewhere (Goss, 1966, 1967) that there are two categories of cells with respect to mitosis: those that cannot divide under any circumstances and those that can but will not unless appropriately stimulated. In the former category are the mitotically static nervous and muscular tissues, as well as many of the differentiated cells of renewing tissues. In addition to neurons and muscle fibers, it is generally accepted that cell division is equally impossible in fully differentiated epidermal cells, erythrocytes, lens fibers, Schwann cells, and spermatozoa, to mention a few. What these cells have in common, besides their inability to divide, are specific end products stored in the cytoplasm, usually as structural proteins. These terminal proteins are exemplified by actin, myosin, rhodopsin, keratin, hemoglobin, crystallin proteins, and so forth. The myelin of Schwann cells and the fat globules in adipose cells are also correlated with the nonproliferative state, although in these cases differentiation can be reversed and mitotic competence recovered.

In contrast, the cells of expanding organs can have it both ways. Despite their fully differentiated condition, cells in the liver, kidney, thyroid, adrenal cortex, pancreas, and other such organs can and will divide when necessary. The versatility of these cells is correlated with the nature of their end products. When fully differentiated, these cells release their products into the extracellular compartment in the form of hormones, enzymes, antibodies, collagen, plasma proteins, and so forth. It is probably no coincidence that secretory cells can divide while nonsecretory cells cannot.

There is yet another category of cell, intermediate between the two kinds already discussed. These are the cells which secrete their end products but do not really get rid of them. Chondrocytes and osteocytes, for example, which elaborate matrices of chondroitin and osteoid, become trapped within their own end products. Similarly, ameloblasts secrete enamel, but remain in intimate contact with the growing prisms. In these cases the mature cell types usually do not divide, and proliferation is largely restricted to zones devoid of end product. This is not to say that mature bone and cartilage cells cannot divide under special circumstances, but they do not ordinarily do so unless released from their enveloping matrices.

One is tempted to conclude that the end products of a cell may operate in a negative feedback circuit to turn off division. As illustrated diagrammatically in Figure 5, when the products are retained within the cell, their inhibitory influences are most effective. When dissolved in extracellular fluids, their efficacy is diminished to the extent that they become diluted and removed from their sites of production. But when allowed to accumulate in the microenvironment of the parent cells, their antimitotic influences are almost as effective as ever.

As was recently pointed out by Hermann et al. (1967), the supposed incompatibility between differentiation and mitosis is not so clear cut as it is sometimes assumed to be. This is because some cells can continue to divide after they have commenced the synthesis of specific proteins. Hence, it seems that it is not the *activation* of specific synthesis by a cell which is incompatible with mitosis, but the *degree* to which such end products accumulate. Different kinds of cells appear to vary in the extent to which their division is susceptible to inhibition by specific

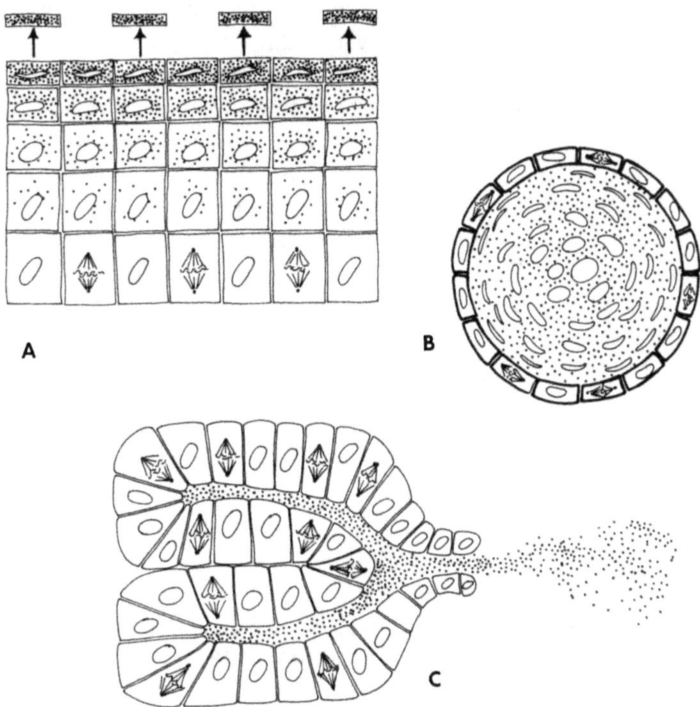

Fig. 5. End product feedback on mitosis. (A) As in epidermis, division is inhibited when product (keratin) accumulates inside cell. (B) When cells become embedded in a matrix of their end products, division occurs largely at the periphery, as in skeletal tissues. (C) Glandular tissues remain mitotically competent because their end products are eliminated by secretion.

proteins. Fibroblasts, for example, can continue to make collagen simultaneously with DNA synthesis (Davies et al., 1968). Red blood cells can divide for some time after the onset of hemoglobin synthesis. Muscle fibers, on the other hand, abruptly cease to multiply when myosin production begins. Thus, comparative cytological evidence points to a mechanism by which the end products of specific syntheses usually rise to a concentration which turns off the capacity for division.

These constraints on the proliferative activities of cells may explain why the static, renewing, and expanding tissues of the body have adopted different modes of growth. Mitotically static tissues are compelled to grow by cell hypertrophy once past their hyperplastic phase early in ontogeny. Since neuroblasts and myoblasts are used up prior to maturity there can be no provision for turnover at the cellular level of organization in these tissues. This is probably why they excel at molecular renewal. Even so, the inexorable loss of neurons and muscle fibers which takes place as an animal ages can only be compensated for by cell enlargement, an adaptive reaction of rather limited adequacy. Why it is that these vitally important tissues have evolved such seemingly inefficient means of growth and maintenance is an interesting field for speculation (cf. Payling Wright, 1963; Goss, 1965, 1966, 1967).

The cells of expanding organs resemble static tissues in only one respect, namely, their lifelong survival. Since they can differentiate and still divide, there is no need to isolate the young from the old. Consequently, when such organs are stimulated to grow, mitoses tend to be randomly distributed throughout the organ.

The cells of renewing tissues, like those of static ones, are incapable of dividing in the fully differentiated state. The chief difference between static and renewing cells is their lifespans, which in static tissues are potentially as long as the animal lives but are considerably shorter in renewing tissues. This necessitates, in the latter case, a system of continual renovation to replace the lost cells in order to maintain the population at a constant level. Renewing tissues, therefore, must grow by cell proliferation, and since they cannot divide and differentiate at the same time, their sites of renewal must be segregated from the region of cell differentiation. This sets up a growth zone composed of immature cells still capable of division. It is from this germinative region that cells are recruited for the zone of differentiation. Thus, the spatial separation of these two populations of cells reflects the temporal separation of the young from the old in the life history of each cell.

Turnover of Red Blood Cells

Of all the renewing tissues in the body, few are so important as blood, and none has been so intensively studied. Blood lends itself especially well to the analysis of turnover because it can be easily sampled from time to time in the course of an experiment without upsetting what is under study. Also, the entire technical armamentarium of the hematologist can be used to provide data on a variety of parameters concurrently. Moreover, a multitude of pathological conditions yield further information concerning the renewal of blood cells and what happens when that renewal is not properly regulated.

The erythron is an object lesson in steady-state control. In the human, for example, erythrocyte production (and destruction) goes on at the rate of about two million cells per second. This phenomenal turnover demands the operation of regulatory mechanisms of great precision if the concentration of circulating red cells is to be held constant. Yet it is not enough just to equate the birth rate with the death rate, for the erythron must also be able to recuperate following the accidental loss of substantial quantities of blood. Regeneration can be achieved only by the stepped-up production of erythrocytes, at least until the normal red cell count has been reestablished. This requires that the number of circulating cells be monitored in order that the marrow can adapt its rate of erythropoiesis to the deficit, a deficit of perpetual duration because red cells never stop dying.

Such are the problems of adjustment confronting the hemostat in its job of maintaining the erythron in equilibrium. Poised between the insufficiencies of anemia and the superabundance of polycythemia, a population of red cells must on the one hand remain constant, but must also be capable of going up or down as circumstances dictate. Hence, what would be anemic under some conditions could be normal under others. Unquestionably, the most extreme case of "normal anemia" occurs in the ice fish (Ruud, 1965), an inhabitant of the supercooled waters surrounding the Antarctic continent. Lacking both hemoglobin and erythrocytes, this fish survives by virtue of a metabolic rate low enough to subsist on what oxygen can be delivered to the tissues in physical solution in the plasma.

Although the ice fish is the only vertebrate known to maintain its population of red cells normally at zero, there are others which surprisingly do not depend for their survival on the erythrocytes they possess. Flores and Frieden (1968) have discovered that bullfrogs and their tadpoles can survive in good condition even after their circulating red cells have been virtually wiped out by phenylhydrazine-induced hemolysis. Grasso and Shephard (1968) have confirmed this in the adult urodele amphibian, *Triturus*. Since it may take many weeks or months before the destroyed cells are replaced, there is a very wide margin of safety in these creatures between the minimal and the optimal numbers of peripheral erythrocytes.

Warm-blooded vertebrates, of course, would die of asphyxiation if deprived of their erythrocytes. Yet, here too the pathology of the erythron is relative. What would be anemic at high altitudes may be normal at sea level; and what is polycythemic at sea level would be normal at high altitudes. It is not the absolute numbers or concentration of red cells that counts, but whether or not they can function adequately. Anemia and polycythemia, therefore, must be defined only in terms of the erythron's ability to supply enough oxygen to meet the demands of the tissues.

There is nothing sacred about the so-called normal hematocrit, or red cell count, or hemoglobin concentration in the blood. Erythropoiesis is not intended to maintain any such predetermined parameters. It is intended, however, to meet the body's demands for oxygen with an adequate supply. To fulfill this responsibility, the erythron must react to respiratory imbalances in the body and adjust its output of cells accordingly. There must be a sensing device, therefore, to sample the metabolic state of the body and to communicate this information to the effector organs, in this case the spleen and bone marrow.

The simplest way to achieve this would be for the hemopoietic stem cells to monitor the oxygen supply and make more red cells whenever they begin to feel "out of breath," as it were. But this is not how it is done. It would be a poorly designed system that required the greatest effort just when the energy supply was minimal. Perhaps for this reason the sensory and effector components that regulate erythron size are anatomically separated. The former is located in the kidneys and the latter in the marrow (or in the spleen in animals too small to have adequate marrow cavities in their bones).

Why the kidneys should be the organs chosen to detect respiratory shortcomings is a matter for conjecture. Perhaps it is an atavistic reflection of the renal source of blood cells in lower vertebrates (although other visceral organs might have qualified as well in that respect). Whatever the explanation, the fact remains that the kidneys must relay information to the bone marrow, information in the form of a humoral agent capable of stimulating erythropoiesis. This agent is erythropoietin.

ERYTHROPOIETIN. The existence of erythropoietin can be demonstrated in a variety of ways. It is produced whenever the number of circulating red cells is reduced, either by bleeding or by hemolysis caused by phenylhydrazine (Goldwasser et al., 1957, 1958). Its concentration in the plasma rises under conditions of hypoxia, as when the percentage of oxygen in the air is lowered, or the atmospheric pressure reduced, to simulate high altitudes (Reynafarje et al., 1964). Erythropoietin is also secreted by the kidneys in response to treatment with cobaltous chloride (Goldwasser et al., 1957, 1958; Fisher, 1961), testosterone (Mirand and Gordon, 1965; Naets and Wittek, 1966), ACTH (Osnes, 1960), and by electrical stimulation of the hypothalamus (Halvorsen, 1961). A particularly rich source of erythropoietin is the urine or plasma of patients suffering from various kinds of anemia, but it can also be detected in fetal plasma, the plasma of pregnant females, and even in normal adult plasma if sufficiently concentrated.

In view of the renal source of erythropoietin (Naets, 1958), its output can be enhanced by a variety of experimental or pathological interventions. For example, Wilms' tumor in the kidney is often accompanied by polycythemia owing to the exaggerated production of erythropoietin (Murphy et al., 1967). Renal infarcts, experimentally induced by the injection of plastic microspheres to occlude the renal arterioles in dogs, also heighten the secretion of erythropoietin (Abbrecht and Malvin, 1966; Abbrecht et al., 1966). Partial constriction of the renal artery causes similar effects (Hansen, 1964; Fisher and Samuels, 1967). Perfusion of the dog kidney with blood containing cobalt (Fisher, 1961) or venous blood shunted from the right atrium of the heart (Pavlovic-Kentera et al., 1965) likewise stimulates the kidney to augment its output of erythropoietin. Evidently anything that brings about a local shortage of oxygen in the kidney will cause it to produce factors that can promote erythropoiesis.

Conversely, the production of erythropoietin can be curtailed by anything that tends to satisfy the body's need for oxygen (Jacobson et al., 1957a). There are two ways of doing this. One is to decrease the demand for oxygen, as by starvation or hypophysectomy. The other is to increase the supply of oxygen. This can be done directly by means of hyperoxia, a condition which effectively reduces the output of erythropoietin (Goldfarb and Tobian, 1963). The same effect can also be achieved by increasing the efficiency with which oxygen is transported to the tissues. If extra red cells are pumped into the circulation, the resulting polycythemia is responsible for delivering a superabundance of oxygen to the tissues. Consequently, the percentage of reticulocytes in the peripheral blood declines drastically (Niece et al., 1963), the rate at which ^{59}Fe is incorporated into hemoglobin diminishes (Elmlinger et al., 1952), and the activity of the renal erythropoietic factor is decreased (Gordon et al., 1966).

Actually, recent evidence suggests that what the kidney produces cannot directly stimulate red cell production. Extracts of kidney homogenates, for example, generally do not affect erythropoiesis (Kuratowska, 1965). Kidneys perfused with hypoxic blood, however, do produce an erythropoietic-stimulating factor, but it is only effective when allowed to interact with plasma, specifically with the alpha globulin fraction (Kuratowska et al., 1964; Contrera and Gordon, 1966; Moores et al., 1966). Either the renal factor is an enzyme capable of catalyzing the production of erythropoietin from a serum protein, or it requires something in the serum to combine with or to be otherwise acted upon before it can become erythropoietin (Contrera et al., 1966; Zanjani et al., 1967).

The velocity of a process can be more carefully regulated if, in addition to an accelerator, there are brakes that can be applied. Now it has been found that erythropoiesis may be subject to constraints as well as stimuli. Krzymowski and Krzymowska (1962) showed that the plasma from sheep rendered polycythemic by hypertransfusion depresses erythropoiesis when injected into rats or rabbits. This important discovery was later confirmed by Whitcomb and Moore (1965), who inhibited the incorporation of iron into the red cells of hypoxic mice by treating them with plasma from hypertransfused sheep, rabbits, or mice. Thus, excessive erythropoiesis is avoided not only by reducing the production of erythropoietin but also by the active production of an inhibitory substance. Whether or not the latter material is produced in kidneys, as is erythropoietin (or its precursor), and if so by what cells, is for the future to discover.

Erythropoiesis in Immature Animals. The admirable efficiency with which erythropoiesis is controlled in the adult is all the more intriguing when compared with embryonic, fetal, and newborn systems (Fig. 6). Control mechanisms, like the tissues whose growth they regulate, must also develop. It should come as no surprise, therefore, that immature animals do not always react the way adults do, erythropoietically, that is.

In the chick embryo, Jalavisto et al. (1965) and Jalavisto (1967) have demonstrated that hemopoiesis does not respond to variations in oxygen, at least during the first 2 weeks of incubation. After the fourteenth day, however, the embryo acquires the capability to respond to low or high oxygen tensions. Thus, the hemoglobin concentration in embryonic chick blood rises when exposed to 15 percent oxygen and decreases at 40 percent oxygen. This maturation of the erythropoietic control mechanism after the second week of incubation coincides with when hemo-

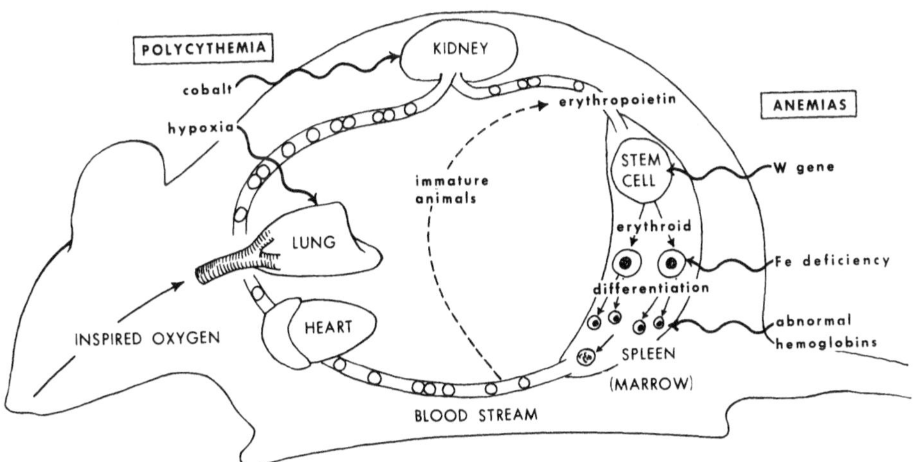

Fig. 6. Summary diagram of how red blood cell production is normally controlled in the mouse. Wavy lines indicate targets of pathological or experimental conditions. Broken line signifies parts of adult system bypassed in immature animals. (Adapted from Keighley et al., 1966.)

poiesis is transferred from the visceral regions to the bone marrow. It is also when the pituitary begins to function and when the metanephros takes over from the mesonephros. If the metanephros is the source of erythropoietin in the chick, it could hardly be expected to affect red cell production during its early stages of development. Perhaps this may explain why cobalt, which stimulates erythropoiesis in the adult, fails to affect hemoglobin synthesis in embryos.

A comparable situation obtains in the infant rat. Deprived of oxygen, young rats refuse to alter their rates of erythropoiesis. During the first 2 weeks or so after birth, rats exhibit no erythropoietic response to exposure to 0.4 atmospheres for 6 hours per day (Carmena et al., 1966). Beginning about the seventeenth day they begin to show very slight reactions and gradually reach the adult condition at about fifty days of age (Garcia, 1957). Even more surprising is the fact that bilateral nephrectomy does not interfere with the normal course of erythropoiesis in rats up to 40 days old (Lucarelli et al., 1964b). Beyond this age erythropoiesis becomes dependent on the kidneys.

If red cell production in young animals is affected by neither hypoxia nor nephrectomy, both of which profoundly influence adults, how is neonatal erythropoiesis regulated? Various observations indicate that it is somehow affected by changes in the concentration of circulating red cells, for erythropoiesis is depressed in 13 to 17-day-old rats made polycythemic by hypertransfusion (Lucarelli, 1964a). It is also stimulated in humans born with severe hemolytic disease (Crawford et al., 1953) and in armadillo embryos suffering from thalidomide-induced hemorrhaging (Marin-Padilla and Benirschke, 1963, 1965). Thus, even immature animals of various species are able to adjust their output of red cells to some predetermined standard.

The next question must obviously be whether fetal or newborn red cell production is influenced by erythropoietin. This has been tested in 18-day-old rats by Lucarelli et al. (1964a). The young assay animals were first made polycythemic in order to suppress their normal erythrocyte production. When subsequently given erythropoietin, their red cell production was stimulated. This important discovery is consistent with other findings reporting the presence of erythropoietin in the fetal blood of humans, sheep, and dogs (Bonsdorff, 1949). Clearly, erythropoietin is present, and presumably active, at a stage of development when the kidneys are not required to sustain normal rates of erythropoiesis. The conclusion is inescapable, therefore, that there must be some extrarenal source of erythropoietin in young animals. Even in adults there is evidence that the kidneys are not the only source of erythropoietic stimulating factor. Human patients without kidneys, for example, can respond to hypoxia by increasing their rates of erythropoiesis (Nathan et al., 1964). Furthermore, de Franciscis et al. (1966) have shown that cell-free spleen extracts cause a marked elevation in the reticulocyte percentage when injected intraperitoneally into mice.

Prenatally, it is conceivable that fetal erythropoiesis could be subject to maternal supervision. A few venturesome experiments have been directed toward this possibility, but their results are not entirely consistent with each other. Jacobson et al. (1959) made pregnant rats polycythemic by hypertransfusion. This effectively suppressed maternal erythropoiesis, but did not affect it in the fetuses. Lucarelli et al. (1964b), in a similar investigation, found normal reticulocyte percentages in the offspring of polycythemic mothers. However, Butturini and Lucarelli (1963) reported that the mitotic activity in the fetal livers of such animals was depressed 50 percent. Since considerable hemopoiesis goes on in fetal livers, maybe the prenatal mammal is not so independent after all. Further evidence along these lines is provided by the investigations by Johnson and Roofe (1960). They kept pregnant rats at a simulated altitude of 18,000 feet and found the hemoglobin and red cells of the offspring to have risen 39 percent and the hematocrits by as much as 20 percent. Since newborn rats do not accelerate their erythropoiesis under hypoxic conditions, this prenatal response may have been mediated indirectly by way of maternal influences.

ANEMIA. In a complex phenomenon such as erythropoiesis there are many things that can go wrong. Each step in the pathway leading to the production of a red blood cell is the site for a potential disease. Hence, pathological conditions provide clues which often give away the secrets of the normal functioning of a process. This is why certain kinds of anemias have proved very useful in revealing where lesions can occur in the process of erythropoiesis.

Hereditary anemia is a case in point. Some of these conditions, long known to afflict mankind, have fortunately been reproduced (or at least simulated) in mice where they are more amenable to experimental manipulation. Two genetic conditions in mice have been particularly rewarding, both overt manifestations of macrocytic anemia but each the result of very different inborn aberrations in the physiology of erythropoiesis.

One strain of mouse carries the dominant W-series genes (W/W^v) and the other the so-called steel-series genes (Sl/Sl^d). The pioneering investigations of Bernstein and Russell (1959), and Keighley et al. (1962, 1966) have revealed how different these two strains of mice really are, despite their anemic similarities. These differences are most strikingly demonstrated by grafting W/W^v and Sl/Sl^d mice together in parabiosis This operation cures both partners of their anemias, an outcome taken to indicate that each may have originally lacked something that the other possessed. When such parabiotic pairs are later separated, it turns out that the W/W^v mouse remains permanently cured of its anemia. The Sl/Sl^d mouse, however, reverts to its former anemic state (Fig. 7).

The explanation of these results lies in the ability of at least some cross-circulating blood cells to settle down in the spleen or marrow and give rise to hemopoietic stem cells. This is exactly what happens when the Sl/Sl^d blood cells are carried across to the W/W^v partner. They colonize the latter's hemopoietic tissues (which are only sparsely populated with host cells) and produce red blood cells at normal rates. Clearly, the W/W^v mouse is not anemic because of any systemic deficiency in erythropoietin production, but because its cells are abnormal. Consequently, its cells cannot compete effectively with normal ones, even if the latter are derived from a mouse that is also anemic, but anemic for different reasons. The Sl/Sl^d anemic mouse in parabiotic partnership with a W/W^v mouse is cured of

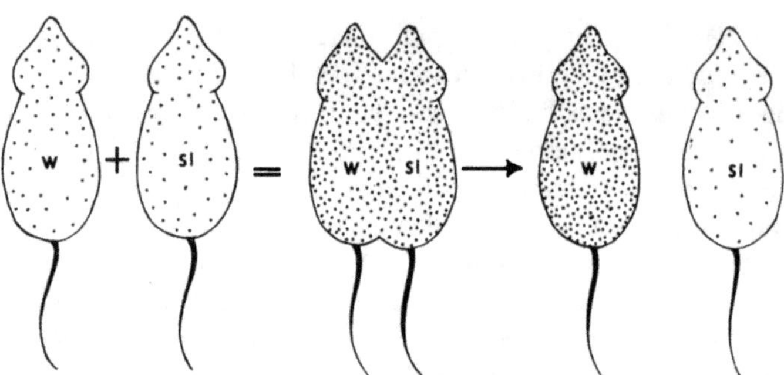

FIG. 7. When W and Sl anemic mice are joined in parabiosis, the blood of both becomes normal. When separated again, the W mouse remains normal because it has received normal stem cells from the Sl partner; the Sl mouse, however, reverts to its former anemic condition owing to some systemic lesion in its erythropoietic regulating mechanism.

its anemia only in the passive sense. Its marrow cells, though present in normal quantities, remain unstimulated while the peripheral blood picture improves. Evidently the anemia is corrected by virtue of the transfusion of blood cells produced in the other mouse, a situation that can be maintained only as long as the parasitic relationship lasts. Hence, the Sl/Sl^d mouse appears to be anemic because its otherwise normal hemopoietic cells are not properly stimulated to divide and differentiate.

How do these strains react to factors that normally stimulate erythropoiesis in healthy mice? When erythropoietin is injected into them, they do not respond normally. W/W^v mice require 150 times as much erythropoietin as do normal mice to elicit equivalent increases in their rates of erythropoiesis. Sl/Sl^d mice are even less responsive, since 1,000 times the normal effective dose of erythropoietin fails to raise the rate of red cell production to normal levels. Yet despite the inability of such heroic doses of erythropoietin to stimulate erythropoiesis, both strains of mice respond to hypoxia, though W/W^v mice do so better than Sl/Sl^d mice. In W/W^v mice, therefore, lowered oxygen tension promotes red cell production (Fig. 8), but it does not do so by elevating the levels of erythropoietin in the plasma. To be sure, W/W^v mice can and do make extra erythropoietin under hypoxic conditions. Plasma from such mice enhances the uptake of ^{59}Fe into hemoglobin when injected into normal assay animals, but it fails to stimulate W/W^v recipients. Hence, the surplus erythropoietin synthesized by hypoxic W/W^v mice is not enough to account for the observed erythropoietic response. One can only conclude, therefore, that there must be another pathway, bypassing erythropoietin, by which red cell production is stimulated. It is conceivable that the erythropoietic inhibitor (see above) might be produced or activated only under sufficiently hyperoxic conditions, and that hypoxia might cause its inactivation thereby releasing the hemopoietic tissues from inhibition.

In any event, anemia is a disease that is caused by a lesion in the control mechanism of red cell production. Yet even anemic animals produce erythrocytes. They simply do so at too slow a pace. We have learned a lot about how the erythropoietic control system manages to hold the population of red cells constant, but

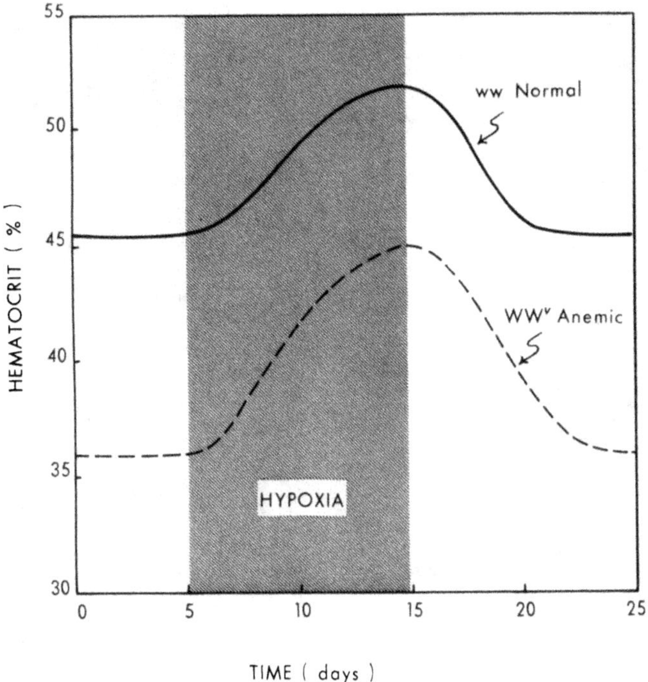

FIG. 8. Effect of hypoxia on normal and anemic mice. W strain mice can still increase their red cell output in response to reduced oxygen, but this is not mediated by erythropoietin. (Adapted from Keighley et al., 1962.)

we know very little about how the mass of the erythron is determined in the first place. There is much to be learned, therefore, by studying anemic mice in which the hemostat is set too low.

Pursuant to the foregoing research on hereditary anemias, some interesting experiments have been carried out on the deleterious effects of x-rays on erythropoiesis and how this condition can be cured. Like all rapidly-dividing cells, hemopoietic stem cells are very sensitive to radiation, with the result that anemia is one of the expected symptoms of radiation sickness. However, it is possible to alleviate this condition by injecting unirradiated marrow into irradiated mice. This repopulates the host's hemopoietic tissue with healthy cells. These radiation chimeras are not only cured of their impending anemia, but can survive otherwise lethal doses of x-rays for almost as long as untreated controls (Bernstein, 1963).

It will be recalled that in mice, the lion's share of erythropoiesis goes on in the spleen, an organ which is therefore very sensitive to radiation. Following exposure to x-rays, the spleen soon becomes depleted of its hemopoietic cells. If healthy cells are injected into an irradiated mouse, they not only recolonize the spleen, but do so in an unexpected manner. Till and McCulloch (1961) showed that the spleens of heavily irradiated mice injected with normal marrow developed erythropoietic tissues in definitive loci. These spleen colonies, appearing as grossly visible

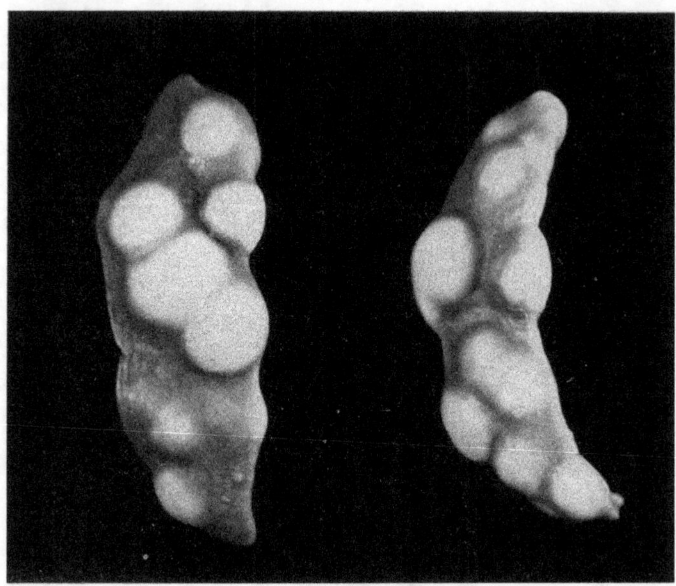

FIG. 9. Hemopoietic spleen colonies. If a heavily irradiated mouse is injected with normal bone marrow cells, some of the latter settle in the spleen. In only 10 days their descendants may proliferate into tumescent colonies grossly visible on the surface of the spleen. (From McCulloch, 1966. *In* The Physiological Basis of Medical Practice, 8th ed., Best, C. H., and Taylor, N. B., eds., pg. 571. Courtesy of the Williams and Wilkins Co., Baltimore.)

lumps under the splenic capsule (Fig. 9), could be counted and correlated with the dose of cells originally injected. They calculated that about one erythroid colony formed for every 10^4 cells injected (Fig. 10). By means of chromosomal markers, Becker et al. (1963) later demonstrated that these spleen colonies are almost certainly clonal in nature.

This exciting technique can also be applied to the study of genetic anemias. Thus, if normal marrow cells are injected into W/W^v mice, their spleens, which are ordinarily poor in hemopoietic cells, become colonized by the healthy cells (McCulloch et al., 1964). Even anemic cells from Sl/Sl^d mice will successfully form spleen colonies in W/W^v mice (McCulloch et al., 1965), but normal marrow injected into irradiated Sl/Sl^d mice proliferates very poorly in host spleens. These and other experiments indicate that it is not enough to have normal blood-forming cells. It is just as important that these cells be in a favorable environment, an internal milieu replete with the appropriate stimulatory (and inhibitory) factors required to maintain normal rates of proliferation and differentiation.

HOW ERYTHROPOIETIN WORKS. Perhaps the most important problem of hemopoiesis is to discover how erythropoietin acts; whether it stimulates proliferation or differentiation, or both. If the unitarian theory of hemopoiesis is true, as seems to be the case, then all kinds of blood cells are ultimately descended from a single

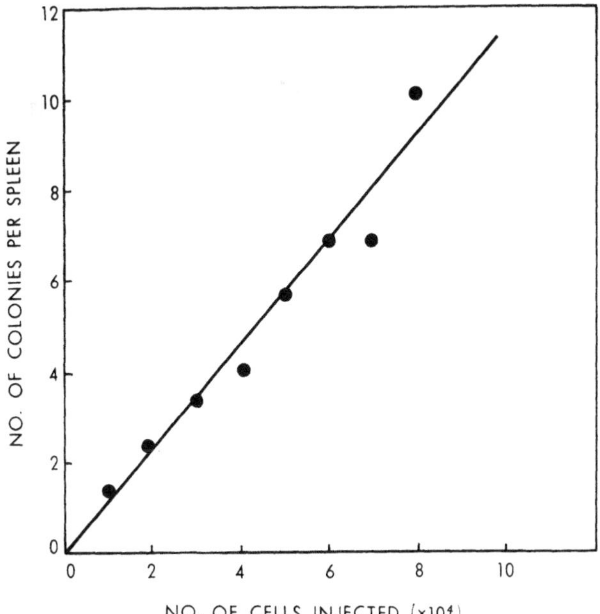

FIG. 10. Linear relationship between number of cells injected and number of hemopoietic colonies in the mouse spleen. (After Till and McCulloch, 1961.)

kind of pluripotent stem cell. With as many as seven pathways open to them, hemocytoblasts must be steered in one direction of differentiation or another, to the exclusion of all alternatives. By shifting the emphasis from one channel to another, it is possible to modulate the populations of various types of blood cells and thereby make up for deficits in the peripheral circulation. Such quality control is important, but it has the disadvantage of forcing each kind of blood cell to compete with the others for a limited supply of stem cells. This undesirable feature can be avoided by adding some sort of quantitative control to the system. This is the only way to make the advantages of differential augmentation possible without creating shortages in other directions. Since the process of erythropoiesis is obviously controlled both qualitatively and quantitatively, we have much to learn from this system about the fundamental problems of cellular differentiation and proliferation (see reviews by Boggs, 1966; Goldwasser, 1966).

From the very outset, most investigators have favored the view that the primary action of erythropoietin was to stimulate the differentiation, rather than the proliferation, of red cells, and that this stimulus was directed toward the early precursor cells. Erslev (1959) showed that if blood was removed from rabbits for 20 hours before being returned to the circulation, this pulsed anemia stimulated the production of new red cells which did not show up in the peripheral circulation as reticulocytes until several days later (Fig. 11). He concluded that the stimulus had to be directed at the stem cells in the marrow, inducing them to become pro-

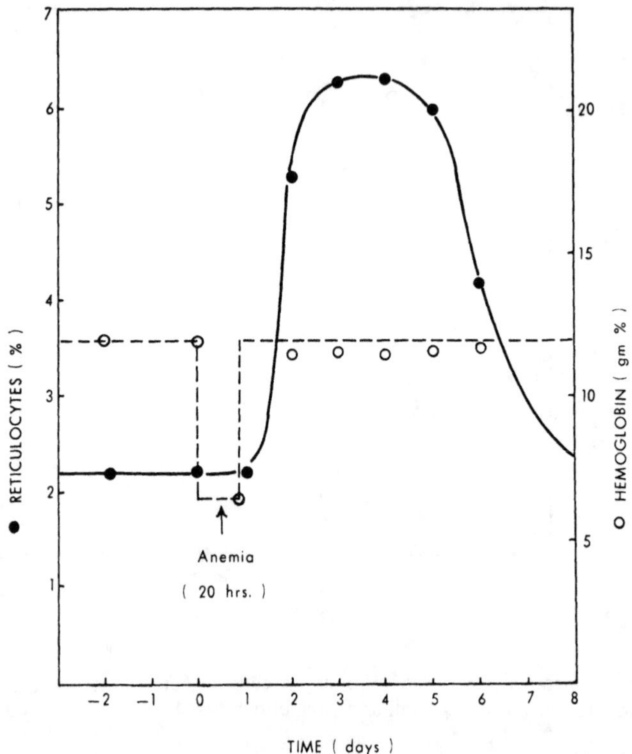

FIG. 11. Effect of temporary anemia on erythropoiesis in the rabbit. When 20 ml of blood per kg is withdrawn and then replaced after 20 hr, reticulocyte percent (●) rises on second day even though hemoglobin levels (○) are normal again. Once it is stimulated, therefore, red cell production goes to completion regardless of circumstances. (Adapted from Erslev, 1959.)

normoblasts. The subsequent maturation of these cells, including several rounds of mitosis, then goes on at the normal pace (Erslev, 1961). Hence, according to this view, more red cells are produced by channeling extra stem cells in the erythroid direction.

Filmanowicz and Gurney (1961) and Bruce and McCulloch (1964) have confirmed the delay between erythropoietic stimulation and the appearance of the end products in the circulation. The maximum number of proerythroblasts appears in mouse spleens 24 hours after treatment with erythropoietin. The rate of hemoglobin synthesis is accelerated during the second and third days, and the normoblast population reaches a peak at 72 hours. Not until after 3½ to 4 days do reticulocytes finally make their appearance in the peripheral blood. If erythropoietin were to stimulate red cell production by accelerating the rate of erythroblast maturation, or by promoting their multiplication, then a more prompt peripheral response might have been expected. Since this does not occur, there can be little doubt that erythropoietin works very early in the process of erythropoiesis (Filmanowicz and Gurney, 1959).

More precisely, the consensus is that hemopoietic stem cells are the targets of the hormone, erythropoietin (Jacobson et al., 1957b; Alpen and Cranmore, 1959; Erslev, 1959; Stohlman, 1960, 1961). As early as 12 hours after hypoxic stimulation, Orlic et al. (1966) have detected proerythroblasts in the mouse spleen by electron microscopic examination. Using biochemical techniques, Goldwasser (1966) and his colleagues have concentrated on the role of erythropoietin in inducing hemoglobin synthesis in rat marrow cells *in vitro*. These important investigations have revealed that the hemoglobin synthesis thus induced is inhibited by actinomycin D if the latter is added to the medium soon after the erythropoietin (Gallien-Lartigue and Goldwasser, 1965). However, once it has started to work, the action of erythropoietin, as measured by the uptake of ^{14}C-glucosamine into hemoglobin, continues for 3 to 4 hours after the removal of the hormone or the addition of actinomycin D (Dukes and Goldwasser, 1965). These observations, coupled with those of Krantz and Goldwasser (1965), who showed that the rate of RNA synthesis begins to rise as early as 15 minutes after the addition of erythropoietin to cultures of rat marrow, suggest that the hormone may induce the production of m-RNA specific for hemoglobin synthesis. Like so many other hormones, it appears that erythropoietin may work by depressing the genes for the synthesis of those proteins by which a differentiating cell expresses its individuality.

This, however, does not explain how the rate of proliferation may be affected by erythropoietin. Lord (1967) has presented evidence that in bled rats the population of maturing erythroid cells in the spleen and marrow is enhanced not only by stimulating more stem cells to differentiate in that pathway, but also by increasing the frequency with which the erythroblasts divide. Be this as it may, it is equally important that the pool of stem cells not be depleted by excessive recruitment for erythroid differentiation. Stohlman (1960, 1961) proposed that erythropoietin might act exclusively on the mechanism of cellular differentiation, while the ensuing proliferation of stem cells might be stimulated directly by the resultant depopulation. Such an interpretation raises the additional problem of how the stem cell compartment detects and reacts to deficiencies in its own population. Hanna (1967) has suggested that the erythron size might be regulated by a direct feedback to the proliferation of stem cells, which begin to multiply in the rat within three hours after blood loss.

Hence, the question is whether the cell which is induced to differentiate by erythropoietin is at the same time stimulated to divide. Till et al. (1967) think not, at least if one equates their spleen colony-forming cells with hemopoietic stem cells. When W/W^v mice are injected with suspensions of normal marrow cells, the latter form erythropoietic colonies in the host spleens. But when the anemic hosts are made polycythemic by hypertransfusion, erythropoietic spleen colonies do not develop from the injected marrow. The colony-forming cells, instead of producing erythropoietic descendants, now proliferate into granulocytes. In the absence of erythropoietin, it seems that the division of stem cells is not prevented but their erythroid differentiation is.

What emerges from these many observations is the notion that primitive stem cells may be up for hire to whichever kind of hemopoietin (e.g., erythro-, leuko-, thrombo-) competes most successfully for their attentions. Once such a cell is engaged by a specific kind of "poietin," it is automatically taken off the market. Thenceforth it differentiates exclusively in the direction dictated by the specific combination of genes that are derepressed by the poietin in charge. It is quite possible that gene activation triggers DNA replication. If so, then erythropoietin, for example, would stimulate proliferation as a consequence of inducing erythroid differentiation. Not until the affected cell and its progeny have matured to the normoblast stage does the differentiated state finally become incompatible with mitosis.

To the extent that poietins of various kinds drain stem cells from the hemocytoblast pool, the latter must make up for the loss. It is conceivable that some cells which have already set out on the pathway of differentiation may occasionally revert to their original primitive conditions. If only one descendant of each induced cell returned to the progenitor pool, then the stem cell population could be maintained. Alternatively, the stem cells might maintain themselves by dividing directly into more of the same. Yet this possibility has the disadvantage of requiring still another stimulatory factor (which is perhaps blastopoietin) to regulate the rate of hemocytoblast proliferation. The most logical mechanism, it seems, would be a compromise between these two models. That is, if each stimulated stem cell should give rise to one daughter cell which would go on to differentiate and another which would remain as a pluripotent progenitor, then the generative compartment could never be depleted because every cell induced to differentiate would first have to provide for its own replacement in the stem cell pool (Fig. 12). Whether or not the logic of this argument is correct we do not know. The verdict must wait until we learn how nature has solved the problem.

Leukopoiesis

It is reasonable to expect that the populations of white cells in the blood might be regulated, as in the case of the red cells, by humoral agents. If so, it

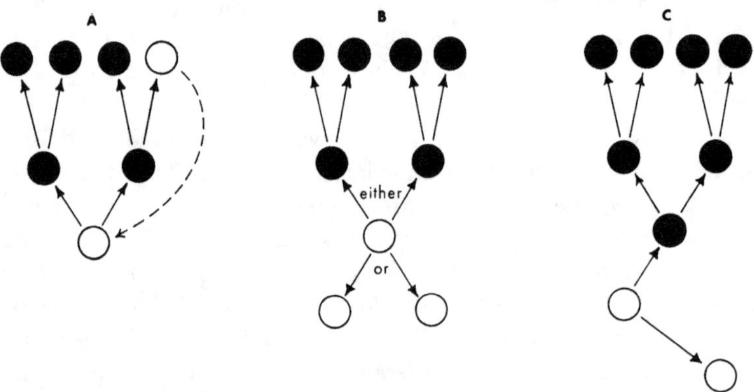

FIG. 12. Alternative hypothetical relationships between hemocytoblast differentiation (●) and stem cell (○) renewal. (A) Stem cells replenished by dedifferentiation. (B) Daughter cells of hemocytoblast both differentiate or both remain undifferentiated. (C) One daughter cell differentiates, the other does not. Combinations of these pathways are also possible.

might be predicted that there should be as many as five different leukopoietins, one for each kind of white cell. In view of the technical difficulties inherent in studying a single kind of leukocyte to the exclusion of all others, the distinction between their mechanisms of growth regulation is not always as clear-cut as one might prefer. Nevertheless, considerable progress has been made along several lines of investigation.

The most direct approach to the problem of leukopoiesis is to remove white cells from the peripheral circulation (leukopheresis) and study the mechanism of their replacement. In the human, the rate of leukopoiesis is increased as soon as the second day (Bierman, 1961), and in the dog by 3 to 4 days (Craddock et al., 1956) after leukopheresis. It is possible to deplete the number of circulating neutrophils by intraperitoneal injections of saline solutions (Gordon et al., 1960) or a 24 percent glucose solution in 0.01N HCl (Estes et al., 1958). These treatments attract quantities of neutrophils into the peritoneal cavity within several hours. By repeatedly harvesting them it is possible to make the animal severely neutropenic. Similar results can be achieved by treating animals with antineutrophilic serums (Patt et al., 1957; Lawrence et al., 1967). Eosinophils can be removed from the peritoneal cavities of rats injected with asbestos (Cohen and Gordon, 1961) and leukopenia can be induced in rabbits by injecting a two percent starch solution intravenously (Gidáli and Fehér, 1964; Fehér and Gidáli, 1965).

In general, the initial reaction to the depletion of leukocytes is their replacement within a few hours by mature cells normally held in reserve. Only by repeated withdrawals can true leukopenia be achieved because the reservoir of differentiated but noncirculating leukocytes is a very extensive one. Craddock et al. (1960) estimated the number of granulocytes sequestered in human marrow to be 20 to 25 times greater than the pool of circulating cells. Likewise in humans, Bierman et al. (1961) concluded that the population of leukocytes in reserve was 59 times that in the circulation. In rats, Cohen and Gordon (1961) were able to remove 134×10^6 eosinophils over a five-day period, or about 45 times the complement of eosinophils normally circulating in the blood at once. Hence, only a fraction of the total number of leukocytes actually circulates at any one time. Hollingsworth et al. (1957) replaced the normal blood of rats with that from which leukocytes had been removed, but found that normal white cell counts were restored in only 30 minutes. This prompt mobilization of mature leukocytes was only partially impeded by splenectomy, suggesting that reserve cells are present in hemopoietic tissues other than the spleen. In germfree mice, there are fewer circulating neutrophils than in normal animals, but just as many in the marrow. When such animals are removed from their sterile environments, however, the mature neutrophils in reserve are promptly released into the blood stream and new ones begin to be produced in the marrow (Boggs et al., 1967). This rapid response may be triggered by a factor which expels leukocytes from the marrow and lymphoid organs (Patt et al., 1957; Steinberg, 1958; Fukuda et al., 1960). The resulting emigration of leukocytes may then be responsible for the compensatory leukopoiesis that ensues.

Despite these obstacles, experimental attempts to detect leukopoietic stimulating factors in the plasma have not been altogether unsuccessful. One way to elicit the production of such factors is by leukopheresis. In the rat, for example, plasma from animals previously depleted of their leukocytes will stimulate the production of white blood cells when injected into normal assay animals (Gordon et al., 1960). The capacity to produce and respond to this leukocytosis-inducing factor is unaffected by hypophysectomy or adrenalectomy (Handler and Gordon, 1962), although glucocorticoids administered to normal animals tend to depress the production of lymphocytes and eosinophils while stimulating neutrophil production (Quittner et al., 1951; Gordon, 1955; Shen and Hoshino, 1961). Other methods of inducing leukopenia, such as the intravenous injection of starch solution (Gidáli and Fehér, 1964; Fehér and Gidáli, 1965) or x-irradiation (Sodicoff and Binhammer, 1968), also give rise to plasma factors which can stimulate leukopoiesis in the marrow.

In addition to such endogenously produced leukopoietins, there are numerous exogenous agents capable of eliciting leukocytosis. Rats injected with plasma from humans (Bierman et al., 1962) or cattle (Bierman, 1964) undergo granulocytosis. Typhoid-paratyphoid vaccine likewise induces leukocytosis in rabbits (Fukuda et al., 1960) and rats (Gordon et al., 1964). Extracts of bovine bone marrow stimulate neutrophil production when injected into rats intraperitoneally (Wasastjerna et al., 1958), as do homogenates of various bovine organs (Bierman, 1964). Even extracts of the thymus and of lymphoblasts cultured *in vitro,* promote the production of lymphocytes (Duplan et al., 1962). Also, the possibility that the humoral factor by which the neonatal thymus induces differentiation in lymphoid organs (Osoba and Miller, 1964) may be a lymphopoietin is more than idle speculation. In view of the variety of substances capable of initiating leukopoiesis, considerable caution must be exercised in interpreting their effects, for they may have little or nothing to do with the *normal* mechanisms by which the body adjusts the output of white cells to the transient population of leukocytes in the peripheral circulation.

Nevertheless, it is pertinent to any serious discussion of leukocyte renewal to explore the ways in which lymphocytes in particular react to various exogenous agents. Basically, these cells respond to antigenic stimulation by proliferating and giving rise to specific antibodies. This they do even in tissue culture, provided they are derived from donors previously sensitized to the antigen (Pearmain et al., 1963; Elves et al., 1963; Gräsbeck et al., 1963; Lycette and Pearmain, 1963). Sometimes prior exposure is not even necessary to elicit this reaction. Bain and Lowenstein (1964) discovered that if lymphocytes from two different individuals are cultured together, they transform into lymphoblasts, synthesize DNA, and proliferate. Dutton and Mishell (1966) reported similar reactions when splenic lymphoid cells from different strains of mice are mixed in culture, provided the two donors differ in their histocompatibility antigens. It was further noted that the *in vitro* response is intensified if the mice have previously swapped skin grafts and have thereby become sensitized against each other. Clearly, the blastogenesis and proliferation of lymphocytes is an immunological response dependent upon exposure to foreign cells in culture. More precisely, it is a reaction to the culture medium itself, for Kasakura and Lowenstein (1965) showed that cell-free medium in which homologous leukocytes had been cultured would also elicit blastogenesis. Apparently cells produce something against which genetically different lymphocytes will react. Since neither devitalized cells nor medium from cultures prevented from synthesizing proteins provoke this response (Gordon and MacLean, 1965), it can be concluded that the cells of one individual synthesize and secrete a proteinaceous antigen which activates the lymphocytes from another individual. One wonders if the antibodies produced by each kind of lymphocyte (see below) might not act as reciprocal antigenic stimuli, especially in view of the mitogenic effects on lymphocytes *in vitro* of antilymphocyte serum (Gräsbeck et al., 1963; Ling et al., 1967).

Indeed, Elkins and Guttmann (1968) have recently reported evidence that the rejection of organ allografts may be elicited not so much by the antigenicity of the transplanted parenchymal cells themselves as by the lymphoid cells which inadvertently accompany them. When exposed to host leukocytes, these passenger lymphocytes are apparently stimulated to undergo the kind of blastogenic transformation and proliferation otherwise observed in tissue cultures. It may be the resultant interaction between donor and host leukocytes within the organ graft that brings about the nonspecific destruction of the transplant. Hence, the donor may be more to blame for the rejection of a graft than the host, simply because the lymphoid cells included in the transplant react immunogenically against the recipient's leukocytes.

Thus it is that in lymphocytes, as in so many other kinds of cells, the stimulus to function is also a mandate to grow and divide. This is one reason why the reactions of lymphocytes to certain agglutinins of plant origin are so important. In 1960, Nowell reported that phytohemagglutinin (PHA) derived from kidney beans (*Phaseolus vulgaris*) would stimulate mitosis among human leukocytes *in vitro*. Similar properties were subsequently discovered in the pokeweed mitogen (PWM) extracted from *Phytolacca americana* (Farnes et al., 1964; Barker et al., 1964; Chessin et al., 1966; Barker and Farnes, 1967a) and even in extracts of the seeds of *Wistaria floribunda* (Barker and Farnes, 1967b). These three substances are chemically different but cause similar (though not identical) effects in cultures of human lymphocytes.

In general, they agglutinate cells and stimulate lymphocytes to enlarge, produce antibodies, and multiply. Presumably the same agent that agglutinates leukocytes (but not red cells) also exerts the mitogenic effects (Börjeson et al., 1966). When cultured in the presence of these phytomitogens, human lymphocytes transform into lymphoblasts (Fig. 13) as they synthesize antibodies. This is accompanied by the rapid uptake of ^{3}H-uridine into RNA (Cooper and Rubin, 1965; Hayhoe and Quaglino, 1965) within the first few hours (Darzynkiewicz et al., 1965), a reaction which is evident first in the nuclei, and after 12 hours or so, in the cytoplasm (Winter and Yoffey, 1965). In the presence of PHA, lymphocytes begin to synthesize protein in excess of control cultures after only two hours (Bach and Hirschhorn, 1963; Hirschhorn et al., 1963). Hence, the production of antibodies and RNA both go on concomitantly from the very beginning. Later on, the activated cells enlarge and develop granules, often identified as lysosomes (Hirschhorn et al., 1964; Hirschhorn and Hirschhorn, 1965; Hirschhorn et al., 1965; Diengdoh and Turk, 1965; Hirschhorn et al., 1967). The appearance of lysosomes precedes mitosis, and may be causally related to the onset of the division mechanism.

Whatever the initial stimulus, the cellular response is one of functional activity, an activity which, in the case of lymphocytes, involves the synthesis of antibodies. Secondarily, it leads to cell division. When the immunological reaction occurs in

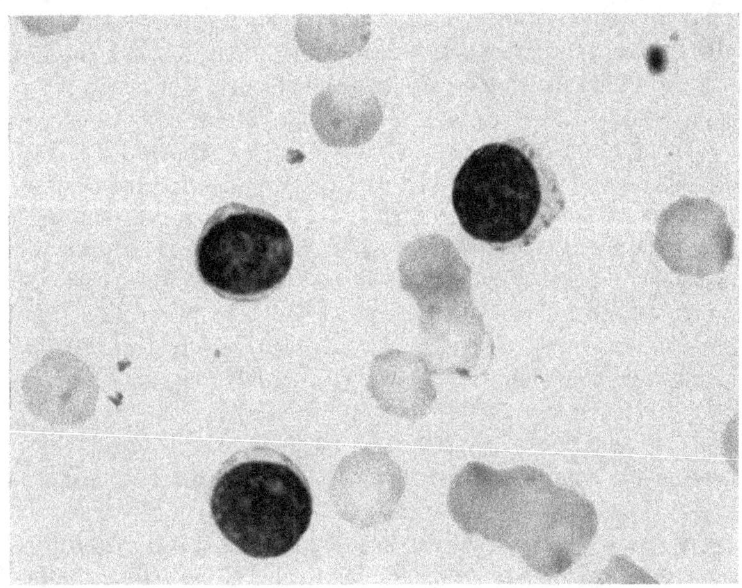

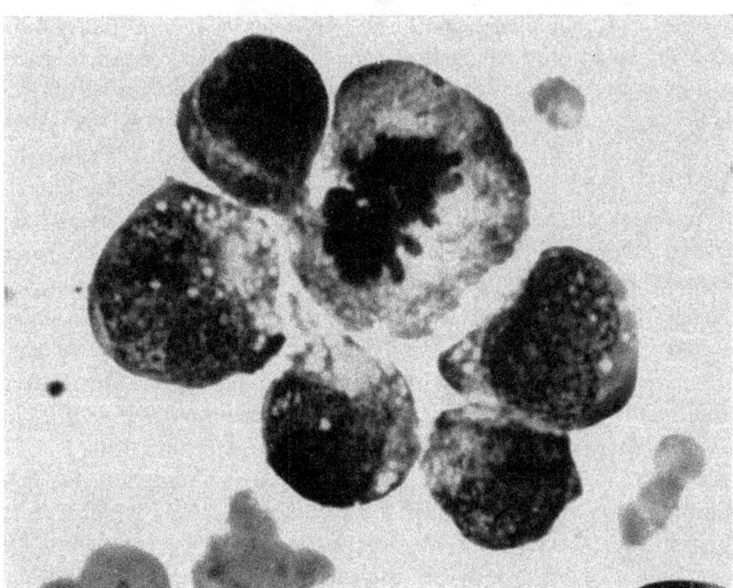

FIG. 13. Above, human peripheral blood lymphocytes cultured for 72 hr in NCTC 109 and 20 percent autologous plasma. The morphology of these cells closely resembles that of fresh human peripheral blood lymphocytes, with the exception of the numerous azurophilic granules seen in the lymphocyte at the upper right. This increase in granulation reflects an increase in lysosomes known to occur *in vitro*. Below, large primitive, basophilic, vacuolated cells with many mitoses from a 72-hr lymphocyte culture from the same source but cultured in the presence of pokeweed mitogen. Wright-Giemsa stain, × 432. (Courtesy of Dr. Barbara E. Barker.)

response to antigens to which the lymphocytes had previously been sensitized, they then give rise to antigen-specific antibodies. But when cells are stimulated by phytomitogens, to which they have never before been exposed, what kind of antibody must they produce? We do not know if these antibodies are specific for PHA, for example, or even why PHA stimulates almost all cells in a culture instead of just a fraction of them as other antigens commonly do. It has been suggested that PHA might be a universal antigen, but available evidence argues against this interpretation (Holland and Holland, 1965; Caron and Sarkany, 1966). This does not mean, however, that PHA is not antigenic. It is a heat-stable (Holland and Holland, 1965) mucoprotein with a molecular weight of 128,000 (Rigas and Johnson, 1964). An antiserum against it has been found to inhibit the lymphocytic proliferation which normally occurs in response to PHA in culture (Byrd et al., 1967). Along these lines, Mellman and Rawnsley (1966) showed that antiPHA serum added to lymphocytes during the first 6 hours of culture in the presence of PHA totally abolishes the mitotic response. The inhibition declines thereafter up to 36 hours, by which time DNA synthesis is going on. Exposure to antiserum after 36 hours permits mitosis to occur as usual between 48 and 72 hours.

In view of the profusion of substances by which lymphocyte blastogenesis can be stimulated, it is hardly unexpected that inhibitors should exist in the plasma to prevent superfluous stimulation *in vivo*. Such an inhibitor has been discovered in the alpha globulin fraction of normal plasma which not only suppresses immunological reactions (Karmin, 1959; Mowbray, 1963) but also prevents PHA-induced blastogenesis and replication *in vitro* (Cooperband et al., 1968). As in the case of erythropoiesis, then, the reproduction of lymphocytes seems to be subject to the antagonistic influences of inhibitors as well as stimulators.

We are still a long way from knowing how antigens and phytomitogens work, but the important thing is that these agents stimulate mitosis at the same time they induce specific physiological activity in the lymphocytes to which they are exposed. As additional facts are uncovered in the future we are destined to learn much about the connection between cell function and cell division. When we do find out how these two activities are related, it may well be in the lymphocyte that the discovery will be made.

Thrombopoiesis

No consideration of blood cell turnover would be complete without a discussion of the renewal of platelets, if only because the questions about this fascinating and important aspect of hemopoiesis far outnumber the answers. As in the case of blood cells, the turnover of platelets is a function of their longevity as well as of their rates of production and destruction. Also, since they are produced by fragmentation from megakaryocyte cytoplasm, the control of megakaryocyte turnover is also relevant to the subject of thrombopoiesis.

The lifespan of platelets varies with the species (and with the method of determination). The usual procedure is to tag platelets with a radioactive label, for which they have an affinity either *in vivo* or supravitally, and then return them to the circulation to find out how long they survive. Their affinity for diisopropylfluorophosphate (DFP^{32}) or $Na_2Cr^{51}O_4$ makes labeling and subsequent detection technically feasible. Using these and other methods, it has been learned that platelets survive in the rat for 4 to 5 days (Odell et al., 1955; Cronkite et al., 1957; Hjort and Paputchis, 1960) and in the human about twice as long (Hirsch and Gardner, 1951; Leeksma and Cohen, 1956; Aas and Gardner, 1958; Gardner et al., 1958). The concentration of platelets in the peripheral circulation averages around several hundred thousand per cubic millimeter, but tends to be higher in smaller animals than in larger ones.

The most direct approach to the problem of platelet population control is to reduce their numbers in the blood and to study their subsequent replacement. Platelets can be depleted by repeatedly removing blood, allowing the plasma to clot, and reinjecting the reconstituted defibrinated blood. When Duke did this in dogs in 1911, he found that the platelets were regenerated in about five days. This is in agreement with the results of Craddock et al. (1955) who reduced the platelet counts in dogs to about 10 percent of the normal number and noted that the thrombocytopenia lasted only 3 to 4 days. In the rat, Matter et al. (1960) found that the rate of platelet production increased 2 days after peripheral depletion and thus restored platelet levels from 10 percent to normal after 7 days. Tocantins (1936) produced thrombopenias down to 10 percent of normal levels by treating dogs with antiplatelet serum. In 8 to 9 days, however, such animals had regenerated completely.

The opposite approach is to elevate the number of circulating platelets to learn if platelet production will decrease. When the platelet levels in rats were doubled or tripled by hypertransfusion, Odell et al. (1965) reported that the excess had disappeared in less than a week. The megakaryocyte concentration, however, did not change, a finding confirmed by Ebbe's (1966) observation that under similar experimental conditions the rate of thrombopoiesis was not decreased, nor was the rate of megakaryocyte maturation altered (Ebbe et al., 1966). Rolovic et al. (1966) increased the platelet counts in rats by splenectomy, but could detect no reduction in the incorporation of tritated thymidine by the megakaryocytes of such animals. This apparent lack of a megakaryocyte depletion in response to excessively high platelet counts in the blood is very interesting because it is not what might have been expected from studies on experimental animals made thrombopenic by exchange transfusion with defibrinated blood (Craddock et al., 1955; Matter et al., 1960). Under the latter conditions the number of megakaryocytes in the marrow increased. In 1922, Firket and Campos destroyed the platelets in rabbits by injecting the glycoside, saponin. This treatment also stimulated megakaryocyte production during the course of platelet regeneration. Ebbe et al. (1968a, b) have recently confirmed that thrombocytopenia is followed by elevated platelet production rates, accompanied by an acceleration of megakaryocyte maturation. Clearly, the rate of platelet production is in part controlled by alterations in the rates of megakaryocyte turnover in the marrow, but the latter are evidently more sensitive to depletions than to excesses of circulating platelets. It is too early to predict that thrombopoietin (as yet hypothetical) really exists, and if so, whether it operates at the level of megakaryocyte production or directly on platelet formation. Considerable insight into such problems, however, has grown out of pathological studies, notably on an interesting disease called idiopathic thrombocytopenic purpura (ITP).

ITP is a disease in which the paucity of platelets is responsible for serious disturbances in the clotting mechanism. The abnormally low platelet count is, in most cases, due to shortened survival times of a day or two, compared with more than a week in normal individuals (Baldini, 1966a). Apparently the platelets themselves are all right in ITP, but the plasma is not. As shown by the pioneering investigations of Harrington et al. (1951, 1953), when normal subjects are injected with plasma from patients with ITP there may be an abrupt fall in their platelet counts and an interference with their clotting mechanisms. Recovery occurs in about 8 days. Thus, some forms of ITP can be temporarily mimicked by transfusing a plasma factor, subsequently identified as an antibody (Harrington et al., 1956; Shulman et al., 1964). Not unexpectedly, if normal platelets are transfused into ITP patients, they are also destroyed (Stefanini and Chatterjea, 1951; Gardner et al., 1958). Most evidence points to the autoimmune nature of ITP, a condition in which the individual makes antibodies against his own platelets, just as a healthy one will become immunized against isologous platelets and thereby destroy them with progressively less delay after each successive round of transfusion, as shown in Figure 14 (Baldini, 1966b).

The spleen plays an especially important role in platelet economy. In normal rats, for example, splenectomy causes the circulating platelet counts to increase 10 to 20 percent (Hjort and Paputchis, 1960; Matter et al., 1960; Rolovic et al., 1966). In patients with ITP, splenectomy also increases the number of platelets (Tocantins, 1938; Harrington et al., 1953). Such remissions of the disease, though not permanent, may be attributed partly to the release of more platelets into the blood stream (Cohen et al., 1961), but mostly to their increased lifespans (Gardner et al., 1958; Baldini, 1966b). The implication is that the spleen participates in the destruction of platelets by sequestering the ones sensitized to antibodies. Splenectomy in the case of ITP is almost like prolonging life by removing the graveyard; almost, because the real cause of death is the antiplatelet gamma globulin which still persists after splenectomy.

The opposite approach can be studied by administering spleen extracts. Early studies by Troland and Lee (1938), Otenasek and Lee (1941), and Paul (1942) suggested that normal spleen extracts exert no effects on the platelet levels of the rabbits into which they are injected. Cronkite (1944) and Plunket et al. (1965), however, reported drops in platelet counts in animals treated with splenic preparations. In view of such inconsistencies, the effects of normal spleen extracts on platelets remain problematical. There is nothing equivocal, however, about the effects of spleen extracts from cases of ITP. When these are injected into assay animals, the number of circulating platelets promptly drops, but rises to normal within a day (Troland and Lee, 1938; Otenasek and Lee, 1941; Paul, 1942; Uihlein, 1942).

Another way to augment the spleen effect is to inject methyl cellulose intraperitoneally into an animal. Following this sort of treatment for 15 weeks in the rat, Palmer et al. (1953) reported a threefold increase in the relative weight of the spleen, together with miscellaneous hematologic disorders, including slightly lowered platelet counts. Hjort and Paputchis (1960) found that rats made hyper-

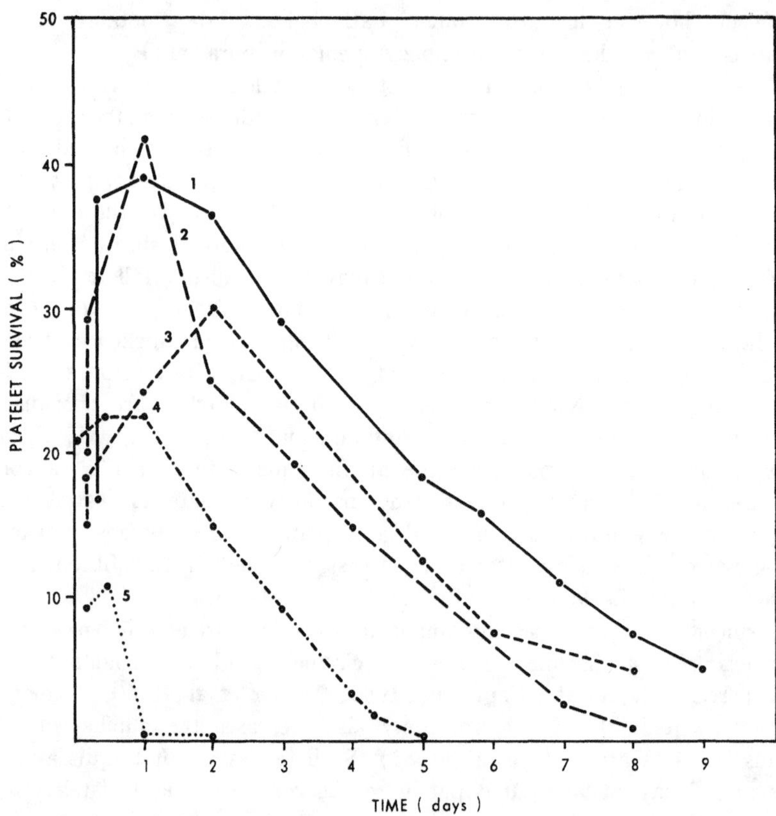

FIG. 14. Patterns of human platelet survival following successive transfusions of isologous platelets. First transfusion (1) represents normal survival curve before sensitization. Subsequently transfused platelets (2 to 5) undergo increasingly rapid destruction as recipient becomes immunized. (Adapted from Baldini, 1966b.)

splenic by methyl cellulose had platelet levels 35 percent below normal, and Matter et al. (1960) reported a 64 percent reduction in circulating platelets in similarly treated rats. In general, since splenectomy elevates platelet counts while enlarged spleens lower them, it seems safe to conclude that the spleen is engaged in disposing of platelets in both health and disease, but more so in disease.

Epidermal Turnover

From the functional point of view there are two categories of renewing tissues. One kind does business with the body as a whole; the other is involved locally. Blood cells are of the former type. Distributed throughout the body, their rates of turnover are systemically controlled. What affects one hemopoietic stem cell affects them all. Epidermis and other tissues are not so physiologically meaningful at the organismic level. Their spheres of influence are physiologically circumscribed. Responsible only unto themselves and their near neighbors, these cells have growth rates controlled by strictly local factors. What affects the epidermis in one region does not necessarily affect the same tissues elsewhere. This is not to say that agents have not been discovered that can exert profound effects on epidermal growth when administered systemically. Important as such agents may be, however, there is reason to believe that they are secondary to the operation of local influences, influences which are responsible for that differential growth by which epidermal and other renewing tissues are distinguished from hemopoietic cells. Our task is to decipher the messages to which the epidermal cells are reacting when their rates of proliferation go up or down.

The epidermis is a divided population of cells. The germinative portion is specialized for proliferation and feeds cells into the differentiated compartment. Typically, these two subpopulations are spatially segregated; the growth zone is located basally and cell maturation progresses toward the upper surface. Such a division of labor, however, is not universal. In the epidermis of the goldfish, for example, Henrikson (1967) has shown that there is an "indiscriminate distribution" of tritiated thymidine labeled cells in the epidermis, some of them preparing for division even in the outermost layer. A comparable situation obtains in the chick embryo. Here, before 13 days of incubation, Wessells (1963, 1967) has demonstrated DNA synthesis at all levels in the epidermis. On the thirteenth day, however, the pattern shifts to that of the adult, a change which coincides with the onset of specific protein synthesis in the strata overlying the basal layer of dividing cells.

As in the case of hemopoiesis, the problem of how differentiated descendants are produced without depleting the stem cell population is an important aspect of epidermal turnover. The method used to resolve this problem involved labeling with tritiated thymidine long enough before sacrifice to allow the tagged cells to complete their next scheduled divisions (Marques Pereira and Leblond, 1963; Leblond, 1965). This technique makes it possible to identify isolated pairs of labeled daughter cells in the epidermis with a view to analyzing their postmitotic fates. It has been found that in some cases both daughter cells remain in the germinative stratum. In other cases they both migrate into the layers above. Sometimes one moves up and the other stays behind. Evidently cells are squeezed out of the basal layer on a hit-or-miss basis.

The apparent incompatibility between cell division and differentiation in the epidermis has dominated many attempts to explain how the skin grows. But whether cells cease to divide because they begin to differentiate, or vice versa, is an elusive question. Like the riddle of whether the chicken or the egg came first, perhaps the answer is neither, in the sense that cause and effect cannot logically be applied to simultaneous events. If it is true, however, that mitosis and differentiation are mutually exclusive events in epidermal cells, then the corollary must also be true, namely, that mitotic competence should persist in the absence of differentiation. This is apparently what happens in the case of certain epidermal cancers. Rothberg and Van Scott (1964) showed that some of the proteins normally present in the epidermis are absent, or markedly reduced, in basal cell tumors; it is probably no coincidence that such nondifferentiating cells multiply out of control.

Psoriasis is still another case in point. Characterized by an excessive thickening of the epidermis, this condition is caused not so much by an increase in the percentage of dividing cells as by an enlargement of the germinative zone. Van Scott and Ekel (1963) and Van Scott (1965) calculated that the dermal-epidermal interface increased threefold in area due to papillary expansion, and that the germinative zone was extended to three cell layers from the usual single layer of basal cells. These two factors combine to yield a ninefold rise in the absolute number of dividing cells per unit area of epidermis. This is in keeping with the observations of Rothberg et al. (1961) on the rate of turnover of human epidermal cells. According to their studies of glycine-^{14}C labeled epidermis, normal cells spend about four weeks in transit from the basal layer to the surface of the stratum corneum. In psoriasis, however, the turnover time is accelerated about nine times to only three or four days, not necessarily because the rate of proliferation is elevated, but because of the expanded population of cells in the germinative compartment.

Rates of epidermal renewal therefore vary considerably from normal to pathological conditions. They also vary in different regions of the body, especially in epidermal appendages where rates of linear growth are most conveniently measured. Human fingernails, for example, grow about 0.12 mm per day and thus renew themselves completely about every four months. Hairs grow even faster and farther. In the chinchilla, Lyne (1965) found that the lateral hairs grow 0.77 mm per day, while hairs in the rat may elongate at the rate of 1.2 mm per day (Johnson, 1965). This is exceeded by the remarkable velocity with which feathers regenerate in birds, which is in turn surpassed by the growth of horns in some ungulates. Dorset rams, for example, begin to grow their horns within days after they are born, and for the first year or so they add on about 5 cm per month, or 1.66 mm per day. Not even the bighorn sheep, which can lay down as much as 28 cm of horn in a seven-month growing season, surpasses this.

Notwithstanding these impressive kinetics of keratinization, the most pressing problem relates to the *control* of cell renewal in epidermal tissues. Are cells lost at the surface because they are constantly produced below, or are they replenished in the basal layer to compensate for the constant desquamation from the stratum corneum? The answer seems to be "yes" on both counts. It is in the nature of postmitotic epidermal cells to die in the process of differentiation, an attribute which makes it imperative that their progenitor cells should multiply constantly, if not automatically, lest the epidermal species of cell become extinct. Yet we know from the classic experiments of Pinkus (1951, 1952) that the rate of basal proliferation can be accelerated by stripping off outer layers of epidermal cells prematurely. This may be taken to indicate that basal growth is at least indirectly sensitive to epidermal thickness. But how the message is communicated we do not know.

Fortunately, however, there are some interesting theories along these lines, not the least of which is that proposed by Bullough (1965) and Bullough and Laurence (1966, 1967). These investigators have taken the view that, other things being equal, basal epidermal cells tend to divide spontaneously. Consequently such cells supposedly do not depend upon mitogenic stimulation. Rather, it has been proposed that their mitotic control is under the influence of inhibitory factors produced by epidermal cells themselves. These factors have been called "chalones" because, unlike hormones, they inhibit growth rather than stimulate it. Evidence for the existence of such chalones rests upon the capacity of epidermal extracts to arrest mitotic activity in epidermis cultured *in vitro*. The resulting depression of cell division occurs within several hours after explantation of pieces of mouse ear skin to a phosphate-buffered saline medium containing an aqueous extract of macerated epidermis (Bullough and Laurence, 1964). Although controls undergo a moderate decrease in proliferative activity, those cultured with chalone are even more depressed (Fig. 15). Furthermore, it has been discovered that the antimitotic action of epidermal extracts was potentiated by the presence of adrenalin, an observation in keeping with the observed *in vivo* inhibition of mitosis by this hormone. The conclusion is that chalone is a tissue-specific antimitogenic agent, while adrenalin acts as a nonspecific cofactor.

It would be premature to speculate on whether the epidermal chalone serves only to inhibit mitosis, or perhaps does so incidental to other physiological activities of primary importance. Since the growth of so many other tissues and organs is linked to functional demands, it would be a curious thing if epidermal growth were related to some genetically predetermined thickness. Yet within certain limits this is indeed the case, for the variations that exist from one region of the body to another tend to anticipate the jobs the epidermis is normally called upon to

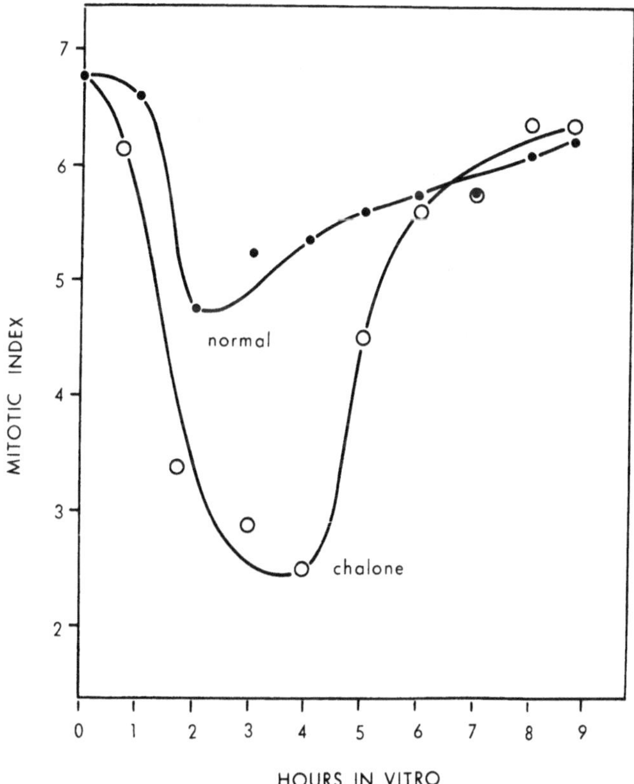

FIG. 15. Depression of mitotic activity in mouse ear epidermis cultured with (○) or without (●) chalone prepared as a water-soluble extract of epidermis. (Adapted from Bullough and Laurence, 1964.)

perform. At the same time, however, the rates of proliferation can rise or fall according to physiological conditions which require thicker or thinner layers of stratum corneum. Clearly, whatever factors mediate epidermal growth must logically operate in response to functional demands, but where nature leaves off and nurture begins is not an easy question to answer.

In view of the important role of the epidermis in preventing desiccation, it would seem reasonable that proliferation and differentiation might be regulated in response to the permeability of the epidermis to water (Aschheim, 1968). An interesting approach to this aspect of the problem is afforded by the reaction of the skin to dietary deficiencies in essential fatty acids. Under these conditions, the skin of mice and rats have been shown to become increasingly permeable to water (Ramalingaswami and Sinclair, 1953; Basnayake and Sinclair, 1954), a condition accompanied by its excessive growth. More recently, Menton (1968) has explored this interesting problem in mice rendered deficient in essential fatty acids. He has shown that the epidermis of such animals becomes markedly thickened after several weeks, a reaction brought on by heightened rates of cell division in the basal layer (Fig. 16). On the basis of the theory that these responses might be attributable to excessive transepidermal water loss, Menton maintained deficient mice under conditions of high water vapor pressure. The results were as expected, for the rates of epidermal proliferation in such animals were lower than in deficient controls kept under normal conditions of ambient humidity. These experiments yield compelling evidence, therefore, in favor of the role of epidermal permeability as a regulator of basal proliferation and superficial cornification. One might suppose that the more hydrated the epidermis becomes, the more concentrated would be the water-soluble chalones which turn off cell division.

This cannot be the whole story, however, because the epidermis, and especially its appendages, often grow intermittently without apparent reference to permeability phenomena. Moreover, various systemic factors, notably hormones, modulate the rates of epidermal growth also. Of great interest, but uncertain significance, is the remarkable capacity of salivary gland extracts to stimulate epidermal growth. In 1962, Cohen discovered that when a protein derived from mouse submaxillary glands was injected into newborn mice, their rates of epidermal maturation were accelerated. Not only did the epidermis become thicker, but due to its precocious keratinization the eyelids opened at an earlier age and the incisors erupted prematurely. Further investigations (Cohen, 1965) demonstrated that the epidermal growth factor also works *in vitro*; cultured explants of chick embryo skin become thickened in its presence due to enhanced rates of proliferation and keratinization. We do not know how this interesting effect is caused, much less its meaning, if any, in the normal development of the organism. Yet sometimes it is just such an empirical curiosity that provides the clue necessary to solve a puzzle.

Spermatogenesis

The testis has much in common with the epidermis in that its seminiferous tubules are lined with a stratified epithelium undergoing renewal by virtue of contributions from a basal layer of germinative cells. In both tissues, the differentiating

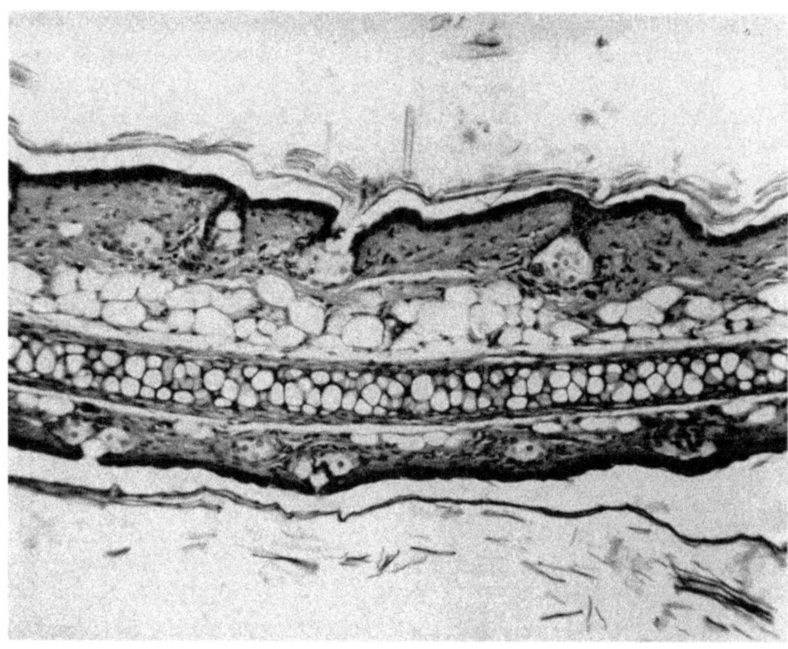

FIG. 16. Effects of essential fatty acid deficiency on mouse ear epidermis. Above, cross section of control ear from mouse on diet supplemented with linoleic acid for 30 days. Below, experimental results in littermate on a diet deficient in essential fatty acids for 30 days. Note thickened ear epidermis and numerous mitoses in basal layer. (Courtesy of Dr. David N. Menton.)

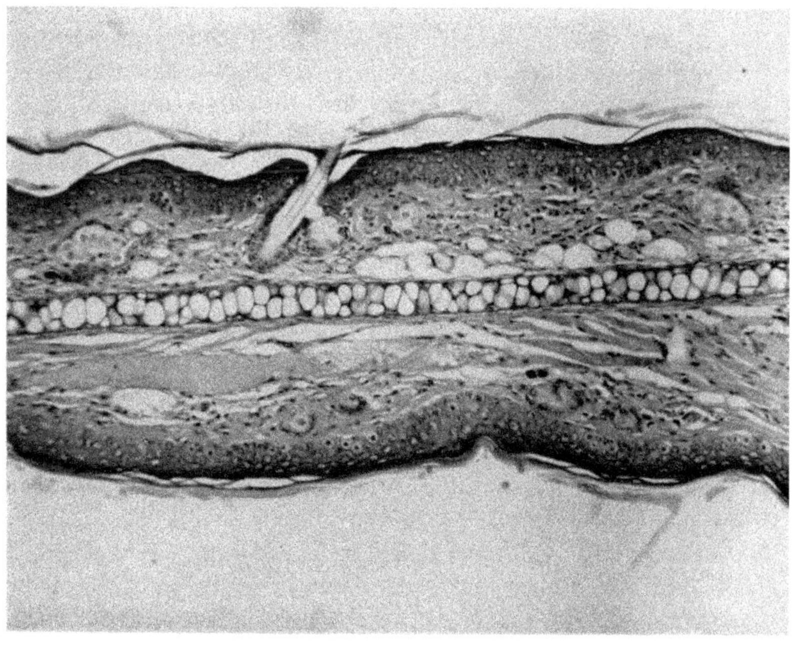

cells move toward the surface, eventually to become detached when mature. Here the similarities end, however. In the epidermis, the histological picture remains essentially the same from place to place and from time to time. Its several stratifications reflect in space what happens to each cell in the course of its life history. In the seminiferous tubule, the cells take a long time to mature and are replaced at infrequent intervals. In other words, the basal layer of spermatogonia wait until their most recent progeny have made a substantial head start along the pathway of differentiation before sending off the next relay. Epidermis, on the other hand, turns out a steady stream of new cells. That is, the interval between successive basal divisions is shorter than that required for the transit of a cell from one stratum to another. For this reason, all stages of differentiation are visible at once in every cross section through the epidermis, while only intermittent steps in the process of spermatogenesis are exemplified at any given locus along the length of a seminiferous tubule.

The rather complicated kinetics of spermatogenesis accounts for the histological variations seen at different levels along a seminiferous tubule. These variations reflect the fact that what is going on at a given moment in time in one place is out of phase from that at other locations. The cycle of spermatogenesis, of course, goes to completion everywhere, but it does not do so synchronously in all places. In the testis of the rat and other small mammals the asynchrony of spermatogenesis is arranged in an orderly manner along the length of the seminiferous tubule. The tubule is thus divided into a series of segments. Within any one segment all cells are in phase, but adjacent segments are composed of cells that are slightly ahead or behind in the cycle. If a tubule is examined longitudinally, therefore, one encounters 14 successive steps in the process of spermatogenesis from beginning to end. This comprises the so-called spermatogenic wave. In the rat, a seminiferous tubule, which is a tightly coiled loop with both ends opening to the rete testis, is made up of about 12 such waves, each of which consists of 14 consecutively arranged segments (Fig. 17). Since the seminiferous epithelium is constantly changing, however, the stage in the spermatogenic cycle represented in any one segment matures continuously. Hence, the map of a given segment in a tubule remains the same, but the degree of differentiation of its cells never stops advancing, and when one cycle is completed the next one follows on its heels.

Thus, the cells in a segment remain in place as they go through spermatogenesis, while the spermatogenic waves sweep along the tubule in the direction of the rete testis. Since this progression occurs in both arms of the tubular loop, there is a point more or less midway between the two ends (site of reversal) where the waves are propagated in opposite directions. In the rat, the average length of each wave is about 2.6 cm, and within each wave there are some 14 segments, each representing a stage in the spermatogenic cycle (Leblond and Clermont, 1952; Clermont and Leblond, 1953; Perey et al., 1961; Clermont, 1962). In the mouse there are 12 such segments (Oakberg, 1956), while 6 have been counted in man (Clermont, 1963; Heller and Clermont, 1964; Clermont, 1966). In the human seminiferous tubule, however, spermatogenic waves cannot be identified since the 6 different phases of the cycle are arranged mosaically and out of order in the walls of the

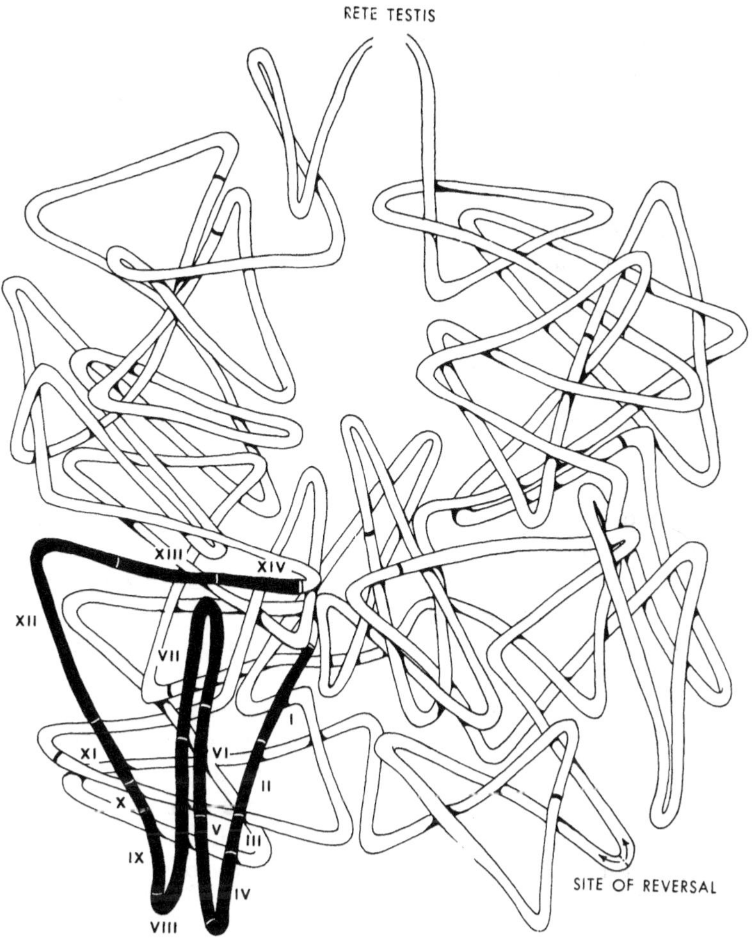

FIG. 17. Diagrammatic illustration of a rat seminiferous tubule. The two ends of the loop open into the rete testis. There are about a dozen spermatogenic waves marked off along its length with their directions of maturation emanating from the site of reversal toward the rete testis. A sample wave subdivided into its constituent segments, or cell associations, is illustrated in black.

tubules. Although this does not interfere with the normal course of spermatogenesis, it does complicate the histological analysis of what is going on.

In the rat testis, which has been so carefully analyzed by Perey et al. (1961), it takes about 48 days for a spermatogonium to differentiate into a family of mature spermatozoa. However, about every 12 days a new group of stem cells embarks on the course of spermatogenesis. Consequently, there is an overlap of four generations within each segment of a spermatogenic wave. In cross section, therefore, one can identify 4 concentrically arranged stages in the process, each of which is

12 days behind those ahead of it. These four stages in the cycle are always found in association with each other, like several generations in a family living under the same roof. Each member of the family (child, young adult, middle aged, oldster), passes through his life cycle in lock step with those in the next generations ahead and behind, and as the oldest ones pass on, new ones are born into the family. Now, if each home on a street contained a family whose members were all a little older than their neighbors on one side, but younger than those on the other, and if in every fourteenth house the people were of the same age, then a row of 14 houses would be comparable to a spermatogenic wave composed of 14 segments. This homely analogy, of course, is not entirely accurate, for each dwelling should actually be an apartment house. In this case, all families living at a given address would be alike in their ages because the segments of a spermatogenic wave contain a number of "units," each of which is descended from a single spermatogonium.

The cycle of spermatogenesis in the rat is diagrammatically illustrated in Figure 18. Reading clockwise, one can visualize the 14 successive cell associations which are radially arranged in the drawing but which in reality are laid out linearly along the length of the seminiferous tubule for a distance of one segment. In progressing from spermatogonium to spermatozoan, each cell moves inward toward the lumen of the tubule, with four generations of cells coexisting from the youngest, on the outside, to the oldest, on the inside.

What is so intriguing about spermatogenesis is that it takes place in a stepwise fashion with cells in several phases of maturation invariably occurring together at any given location along a tubule. The regular occurrence of these cell associations suggests the operation of a control mechanism serving to coordinate the four cycles in phase with one another. Spermatogenesis is in part subject to systematically distributed hormones such as FSH, without which the seminiferous epithelium cannot give rise to mature gametes. Over and above this, however, the timing of spermatogenesis within each segment of the tubule must be independently regulated. The nature of this local control has been investigated by Lacy (1960, 1967) and Lacy and Lofts (1965).

It has long been known that when mature spermatozoa are released into the lumen of the seminiferous tubule, they shed numerous small fragments of cytoplasm, called residual bodies. Laden with lipids, mitochondria, and RNA, these residual bodies find their way to the periphery of the tubule where they are phagocytosed by Sertoli cells. Hence, at stage VIII of the spermatogenic cycle in the rat, when fully differentiated sperm have been produced, these bodies may be found along the luminal border of the seminiferous epithelium. They then migrate centrifugally to become embedded in the Sertoli cells during stages IX through XIV, and disappear thereafter until a new generation of spermatozoa has matured. Once they have been engulfed by Sertoli cells, their contents can no longer be traced except for the persistent lipid bodies.

Lacy believes that the residual bodies may play a role in regulating the cyclical production of sperm cells. Although the evidence for this is circumstantial, the logic of his hypothesis is compelling for a number of reasons. The most opportune time to initiate a new cycle of maturation, for example, would be when the previous

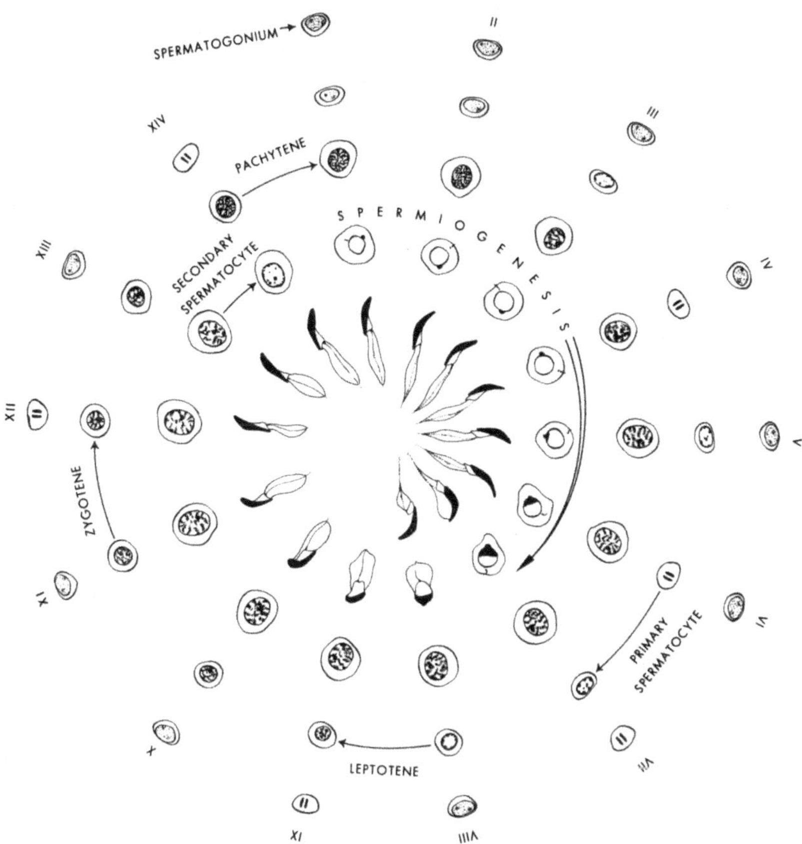

FIG. 18. Spermatogenesis in the rat. Reading clockwise from the top, a spermatogonium may be followed through its successive divisions and differentiation on the way to becoming a mature spermatozoan (center). Each of the 4½ cycles show how far a given cell matures before the next generation begins. The fourteen radially arranged cell associations represent the segments along a spermatogenic wave. See Figure 17. (Adapted from Perey et al., 1961.)

class of spermatozoa has graduated. This is precisely when the residual bodies are released. It may be no accident, therefore, that primary spermatocytes begin to develop coincident with the detachment of sperm. The transfer of material from sperm to Sertoli cell is also a suspicious event, especially in view of the intimate association between the latter cells and the spermatocytes they envelop. It is conceivable that the residual bodies may induce the production of a factor (something like a Sertoli cell hormone) which stimulates the next cycle of spermatogenesis. Although these events are open to wide interpretation, the importance of learning how to control populations of sperm, not to mention the organisms they beget, cannot be exaggerated.

CONCLUSIONS

Space does not permit an examination in depth of all renewing tissues, nor would it necessarily contribute to our understanding to reiterate additional variations on an already overplayed theme. Suffice it to say that most other tissues which constantly renew their constituent cells are epithelial. Perhaps the most thoroughly investigated of these is that lining the intestinal mucosa, for the separation between the germinative zone in the crypts of Lieberkühn and the differentiated cells of the villous epithelium has proved to be convenient for kinetic studies of cellular turnover (Leblond and Messier, 1958; Quastler and Sherman, 1959; Cairnie et al., 1965). In the respiratory tract no such distinct growth zone exists, yet here too there is an endless loss and replacement of cells (Bertalanffy and Leblond, 1953; Rhodin and Dalhamn, 1956; Blenkinsopp, 1967). Turnover of cells likewise occurs in taste buds (Beidler and Smallman, 1965), the transitional epithelium of the urinary bladder (Leblond et al., 1955), and bone (Young, 1962). Although the spatial and temporal patterns of replacement vary from one tissue to another, the problems of how to balance the input and the output are essentially the same throughout.

There is abundant evidence that cell division is controlled by specific extrinsic factors, both inhibitory and stimulatory. We do not know whether or not cells will still multiply in the absence of all such influences. Instances of apparently spontaneous mitotic activity (e.g., in tissue culture) may eventually be explained in terms of extrinsic factors yet to be discovered. In our present state of ignorance, therefore, we can do little more than attempt to attribute as many cases of cellular proliferation as possible to causative influences and assume for the time being that the rest are intrinsically controlled.

The differentiated state of a cell is unquestionably concerned with the regulation of mitosis. Differentiation involves the elaboration of specific end products upon which the functional efficacy of a cell or tissue depends. By virtue of their physiological activities, such end products are of the utmost importance in the negative feedback mechanisms by which the function and the growth of a tissue are governed. It is this end-product inhibition which turns off the output of cells and cell products. And when there are insufficient end products, it is the resulting functional demand which stimulates specific synthesis of more products, and of the cells that produce them. In view of these inevitable relationships, the innumerable tissues of the body can be classified according to how their growth is affected by functional activity (Fig. 19).

In such a scheme, certain structures not known to grow in response to functional demands must tentatively be grouped in an independent category. Hence, the growth and molting of various integumental appendages often exhibits periodicities which may sometimes be correlated with seasonal cues. Other kinds of growth, such as those associated with estrous cycles, seem often to go on autonomously. In either case, it is clear that their patterns of development represent genetic rather than physiological adaptations. Similarly, the mammalian placenta does not enlarge in response to the demands created by the growing fetus, for its dimensions continue

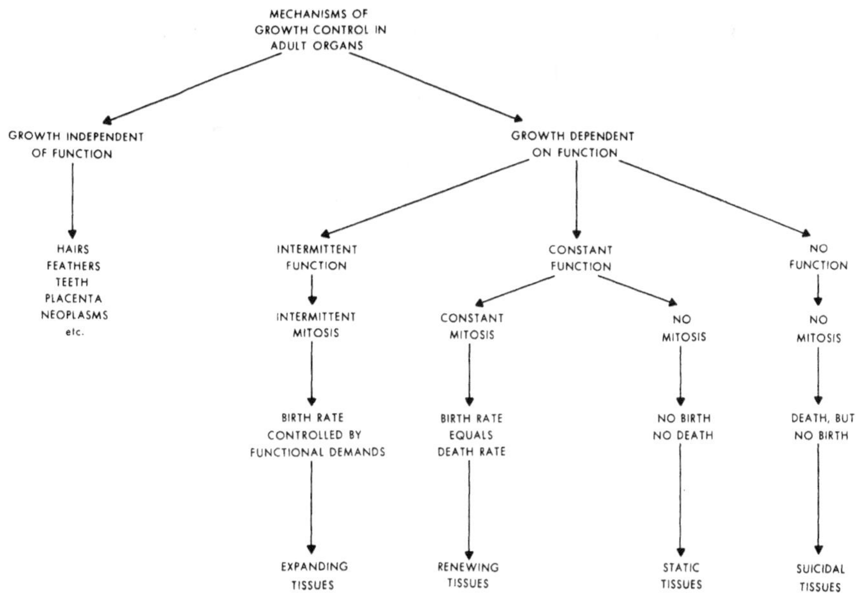

FIG. 19. Relation between function and cell division. Some tissues (left) are not known to grow in response to functional demands. Others, however, proliferate intermittently (expanding tissues) or constantly (renewing tissues) depending upon their work schedules. Static tissues function constantly, do not divide, and die only by accident. Where there is no function there is no mitosis. Such tissues tend to be resorbed.

to increase even after fetectomy in the rat (Prichard and Huggett, 1947) and embryectomy in the Rhesus monkey (Lewis and Hertz, 1966). Perhaps the neoplastic condition may arise when a cell's growth is uncoupled from functional control mechanisms.

In most tissues and organs, however, growth is not entirely autonomous. Within certain limits, their sizes are determined by the work they are called upon to perform. In expanding organs (e.g., glands) the rate of growth accurately reflects the functional demands impinging upon the cells. But since their physiological activities are intermittent and usually below the mitogenic threshold, such organs generally exhibit little or no cell division except in emergency conditions when they undergo compensatory hyperplasia. Expanding organs, therefore, are in little danger of overproducing their cells.

Other tissues, however, must function constantly. If growth rates are still to be linked to physiological demands, they must adopt some strategy to avoid a population explosion at the cellular level. This they achieve in either of two ways. One is to remain mitotically competent, to proliferate incessantly, and to eliminate cells as fast as they are produced. These are the conditions which define renewing tissues, tissues which maintain their morphological integrity not so much by birth control as by death insurance.

The other way to prevent overgrowth in the face of constant function is to give up the capacity for cell division. This is what static tissues have done. Nerves and muscles, therefore, enjoy the advantages of adapting to physiological demands without running the risk of producing too many cells. The price they pay is the limited growth that can occur by cell hypertrophy, and the inability to replace lost cells. But in the theoretical absence of cell death they can forego the inconveniences of cell reproduction and the related problems of birth control.

Thus, in the expanding, renewing, and static tissues of the body there are represented three of the four hypothetical combinations of cell birth and death. Expanding tissues exhibit birth, but not death of cells. Here the control of mitosis must be very tight. Renewing tissues have birth and death with minimal control of cell division. Static tissues avoid the problems of population control by abolishing the birth and death of cells altogether. Such immortal societies as these cannot afford the luxury of reproduction. Finally, the fourth category, in which there is programmed cell death with no provision for renewal, is of more than hypothetical interest. Even suicidal tissues exist, as in the cases of regression of the thymus, tissue resorption in metamorphosing animals, and in certain aspects of embryonic development (Saunders and Fallon, 1966). The demise of these tissues may perhaps be attributed to the withdrawal of all functional demands, thus abolishing not only the need to produce new cells but also their very excuse for existence.

REFERENCES

Aas, K. A., and F. H. Gardner. 1958. Survival of blood platelets labeled with chromium[51]. J. Clin. Invest., 37:1257–1268.

Abbrecht, P. H., and R. L. Malvin. 1966. Renal production of erythropoietin in the dog. Amer. J. Physiol., 210:237–242.

——— R. L. Malvin, and A. J. Vander. 1966. Renal production of erythropoietin and renin after experimental kidney infarction. Nature (London), 211:1318–1319.

Åkeson, Å., G. v. Ehrenstein, S. Hevesy, and H. Theorell. 1960. Life span of myoglobin. Arch. Biochem., 91:310–318.

Alpen, E. L., and D. Cranmore. 1959. Observations on the regulation of erythropoiesis and cellular dynamics by Fe^{59} autoradiography. *In* The Kinetics of Cellular Proliferation, Stohlman, F. Jr., ed., Grune and Stratton, New York, pp. 290–300.

Aschheim, E. 1968. Epidermal homeostasis—a numerical model. Kinetics of epidermal cells. Experientia, 24:94–98.

Auclair, W., and B. W. Siegel. 1966. Cilia regeneration in the sea urchin embryo: Evidence for a pool of ciliary proteins. Science, 154:913–915.

Bach, F., and K. Hirschhorn. 1963. γ-globulin production by human lymphocytes *in vitro.* Exp. Cell Res., 32:592–595.

Bain, B., and L. Lowenstein. 1964. Genetic studies on the mixed leukocyte reaction. Science, 145:1315–1316.

Baldini, M. 1966a. Platelet circulation and survival. Proc. XI Congr. Internat. Soc. Hemat. and Blood Transfusion, Sydney, V. C. N. Blight Publisher, pp. 332–346.

——— 1966b. Idiopathic thrombocytopenic purpura. New Eng. J. Med., 274:1245–1251, 1302–1306, 1360–1367.

Barker, B. E., and P. Farnes. 1967a. Histochemistry of blood cells treated with pokeweed mitogen. Nature (London), 214:787–789.

——— and P. Farnes. 1967b. Mitogenic property of *Wistaria floribunda* seeds. Nature (London), 215:659.

——— P. Farnes, and H. Fanger. 1964. Mitogenic activity in *Phytolacca americana* (pokeweed). Lancet, 1:170.

Basnayake, V., and H. M. Sinclair. 1954. Skin permeability in deficiency of essential fatty acids. J. Physiol. (London), 126:55P–56P.

Becker, A. J., E. A. McCulloch, and J. E. Till. 1963. Cytological demonstration of the clonal nature of spleen colonies derived from transplanted mouse marrow cells. Nature (London), 197:452–454.

Beidler, L. M., and R. L. Smallman. 1965. Renewal of cells within taste buds. J. Cell Biol., 27:263–272.

Beránek, R., P. Hník, L. Vyklicky, and J. Zelená. 1963. Plastic effect of long-term tenotomy upon the spinal monosynaptic response. *In* The Effect of Use and Disuse on Neuromuscular Functions, Gutmann E. and Hník, P., eds., Elsevier Publishing Company, Amsterdam, pp. 305–311.

Berg, B. N. 1956. Muscular dystrophy in aging rats. J. Geront., 11:134–139.

Bernstein, S. 1963. Modification of radiosensitivity of genetically anemic mice by implantation of blood-forming tissue. Radiat. Res., 20:695–702.

——— and E. S. Russell. 1959. Implantation of normal blood forming tissue in genetically anemic mice without X-irradiation of host. Proc. Soc. Exp. Biol. Med., 101:769–773.

Bertalanffy, F. D., and C. P. Leblond. 1953. The continuous renewal of the two types of alveolar cells in the lung of the rat. Anat. Rec., 115:515–541.

Bierman, H. R. 1961. Homeostasis of the blood cell elements. *In* Functions of the Blood, MacFarlane, R. G., and Robb-Smith, A. H. T., eds., New York, Academic Press, Inc., Ch. 8, pp. 349–418.

——— 1964. Characteristics of leukopoietin G in animals and man. Ann. N.Y. Acad. Sci., 113(2):753–765.

——— K. H. Kelly, R. L. Byron, Jr., and G. J. Marshall. 1961. Leukapheresis in man. I. Hematological observations following leukocyte withdrawal in patients with non-hematologic disorders. Brit. J. Haemat., 7:51–63.

——— G. J. Marshal, T. Maekawa, and K. H. Kelly. 1962. Granulocytic activity of human plasma. I. Acta. Haemat., 27:217–228.

Bizzozero, G. 1894. An address on the growth and regeneration of the organism. Brit. Med. J., 1:728–732.

Blenkinsopp, W. K. 1967. Proliferation of respiratory tract epithelium in the rat. Exp. Cell Res., 46:144–154.

Boggs, D. R. 1966. Homeostatic regulatory mechanisms of hematopoiesis. Ann. Rev. Physiol., 28:39–56.

——— P. A. Chervenick, J. C. Marsh, H. I. Pilgrim, G. E. Cartwright, and M. M. Wintrobe. 1967. Granulocytopoiesis in germfree mice. Proc. Soc. Exp. Bio. Med., 125:325–330.

Bonsdorff, E. 1949. On the presence of erythropoietins in the plasma from sheep foetuses during the latter half of gestation. Acta. Physiol. Scand., 18:51–62.

Börjeson, J., R. Reisfeld, L. N. Chessin, P. D. Welsh, and S. D. Douglas. 1966. Studies on human peripheral blood lymphocytes in vitro. I. Biological and physiochemical properties of the pokeweed mitogen. J. Exp. Med., 124:859–872.

Brenner, R. M. 1968. The biology of oviductal cilia. *In* The Mammalian Oviduct, Hafez E. S., and Blandau, R .J., eds., Chicago, University of Chicago Press.

Bruce, W. R., and E. A. McCulloch. 1964. The effect of erythropoietic stimulation on the hemopoietic colony-forming cells of mice. Blood, 23:216–232.

Bullough, W. S. 1965. Mitotic and functional homeostasis: A speculative review. Cancer Res., 25:1683–1727.

——— and E. B. Laurence. 1964. Mitotic control by internal secretion: The role of the chalone-adrenalin complex. Exp. Cell Res., 33:176–194.

——— and E. B. Laurence. 1966. Tissue homeostasis in adult mammals. *In* Advances in Biology of Skin. Carcinogenesis. Montagna W. and Dobson, R. L., eds., New York, Pergamon Press, 7:1–36.

——— and E. B. Laurence. 1967. Epigenetic mitotic control. *In* Control of Cellular Growth in Adult Organisms, Teir, H. and Rytömaa, T., eds., New York, Academic Press, Inc., pp. 28–40.

Butturini, U., and G. Lucarelli. 1963. Influence de la polycythémie par érythrotransfusion chez le rat femelle pleine sur l'érythropoïèse foetale. Proc. 9th Cong. Europ. Soc. Haemat. (Lisbon), Basel and New York, S. Karger, pp. 804–807.

Byrd, W. J., K. Hare, W. Finley and S. Finley. 1967. Inhibition of the mitogenic factor in phytohaemagglutinin by an antiserum. Nature (London), 213:622.

Cairnie, A. B., L. F. Lamberton, and G. G. Steele. 1965. Cell proliferation studies in the intestinal epithelium of the rat. I. Determination of the kinetic parameters. Exp. Cell Res., 39:528–538.

Campbell, J. A. 1932. Hypertrophy of the heart in acclimatization to chronic carbon monoxide poisoning. J. Physiol. (London), 77:8P–9P.

Carmena, A., G. Lucarelli, C. Carnevali, and F. Stohlman, Jr. 1966. Regulation of erythropoiesis. XIX Effect of hypoxia on erythropoiesis in the newborn animal. Proc. Soc. Exp. Biol. Med., 121:652–655.

Carney, J. A., and A. L. Brown, Jr. 1964. Myofilament diameter in the normal and hypertrophic rat myocardium. Amer. J. Path., 44:521–529.

Caron, G. A., and I. Sarkany. 1966. Role of plasma factors in the transformation of peripheral blood lymphocytes into lymphoblasts. Nature (London), 210:314–315.

Chessin, L. N., J. Börjeson, P. D. Welsh, S. D. Douglas, and H. L. Cooper. 1966. Studies on human peripheral blood lymphocytes in vitro. II. Morphological and biochemical studies on the transformation of lymphocytes by pokeweed mitogen. J. Exp. Med., 124:873.

Clermont, Y. 1962. Quantitative analysis of spermatogenesis of the rat: A revised model for the renewal of spermatogonia. Amer. J. Anat., 111:111–129.

——— 1963. The cycle of the seminiferous epithelium in man. Amer. J. Anat., 112:35–51.

——— 1966. Spermatogenesis in man: A study of the spermatogonial population. Fertil. Steril., 17:705.

——— and C. P. Leblond. 1953. Renewal of spermatogonia in the rat. Amer. J. Anat., 93:475–501.

Cohen, N. S., and A. S. Gordon. 1961. Studies on humoral regulation of eosinophil release and production. Amer. Zool., 1:442.

Cohen, P., F. H. Gardner, and G. O. Barnett. 1961. Reclassification of the thrombocytopenias by the Cr^{51}-labeling method for measuring platelet lifespan. New Eng. J. Med., 264:1294–1299, 1350–1355.

Cohen, S. 1962. Isolation of a mouse submaxillary gland protein accelerating incisor eruption and eyelid opening in the newborn animal. J. Biol. Chem., 237:1555–1562.

——— 1965. The stimulation of epidermal proliferation by a specific protein (EGF). Develop. Biol., 12:394–407.

Contrera, J. F., and A. S. Gordon. 1966. Erythropoietin: Production by a particulate fraction of rat kidney. Science, 152:653–654.

——— A. S. Gordon, and A. H. Weintraub. 1966. Extraction of an erythropoietin-producing factor from a particulate fraction of rat kidney. Blood, 28:330–343.

Cooper, H. L., and A. D. Rubin. 1965. RNA metabolism in lymphocytes stimulated by phytohemagglutinin: Initial responses to phytohemagglutinin. Blood, 25:1014–1028.

Cooperband, S. R., H. Bondevik, K. Schmid, and A. Mannick. 1968. Transformation of human lymphocytes: Inhibition by homologous alpha globulin. Science, 159:1243–1244.

Craddock, C. G., Jr., W. S. Adams, S. Perry, and J. S. Lawrence. 1955. The dynamics of platelet production as studied by a depletion technique in normal and irradiated dogs. J. Lab. Clin. Med., 45:906–919.

——— S. Perry, and J. S. Lawrence. 1956. The dynamics of leukopoiesis and leukocytosis, as studied by leukopheresis and isotope technique. J. Clin. Invest., 35:285–296.

——— S. Perry, and J. S. Lawrence. 1960. The dynamics of leukopenia and leukocytosis. Ann. Intern. Med., 52:281–294.

Crawford, H., M. Cutbush, and P. L. Mollison. 1953. Hemolytic disease of the newborn due to anti-A. Blood, 8:620–639.

Cronkite, E. P. 1944. Further studies of platelet reducing substances in splenic extracts. Ann. Intern. Med., 20:52–62.

——— V. P. Bond, J. S. Robertson, and D. E. Paglia. 1957. The survival, distribution and apparent interaction with capillary endothelium of transfused radiosulfate labeled platelets in the rat. J. Clin. Invest., 36:881.

Darzynkiewicz, Z., T. Krassowski, and E. Skopinska. 1965. Effect of phytohemagglutinin on the synthesis of "rapidly labelled" RNA in human lymphocytes. Nature (London), 207:1402–1403.

Davies, L. M., J. H. Priest, and R. E. Priest. 1968. Collagen synthesis by cells synchronously replicating DNA. Science, 159:91–93.

de Franciscis, P., G. De Bella, and S. Cifaldi. 1966. Spleen as a production site for erythropoietin. Science, 150:1831–1833.

Diengdoh, J. V., and J. L. Turk. 1965. Immunological significance of lysosomes within lymphocytes *in vivo*. Nature (London), 207:1405–1406.

Dowling, J. E. 1966. Night blindness. Sci. Amer., 215(4):78–84.

Drachman, D. B. 1964. Atrophy of skeletal muscle in chick embryos treated with Botulinum toxin. Science, 145:719–721.

Drahota, Z., and E. Gutmann. 1962. The effect of age on compensatory and "post-functional hypertrophy" in cross-striated muscle. Gerontologia, 6:81–90.

Drasnin, R., J. T. Hughes, R. F. Krause, and E. J. Van Liere. 1958. Glycogen content in normal and hypertrophied rat heart. Proc. Soc. Exp. Biol. Med., 99:438–439.

Dreyfus, J. C. 1963. Problems in the biochemistry of progressive muscular dystrophy. *In* Research in Muscular Dystrophy, Philadelphia, J. B. Lippincott Co., pp. 127–141.

———J. Kruh, and G. Schapira. 1960. Metabolism of myosin and life time of myofibrils. Biochem. J., 75:574–578.

Droz, B. 1963. Dynamic condition of proteins in the visual cells of rats and mice as shown by radioautography with labeled amino acids. Anat. Rec., 145:157–167.

——— and C. P. Leblond. 1962. Migration of proteins along the axons of the sciatic nerve. Science, 137:1047–1048.

Duke, W. W. 1911. The rate of regeneration of blood platelets. J. Exp. Med., 14:265–273.

Dukes, P. P., and E. Goldwasser. 1965. On the mechanism of erythropoietin-induced differentiation. III. The nature of erythropoietin action on (^{14}C) glucosamine incorporation by marrow cells in culture. Biochim. Biophys. Acta., 108:447–454.

Duplan, J. F., G. V. Foschi, and L. A. Manson. 1962. The lymphocytosis stimulating factor (LSF). Proc. Soc. Exp. Biol. Med., 110:426–429.

Dutton, R. W., and R. I. Mishell. 1966. Lymphocytic proliferation in response to homologous tissue antigens. Fed. Proc., 25:1723–1726.

Ebbe, S. 1966. Platelet survival in rats with transfusion-induced thrombocytosis. J. Lab. Clin. Med., 68:813–823.

——— F. Stohlman, Jr., J. Donovan, and D. Howard. 1966. Platelet survival in the rat as measured with tritium-labeled diisopropylfluorophosphate. J. Lab. Clin. Med., 68:233–243.

——— F. Stohlman, Jr., J. Overcash, J. Donovan, and D. Howard. 1968a. Megakaryotic size in thrombocytopenic and normal rats. Blood, 32:383–392.

——— F. Stohlman, Jr., J. Donovan, and J. Overcash. 1968b. Megakaryocyte maturation rate in thrombocytopenic rats. Blood, 32:787–795.

Elkins, W. L., and R. D. Guttmann. 1968. Pathogenesis of a local graft versus host reaction: Immunogenicity of circulating host leukocytes. Science, 159:1250–1251.

Elmlinger, P. J., R. L. Huff, and J. M. Oda. 1952. Depression of red cell iron turnover by transfusion. Proc. Soc. Exp. Biol. Med., 79:16–19.

Elves, M. W., S. Roath, and M. C. G. Israels. 1963. The response of lymphocytes to antigen challenge *in vitro*. Lancet, 1:806–807.

Erslev, A. J. 1959. The effect of anemic anoxia on the cellular development of nucleated red cells. Blood, 14:386–398.

——— 1961. The effect of environment on proliferation and maturation of nucleated red cells. Angiology, 12:307–309.

Estes, F. L., S. Smith, and J. H. Gast. 1958. A method for obtaining polymorphonuclear leukocytes from intraperitoneal exudates. Blood, 13:1192–1197.

Farnes, P., B. D. Barker, L. E. Brownhill, and H. Fanger. 1964. Mitogenic activity in *Phytolacca americana* (Pokeweed). Lancet, 2:1100–1101.

Fehér, I., and J. Gidáli. 1965. Quantitative changes in the level of the myelopoiesis-stimulating agent in rabbit sera. J. Lab. Clin. Med., 66:272–279.

Feng, T. P., H. W. Jung, and W. Y. Wu. 1963. The contrasting trophic changes of the anterior and posterior latissimus dorsi of the chick following denervation. *In* The Effect of Use and Disuse on Neuromuscular Functions, Gutmann, E. and Hník, P., eds., Amsterdam, Elsevier Publ. Co., pp. 431–441.

Filmanowicz, E. V., and C. W. Gurney. 1959. A study of the kinetics of erythropoiesis. J. Lab. Clin. Med., 54:813–814.

——— and C. W. Gurney. 1961. Studies on erythropoiesis. XVI. Response to a single dose of erythropoietin in polycythemic mice. J. Lab. Clin. Med., 57:65–72.

Firket, J., and E. S. Campos. 1922. Generalized megalocaryocytic reaction to saponin poisoning. Bull. Hopkins Hosp., 33:271–283.

Fischer, E., and V. W. Ramsey. 1946. The effect of daily electrical stimulation of normal and denervated muscles upon their protein content and upon some of the physicochemical properties of the protein. Amer. J. Physiol., 145:583–586.

Fisher, J. W. 1961. The production of an erythropoietic factor by the *in situ* perfused kidney. Acta Haemat., 26:224–232.

——— and A. I. Samuels. 1967. Relationship between renal blood flow and erythropoietin production in dogs. Proc. Soc. Exp. Biol. Med., 125:482–485.

Flores, G., and E. Frieden. 1968. Induction and survival of hemoglobin-less and erythrocyte-less tadpoles and young bullfrogs. Science, 159:101–103.

Fukuda, T., H. Oide, and A. Miyasaka. 1960. Nature of the leucocytosis-inducing factor in plasma. Nature (London), 188:860–861.

Gallien-Lartigue, O., and E. Goldwasser. 1965. On the mechanism of erythropoietin-induced differentiation. I. The effects of specific inhibitors on hemoglobin synthesis. Biochim. Biophys. Acta, 103:319–324.

Garcia, J. F. 1957. Erythropoietic response to hypoxia as a function of age in the normal male rat. Amer. J. Physiol., 190:25–30.

Gardner, F. G., K. A. Aas, P. Cohen, and J. C. Pringle. 1958. Investigations of the mechanisms of thrombocytopenia. Clin. Res., 6:199–200.

Geber, W. F., and T. A. Anderson. 1967. Cardiac hypertrophy due to chronic stress in the rat, *Rattus norvegicus albinus,* and rabbit, *Lepus cuniculus*. Comp. Biochem. Physiol., 21:273–277.

Gemmill, C. L. 1958. Cardiac hypertrophy in rats and mice given 3, 3′, 5-triiodo-L-thyronine orally. Amer. J. Physiol., 195:385–390.

Gidáli, J., and I. Fehér. 1964. Myelopoiesis controlling agents in vitro. Amer. J. Physiol., 206:585–588.

Goldfarb, B., and L. Tobian. 1963. Effect of high O_2 concentration on erythropoietin and renal juxtaglomerular cells. Proc. Soc. Exp. Biol. Med., 113:35–36.

Goldwasser, E. 1966. Biochemical control of erythroid cell development. *In* Current Topics in Developmental Biology, Monroy, A. and Moscona, A., eds., New York, Academic Press, Inc. 1:173–211.

——— L. O. Jacobson, W. Fried, and L. Plzak. 1957. Mechanism of erythropoietic effect of cobalt. Science, 125:1085–1086.

——— L. O. Jacobson, W. Fried, and L. F. Plzak. 1958. Studies on erythropoiesis. V. The effect of cobalt on the production of erythropoietin. Blood, 13:55–60.

Gordon, A. S. 1955. Some aspects of hormonal influences upon the leukocytes. Ann. N. Y. Acad. Sci., 59:907–927.

——— R. O. Neri, G. D. Siegel, B. S. Dornfest, E. S. Handler, J. LoBue, and M. Eisler. 1960. Evidence for a circulating leucocytosis-inducing factor (LIF). Acta. Haemat., 23:323–341.

——— E. S. Handler, C. D. Siegel, B. S. Dornfest, and J. LoBue. 1964. Plasma factors influencing leukocyte release in rats. Ann. N. Y. Acad. Sci., 113(2):766–789.

——— R. Katz, E. D. Zanjani, and E. A. Mirand. 1966. Renal mechanisms underlying actions of androgen and hypoxia on erythropoiesis. Proc. Soc. Exp. Biol. Med., 123: 475–478.

Gordon, J., and L. D. MacLean. 1965. A lymphocyte-stimulating factor produced in vitro. Nature (London), 208:795–796.

Gori, Z., C. Pellegrino, and M. Pollera. 1967. The castration atrophy of the dorsal bulbocavernosus muscle of rat: An electron microscopic study. Exp. Molec. Path., 6:172–198.

Goss, R. J. 1965. Adaptive Growth. New York, Academic Press, Inc.

——— 1966. Hypertrophy versus hyperplasia. Science, 153:1615–1620.

——— 1967. The strategy of growth. *In* Control of Cellular Growth in Adult Organisms, Teir, H. and Rytömaa, T., eds., New York, Academic Press, Inc., pp. 3–27.

Gräsbeck, R., C. Nordman, and A. de la Chapelle. 1963. Mitogenic action of antileucocyte immune serum on peripheral leucocytes in vitro. Lancet, 2:385–386.

Grasso, J. A., and D. C. Shephard. 1968. Experimental production of totally anaemic newts. Nature (London), 218:1274–1276.

Grimm, A. F., R. Kubota, and W. V. Whitehorn. 1966. Ventricular nucleic acid and protein levels with myocardial growth and hypertrophy. Circ. Res., 19:552–558.

Gudbjarnason, S., M. Telerman, and R. J. Bing. 1963. Protein synthesis in myocardial hypertrophy and experimental heart failure. Fed. Proc., 22:345.

——— M. Telerman, C. Chiba, P. L. Wolf, and R. J. Bing. 1964. Myocardial protein synthesis in cardiac hypertrophy. J. Lab. Clin. Med., 63:245–253.

Guth, L., A. A. Zalewski, and W. C. Brown. 1966. Quantitative changes in cholinesterase activity of denervated sole plates following implantation of nerve into muscle. Exp. Neurol., 16:136–147.

Gutmann, E., and J. Zelená. 1962. Morphological changes in the denervated muscle. *In* The Denervated Muscle, Gutmann, E., ed., Prague, Publ. House of the Czechoslovak Acad. Sci., pp. 57–102.

Halvorsen, S. 1961. Plasma erythropoietin levels following hypothalamic stimulation in the rabbit. Scand. J. Clin. Lab. Invest., 13:564–575.

Hamosh, M., M. Lesch, J. Baron, and S. Kaufman. 1967. Enhanced protein synthesis in a cell-free system from hypertrophied skeletal muscle. Science, 157:935–937.

Handler, E., and A. S. Gordon. 1962. Endocrine influence on the leucocytosis-inducing factor (LIF). Amer. Zool., 2:528.

Hanna, I. R. A. 1967. Response of early erythroid precursors to bleeding. Nature (London), 214:355–357.

Hansen, P. 1964. Demonstration of erythropoietin in urine and in kidney extracts from rabbits with experimental constriction of the left renal artery. Acta Path. Microbiol. Scand., 61:514–520.

Harrington, W. J., V. Minnich, and G. Arimura. 1956. The autoimmune thrombocytopenias. *In* Progress in Hematology, Tocantins, L. M., ed., New York, Grune and Stratton, Inc., Vol. 1, pp. 166–192.

——— V. Minnich, J. W. Hollingsworth, and C. V. Moore. 1951. Demonstration of a thrombocytopenic factor in the blood of patients with thrombocytopenic purpura. J. Lab. Clin. Med., 38:1–10.

——— C. C. Sprague, V. Minnich, C. V. Moore, R. C. Ahlvin, and R. Duback. 1953.

Immunologic mechanisms in idiopathic and neonatal thrombocytopenic purpura. Ann. Intern. Med., 38:433–469.

Hayhoe, F. G. J., and D. Quaglino. 1965. Autoradiographic investigations of RNA and DNA metabolism of human leucocyte cultures with phytohemagglutinin; uridine-5-^{3}H as a specific precursor of RNA. Nature (London), 205:151–154.

Helander, E. A. S. 1961. Influence of exercise and restricted activity on the protein composition of skeletal muscle. Biochem. J., 78:478–482.

Heller, C. G., and Y. Clermont. 1964. Kinetics of the germinal epithelium in man. Recent Prog. in Hormone Res., 20:545–575.

Henrikson, R. C. 1967. Incorporation of tritiated thymidine by teleost epidermal cells. Experientia, 23:357–358.

Herrmann, H., A. C. Marchok, and E. F. Baril. 1967. Growth rate and differentiated function of cells. Nat. Canc. Inst. Monogr. No. 26, pp. 303–326.

Hines, H. M., and G. C. Knowlton. 1934. The influence of fasting and environmental temperature upon the atrophy of denervated skeletal muscle. Amer. J. Physiol., 110:8–13.

——— and G. C. Knowlton. 1937. Electrolyte and water changes in muscle during atrophy. Amer. J. Physiol., 120:719–723.

Hirsch, E. O., and F. H. Gardner. 1951. Life span of transfused human blood platelets. J. Clin. Invest. 30:649–650.

Hirschhorn, K., F. Bach, R. Kolodny, I. L. Firschein, and N. Hashem. 1963. Immune response and mitosis of human peripheral blood lymphocytes. Science, 142:1185–1187.

——— and R. Hirschhorn. 1965. Role of lysosomes in the lymphocyte response. Lancet, 1:1046–1047.

Hirschhorn, R., G. Kaplan, K. Hirschhorn, and G. Weissmann. 1964. Appearance of lysosomes before mitosis induced in human lymphocytes by phytohemagglutinin. Clin. Res., 12:449.

——— J. M. Kaplan, A. F. Goldberg, K. Hirschhorn, and G. Weissmann. 1965. Acid phosphatase-rich granules in human lymphocytes induced by phytohemagglutinin. Science, 147:55–57.

——— K. Hirschhorn, and G. Weissmann. 1967. Appearance of hydrolase rich granules in human lymphocytes induced by phytohemagglutinin and antigens. Blood, 30:84–102.

Hjort, P. F., and H. Paputchis. 1960. Platelet life span in normal, splenectomized and hypersplenic rats. Blood, 15:45-51.

Holland, N. H., and P. Holland. 1965. Haemagglutinating, precipitating and lymphocyte-stimulating factors of phytohaemagglutinin. Nature, 207:1307–1308.

Hollingsworth, J. W., J. A. Berend, D. R. Silbert, and S. C. Finch. 1967. Leukocyte mobilization in normal, splenectomized, and leukemic rats after replacement transfusion. J. Lab. Clin. Med., 50:36–44.

Hoyle, T. C., III., R. G. Sumner, and H. D. McIntosh. 1963. Nucleic acid and protein determinations in the study of experimental cardiomegaly. J. Lab. Clin. Med., 62: 632–638.

Hultgren, H. N., and H. Miller. 1965. Right ventricular hypertrophy at high altitude. Ann. N. Y. Acad. Sci., 127:627–631.

Imig, C. J., B. F. Randall, and H. M. Hines. 1953. Effect of immobilization on muscular atrophy and blood flow. Arch. Phys. Med., 34:296–299.

Iwaikawa, Y. 1967. Regeneration of cilia in the sea urchin embryo. Embryologia, 9:287–294.

Jacobson, L. O., E. Goldwasser, W. Fried, and L. Plzak. 1957a. Role of the kidney in erythropoiesis. Nature (London), 179:633–634.

——— E. Goldwasser, L. Plzak, and W. Fried. 1957b. Studies on erythropoiesis. Part IV. Reticulocyte response of hypophysectomized and polycythemic rodents to erythropoietin. Proc. Soc. Exp. Biol. Med., 94:243–249.

——— E. K. Marks, and E. O. Gaston. 1959. Studies on erythropoiesis. XII. The effect of transfusion-induced polycythemia in the mother on the fetus. Blood, 14:644–653.

Jalavisto, E. 1967. The development of responsiveness to erythropoietic stimuli during ontogeny. *In* Control of Cellular Growth in Adult Organisms, Teir, H. and Rytömaa, T., eds., New York, Academic Press, Inc., pp. 139–147.

——— I. Kuorinka, and M. Kyllästinen. 1965. Responsiveness of the erythron to variations of oxygen tension in the chick embryo and young chicken. Acta Physiol. Scand., 63:479–486.

Johnson, D., and P. G. Roofe. 1966. Blood constituents of normal newborn rats and those exposed to low oxygen tension during gestation; weight of newborn and litter size also considered. Anat. Rec., 153:303–309.

Johnson, E. 1965. Inherent rhythms of activity in the hair follicle and their control. *In* Biology of the Skin and Hair Growth, Lyne, A. G. and Short, B. F., eds., Sydney, Angus and Robertson, pp. 491–505.

Kamrin, B. B. 1959. Successful skin homografts in mature non-littermate rats treated with fractions containing alpha-globulins. Proc. Soc. Exp. Biol. Med., 100:58–61.

Kasakura, S., and L. Lowenstein. 1965. A factor stimulating DNA synthesis derived from the medium of leucocyte cultures. Nature (London), 208:794–795.

Keighley, G. H., E. S. Russell, and P. Lowy. 1962. Response of normal and genetically anaemic mice to erythropoietic stimuli. Brit. J. Haemat., 8:429–441.

——— P. Lowy, E. S. Russell, and M.W. Thompson. 1966. Analysis of erythroid homeostatic mechanisms in normal and genetically anaemic mice. Brit. J. Haemat., 12:461–477.

Kleitke, B., and H. Sydow. 1965. Über den Nukleinsäuregehalt des pathologisch hypertrophierten Hundeherzens. Acta Biol. Med. German., 14:447–455.

Kohn, R. R. 1964. Mechanism of protein loss in denervation muscle atrophy. Amer. J. Path., 45:435–447.

Konigsberg, I. R. 1963. Clonal analysis of myogenesis. Science, 140:1273–1284.

Krantz, S. B., and E. Goldwasser. 1965. On the mechanism of erythropoietin-induced differentiation. II. The effect on RNA synthesis. Biochim. Biophys. Acta, 103:325–332.

Kruh, J., J. C. Dreyfus, G. Schapira, and G. Gey. 1960. Abnormalities of muscle protein metabolism in mice with muscular dystrophy. J. Clin. Invest., 39:1180–1184.

Krzymowski, T., and H. Krzymowska. 1962. Studies on the erythropoiesis inhibiting factor in the plasma of animals with transfusion polycythemia. Blood, 19:38–44.

Kuratowska, Z. 1965. On the subcellular localization of the erythropoietic renal factor. Bull. Acad. Pol. Sci. [Biol.], 13:385–390.

——— B. Lewartowski, and B. Lipinski. 1964. Chemical and biologic properties of an erythropoietin-generating substance obtained from perfusates of isolated anoxic kidneys. J. Lab. Clin. Med., 64:226–237.

Lacy, D. 1960. Light and electron microscopy and its use in the study of factors influencing spermatogenesis in the rat. J. Roy. Micr. Soc., 79:209–225.

——— 1967. The seminiferous tubule in mammals. Endeavour, 26:101–108.

——— and B. Lofts. 1965. Studies on the structure and function of the mammalian testis. I. Cytological and histochemical observations after continuous treatment with oestrogenic hormone and the effects of F.S.H. and L.H. Proc. Roy. Soc. (London), B162: 188–197.

Lawrence, J. S., C. G. Craddock, Jr., and T. N. Campbell. 1967. Antineutrophilic serum, its use in studies of white blood cell dynamics. J. Lab. Clin. Med., 69:88–101.

Leblond, C. P. 1965. The time dimension in histology. Amer. J. Anat., 116:1–27.

——— and Y. Clermont. 1952. Definition of the stages of the cycle of the seminiferous epithelium in the rat. Ann. N. Y. Acad. Sci., 55:548–573.

——— and B. Messier. 1958. Renewal of chief cells and goblet cells in the small intestine as shown by radioautography after injection of thymidine-H^3 into mice. Anat. Rec., 132:247–259.

——— M. Vulpe, and F. D. Bertalanffy. 1955. Mitotic activity of epithelium of urinary bladder in albino rat. J. Urol., 73:311–313.

Leeksma, C. H. W., and J. A. Cohen. 1956. Determination of the lifespan of human blood platelets using labeled diisopropylfluorophosphate. J. Clin. Invest., 35:964–969.

Lewis, J. Jr., and R. Hertz. 1966. Effects of early embryectomy and hormonal therapy on the fate of the placenta in pregnant Rhesus monkeys. Proc. Soc. Exp. Biol. Med., 123:805–809.

Ling, N. R., S. Knight, D. Hardy, D. R. Stanworth, and P. J. L. Holt. 1967. Antibody-induced lymphocyte transformation *in vitro*. *In* Antilymphocyte Serum, Wolstenholme, G. E. W., and O'Connor, M., eds., Ciba Foundation Study Group No. 29, Boston, Little, Brown and Co., pp. 41–56.

Lipschütz, A., and A. Audova. 1921. The comparative atrophy of the skeletal muscle after cutting the nerve and after cutting the tendon. J. Physiol. (London), 55:300–304.

Lissak, K., A. Tigyi, G. Hollosi, J. Benedeczky, and P. Juhász. 1963. The role of neural regulation in the nucleic acid metabolism of the striated muscle. *In* The Effect of Use and Disuse on Neuromuscular Functions, Gutmann, E. and Hník, P., eds., Amsterdam, Elsevier Publ. Co., pp. 425–429.

Lord, B. I. 1967. Erythropoietic cell proliferation during recovery from acute haemorrhage. Brit. J. Haemat., 13:160–167.

Lucarelli, G., D. Howard, B. Leventhal, and F. Stohlman, Jr. 1964a. The effect of nephrectomy and hypertransfusion on neonatal erythropoiesis. J. Clin. Invest., 43:1259.

——— D. Howard and F. Stohlman, Jr. 1964b. Regulation of erythropoiesis. XV. Neonatal erythropoiesis and the effect of nephrectomy. J. Clin. Invest., 43:2195–2203.

Lycette, R. R., and G. E. Pearmain. 1963. Further observations on antigen-induced mitosis. Lancet, 2:386.

Lyne, A. G. 1965. The hair cycle in the chinchilla. *In* Biology of the Skin and Hair Growth, Lyne, A. G. and Short, B. F., eds., Sydney, Angus and Robertson, pp. 467–489.

Margulis, L. 1968. Evolutionary criteria in thallophytes: A radical alternative. Science, 161:1020–1022.

Marin-Padilla, M., and K. Benirschke. 1963. Thalidomide induced alterations in the blastocyst and placenta of the armadillo, *Dasypus novencinctus mexicanus*, including achoriocarcinoma. Amer. J. Path., 43:999–1016.

——— and K. Benirschke. 1965. Thalidomide injury to the myocardium of armadillo embryo. J. Embryol. Exp. Morph., 13:235–241.

Marques Pereira, J. P., and C. P. Leblond. 1963. Mode of renewal of epithelial cells in esophagus of the albino rat. Anat. Rec., 145:257.

Matsumoto, S., T. Kishii, Y. Ito, and T. Kobayashi. 1965. Nucleic acid metabolism in experimentally hypertrophied myocardium. Jap. Heart J., 6:443–451.

Matter, M., J. R. Hartmann, J. Kautz, Q. B. DeMarsh, and C. A. Finch. 1960. A study of thrombopoiesis in induced acute thrombocytopenia. Blood, 15:174–185.

McCallister, B. D., and A. L. Brown, Jr. 1965. A quantitative study of myocardial mitochondria in experimental cardiac hypertrophy. Lab. Invest., 14:692–700.

McCulloch, E. A., L. Siminovitch, and J. E. Till. 1964. Spleen-colony formation in anemic mice of genotype WW^v. Science, 144:844–846.

——— L. Siminovitch, J. E. Till, E. S. Russell, and S. E. Bernstein. 1965. The cellular basis of the genetically determined hemopoietic defect in anemic mice of genotype Sl/Sl^d. Blood, 26:399–410.

Meerson, F. Z., T. A. Zaletayeva, S. S. Lagutchev, and M. G. Pshennikova. 1964. Structure and mass of mitochondria in the process of compensatory hyperfunction and hypertrophy of the heart. Exp. Cell Res., 36:568–578.

——— and T. L. Zayats. 1960. Change in the intensity of protein synthesis in the myocardium during compensatory hyperfunction of the heart. Bull. Exp. Biol. Med. USSR, 50:33–36.

Mellman, W. J., and H. M. Rawnsley. 1966. Blastogenesis in peripheral blood lymphocytes in response to phytohemagglutinin and antigens. Fed. Proc., 25:1720–1722.

Menton, D. N. 1968. The effects of essential fatty acid deficiency on the skin of the mouse. Amer. J. Anat., 122:337–356.

Mirand, E. A., and A. S. Gordon. 1965. Action of estrogen in erythropoiesis. Amer. Zool., 5:716.

Moores, R. R., E. Gardner, Jr., C. S. Wright, and J. P. Lewis. 1966. Potentiation of purified erythropoietin with serum proteins. II. Serial dose response relationships. Proc. Soc. Exp. Biol. Med., 123:618–620.

Morpurgo, B. 1897. Über Activitäts-Hypertrophie der willkurlichen Muskeln. Virchow's Arch. Path. Anat., 150:522–554.

Mowbray, J. F. 1963. Effect of large doses of an α_2-glycoprotein fraction on the survival of rat skin homografts. Transplantation, 1:15–20.

Murphy, G. P., E. A. Mirand, G. S. Johnston, R. P. Gibbons, R. L. Jones, and W. W. Scott. 1967. Erythropoietin release associated with Wilms' tumor. Johns Hopkins Med. J. 120:26–32.

Naets, J. P. 1958. The kidney and erythropoiesis. Nature (London), 182:1516–1517.

——— and M. Wittek. 1966. Mechanism of action of androgens on erythropoiesis. Amer. J. Physiol., 210:315–320.

Nathan, D. G., E. Schupak, F. Stohlman, Jr., and J. P. Merrill. 1964. Erythropoiesis in anephric man. J. Clin. Invest., 43:2158–2165.

Niece, R. L., E. C. McFarland, and E. S. Russell. 1963. Erythroid homeostasis in normal and genetically anemic mice: Reaction to induced polycythemia. Science, 142:1468–1469.

Norman, T. D., and R. D. McBroom. 1958. Cardiac hypertrophy in rats with phenylhydrazine anemia. Circ. Res., 6:765–770.

Nowell, P. C. 1960. Phytohemagglutinin: an initiator of mitosis in cultures of normal human leukocytes. Cancer Res., 20:462–466.

Oakberg, E. F. 1956. A description of spermiogenesis in the mouse and its use in analysis of the cycle of the seminiferous epithelium and germ cell renewal. Amer. J. Anat., 99:391–413.

Odell, T. T., Jr., T. P. McDonald, and C. W. Jackson. 1965. Hypertransfusion of blood platelets. Blood, 25:609.

——— F. G. Tausche, and J. Furth. 1955. Platelet lifespan as measured by transfusion of isotopically labeled platelets into rats. Acta Haemat., 13:45–52.

Orlic, D., A. S. Gordon, and J. A. G. Rhodin. 1966. An ultrastructural study of erythropoietin-induced red cell formation in mouse spleen. J. Ultrastruct. Res., 13:516–541.

Osnes, S. 1960. Influence of the pituitary on the erythropoietic principle produced in the kidney. Brit. Med. J., 1:1153–1157.

Osoba, D., and J. F. A. P. Miller. 1964. The lymphoid tissues and immune responses of mice bearing thymus tissue in millipore diffusion chambers. J. Exp. Med., 119:177–194.

Otenasek, F., and F. C. Lee. 1941. Further observations on thrombocytopen. J. Lab. Clin. Med., 26:1266–1273.

Palmer, J. G., E. J. Eichwald, G. E. Cartwright, and M. M. Wintrobe. 1953. The experimental production of splenomegaly, anemia and leukopenia in albino rats. Blood, 8:72–80.

Paplanus, S. H., M. J. Zbar, and J. W. Hays. 1958. Cardiac hypertrophy as a manifestation of chronic anemia. Amer. J. Path., 34:149–159.

Pater, J. L., and R. R. Kohn. 1967. Turnover of structural protein fraction in denervated muscle. Proc. Soc. Exp. Biol. Med., 125:476–481.

Patt, H. M., M. A. Maloney, and E. M. Jackson. 1957. Recovery of blood neutrophils after acute peripheral depletion. Amer. J. Physiol., 188:585–592.

Paul, J. T. 1942. The effect of splenic extracts from cases of essential thrombocytopenic purpura on the platelets and hematopoietic organs of rabbits. J. Lab. Clin. Med., 27:754–762.

Payling Wright, G. 1963. Cellular reactions to injurious agents as seen in muscle. *In*

Symposium on Current Research in Muscular Dystrophy, Philadelphia, J. B. Lippincott Co., pp. 63–72.

Pavlovic-Kentera, V., D. P. Hall, C. Bragosa, and R. D. Lange. 1965. Unilateral renal hypoxia and the production of erythropoietin. J. Lab. Clin. Med., 65:577–588.

Pearmain, G., R. R. Lycette, and P. H. Fitzgerald. 1963. Tuberculin-induced mitosis in peripheral blood leucocytes. Lancet, 1:637–638.

Perey, B., Y. Clermont, and C. P. Leblond. 1961. The wave of the seminiferous epithelium in the rat. Amer. J. Anat., 108:47–77.

Pinkus, H. 1951. Examination of the epidermis by the strip method of removing horny layers. I. Observations on thickness of the horny layer, and on mitotic activity after stripping. J. Invest. Derm., 16:383–386.

——— 1952. Examination of the epidermis by the strip method. II. Biometric data on regeneration of the human epidermis. J. Invest. Derm., 19:431–447.

Plunket, D. C., W. H. Crosby, F. Reiz, and H.-J. Huser. 1965. The effect of orally administered bovine spleen preparations on platelet and leukocyte counts of rats. Blood, 26:757–764.

Pritchard, J. J., and A. St. G. Huggett. 1947. Experimental foetal death in the rat: histological changes in the membranes. J. Anat., 81:212–224.

Quastler, H., and F. G. Sherman. 1959. Cell population kinetics in the intestinal epithelium of the mouse. Exp. Cell. Res., 17:420–438.

Quittner, H. N., N. Ward, L. Sussman, and W. Antopol. 1951. Effect of massive doses of cortisone on peripheral blood and bone marrow of mouse. Blood, 6:513–521.

Ramalingaswami, V., and H. M. Sinclair. 1953. The relation of deficiencies of vitamin A and of essential fatty acids to follicular hyperkeritosis in the rat. Brit. J. Derm., 165:1–22.

Reynafarje, C., J. Ramos, J. Faura, and D. Villavicencio. 1964. Humoral control of erythropoietic activity in man during and after altitude exposure. Proc. Soc. Exp. Biol. Med., 116:649–650.

Rhodin, J., and T. Dalhamn. 1956. Electron microscopy of the tracheal ciliated mucosa in the rat. Z. Zellforsch. Mikroskop. Anat., 44:345–412.

Richter, G. W., and A. Kellner. 1963. Hypertrophy of the human heart at the level of fine structure. An analysis and two postulates. J. Cell Biol., 18:195–206.

Rigas, D. A., and E. A. Johnson. 1964. Studies on the phytohemagglutinin of *Phaseolus vulgaris* and its mitogenicity. Ann. N. Y. Acad. Sci., 113:800–818.

Rolovic, Z., M. Baldini, and W. Dameshek. 1966. Study of megakaryocytopoiesis in splenectomized rats. Fed. Proc., 25:703.

Rosenbaum, J. L., and F. M. Child. 1967. Flagellar regeneration in protozoan flagellates. J. Cell Biol., 34:345–364.

Rossi, G. F., and M. A. Dianzani Mor. 1958. Iprocessi ossidativi del miocardio nell'ipertrofia sperimentale di cuore. Sperimentale, 108:328–337.

Rothberg, S., R. G. Crounse, and J. L. Lee. 1961. Glycine-C^{14} incorporation into the proteins of normal stratum corneum and the abnormal stratum corneum of psoriasis. J. Invest. Derm., 37:497–504.

——— and E. J. Van Scott. 1964. Absence of normal epidermal protein in basal cell tumor. J. Invest. Derm., 42:141–143.

Ruud, J. T. 1965. The ice fish. Sci. Amer., 213(5):108–114.

Sagan, L. 1967. On the origin of mitosing cells. J. Theor. Biol., 14:225–274.

Saunders, H. L., and S. L. Leonard. 1955. The effect of testosterone propionate on muscle glycogen in alloxan diabetic rats. Endocrinology, 57:291–295.

Saunders, J. W., Jr., and J. F. Fallon. 1966. Cell death in morphogenesis. *In* Major Problems in Developmental Biology, Locke, M. ed., The 25th Symposium of the Society for Developmental Biology. New York, Academic Press, Inc., pp. 289–314.

Schapira, G., J. C. Dreyfus, and F. Schapira. 1950. L'anabolisme proteique du muscle atrophie en periode de croissance. C. R. Soc. Biol. (Paris), 144:829–832.

——— J. Kruh, J. C. Dreyfus, and F. Schapira. 1960. The molecular turnover of muscle aldolase. J. Biol. Chem., 235:1738–1741.

Shemin, D., and R. Rittenberg. 1945. The utilization of glycine for the synthesis of a porphyrin. J. Biol. Chem., 159:567–568.

Shen, S. C., and T. Hoshino. 1961. Study of humoral factors regulating the production of leukocytes. I. Demonstration of a "Neutropoietin" in the plasma after administration of triamcinolone to rats. Blood, 17:434–443.

Shulman, N. R., V. J. Marder, and R. S. Weinrach. 1964. Comparison of immunologic and idiopathic thrombocytopenia. Trans. Ass. Amer. Physicians, 77:65–78.

Simon, E. J., C. S. Gross, and I. M. Lessell. 1962. Turnover of muscle and liver proteins in mice with hereditary muscular dystrophy. Arch. Biochem., 96:41–46.

Slack, H. G. B. 1954. Metabolism of limb atrophy in the rat. Clin. Sci., 13:155–163.

Sleigh, M. A. 1962. The biology of cilia and flagella. New York, The MacMillan Company.

Sodicoff, M., and R. T. Binhammer. 1968. Leukocytosis-inducing factor in the blood of X-irradiated rats. Radiat. Res., 33:82–93.

Spraragen, S. C., V. P. Bond, and L. K. Dahl. 1962. DNA synthesizing cells in rabbit heart tissue after cholesterol feeding. Circ. Res., 11:982–986.

Stefanini, M., and J. B. Chatterjea. 1951. Rate of platelet survival in thrombocytopenia. J. Clin. Invest., 30:676.

Steinberg, B. 1958. Mechanism of hematopoiesis: Factors regulating circulatory leukocytes. Arch. Path., 65:237–243.

Stewart, D. M. 1955. Changes in the protein composition of muscles of the rat in hypertrophy and atrophy. Biochem. J., 59:553–558.

——— 1966. Factors affecting the weight of the cremaster muscle. Amer. J. Physiol., 210:1239–1242.

Stohlman, F., Jr. 1960. Observations on the changes in the kinetics of red cell proliferation following irradiation and hypertransfusion. Blood, 16:1777–1787.

——— 1961. Humoral regulation of erythropoiesis. VI. Mechanism of action of erythropoietin in the irradiated animal. Proc. Soc. Exp. Biol. Med., 107:751–754.

Suftin, D. C., J. D. Thomson, and H. M. Hines. 1954. Kinetics of denervation atrophy in the skeletal muscles of the rat. Amer. J. Physiol., 178:535–537.

Sulkin, N. M., and D. F. Sulkin. 1965. An electron microscopic study of the effects of chronic hypoxia on cardiac muscle, hepatic, and autonomic ganglion cells. Lab. Invest., 14:1523–1546.

Summers, T. B., and H. M. Hines. 1951. Effect of immobilization in various positions upon the weight and strength of skeletal muscle. Arch. Phys. Med., 32:142–145.

Sumner, R. G., and H. D. McIntosh. 1963. Nucleic acid studies in experimental cardiomegaly. Circ. Res., 12:170–175.

Tamm, S. 1967. Flagellar development in the protozoan *Peranema trichophorum*. J. Exp. Zool., 164:163–186.

Thomsen, P., and J. V. Luco. 1944. Changes of weight and neuromuscular transmission in muscles of immobilized joints. J. Neurophysiol., 7:245–252.

Till, J. E., and E. A. McCulloch. 1961. A direct measurement of the radiation sensitivity of normal mouse bone marrow cells. Radiat. Res., 14:213–222.

——— Siminovitch, and E. A. McCulloch. 1967. The effect of plethora on growth and differentiation of normal hemopoietic colony-forming cells transplanted in mice of genotype W/W^v. Blood, 29:102–113.

Tobian, L., O. Severseike, and J. Cich. 1960. Do mitochondria participate in general cardiac hypertrophy? Proc. Soc. Exp. Biol. Med., 103:774–777.

Tocantins, L. M. 1936. Experimental thrombopenic purpura: cytolological and physical changes in the blood. Ann. Intern. Med., 9:838–849.

——— 1938. The mammalian blood platelet in health and disease. Medicine, 17:155–260.

Tomita, K. 1966. Studies on myocardial protein metabolism in cardiac hypertrophy. Jap. Heart J., 7:566–589.

Troland, C. E., and F. C. Lee. 1938. Thrombocytopen. A substance in the extract from the spleen of patients with idiopathic thrombocytopenic purpura that reduces the number of blood platelets. J.A.M.A., 111:221–226.

Uihlein, A. 1942. Effect of injection of tissue extracts on the number of blood platelets. J. Lab. Clin. Med., 28:157–162.

Van Scott, E. J. 1965. Replacement kinetics of integumental epithelia. *In* Biology of the Skin and Hair Growth, Lyne, A. G. and Short, B. F., eds., Sydney, Angus and Robertson, pp. 399–408.

——— and T. M. Ekel. 1963. Kinetics of hyperplasia in psoriasis. Arch. Derm., 88:373–381.

Venable, J. H. 1966. Morphology of the cells of normal, testosterone-deprived and testosterone-stimulated levator ani muscles. Amer. J. Anat., 119:271–302.

Wainman, P., and G. C. Shipounoff. 1941. The effects of castration and testosterone propionate on the striated perineal musculature in the rat. Endocrinology, 29:975–978.

Wasastjerna, C., H. Teir, and A. Larmo. 1958. The effect of heterologous bone marrow extracts on the leukocytes of rats. Acta Path. Microbiol. Scand., 44:80–87.

Wessells, N. K. 1963. Effects of extra-epithelial factors on incorporation of thymidine by embryonic epidermis. Exp. Cell Res., 30:36–55.

——— 1967. Differentiation of epidermis and epidermal derivatives. New Eng. J. Med., 277:21–33.

West, W. T. 1963. Muscular dystrophy of vitamin E deficiency. *In* Muscular Dystrophy in Man and Animals, Bourne, G. H. and Golarz, M. N., eds., New York, Hafner Publ. Co., pp. 367–405.

Whitcomb, W. H., and M. Z. Moore. 1965. The inhibitory effect of plasma from hypertransfused animals on erythrocyte iron incorporation in mice. J. Lab. Clin. Med., 66:641–651.

Whitehorn, W. V., and A. F. Grimmenga. 1956. Effect of exercise on properties of the myocardium. J. Lab. Clin. Med., 48:959.

Winter, G. C. B., and J. M. Yoffey. 1965. Cytoplasmic labelling with uridine-5-^{3}H in human lymphocytes cultured with phytohemagglutinin. Nature (London), 208:1018–1019.

Wollenberger, A., and W. Schulze. 1961. Mitochondrial alterations in the myocardium of dogs with aortic stenosis. J. Biophys. Biochem. Cytol., 10:285–288.

Young, R. W. 1962. Cell proliferation and specialization during endochondrial osteogenesis in young rats. J. Cell Biol., 14:357–370.

——— 1967. The renewal of photoreceptor cell outer segments. J. Cell Biol., 33:61–72.

Zanjani, E. D., J. F. Contrera, A. S. Gordon, G. W. Cooper, K. K. Wong, and R. Katz. 1967. The renal erythropoietic factor (REF). III. Enzymatic role in the erythropoietin production. Proc. Soc. Exp. Biol. Med., 125:505–508.

Zelená, J. 1962. The effect of denervation on muscle development. *In* The Denervated Muscle, Gutmann, E., ed., Prague, Publ. House of the Czechoslovak Acad. Sci., pp. 103–126.

Harold P. Rusch

McArdle Laboratory for Cancer Research, Medical Center, University of Wisconsin, Madison, Wisconsin

7

Some Biochemical Events in the Life Cycle of *Physarum polycephalum*

INTRODUCTION

Physarum polycephalum, a myxomycete or true slime mold, has been gaining popularity for studies on the biochemical events associated with growth and differentiation ever since it became possible to culture it in a defined axenic medium. The characteristics which make it an attractive organism for such studies are the large size of the plasmodium, the naturally occurring synchronous mitosis in the plasmodium, and the fact that growth and differentiation occur in readily recognizable stages, which are separate and distinct from each other (Rusch, 1962). It is also ideal for studying the primitive motive forces involved in protoplasmic movement, for investigating the effect of growth inhibitors (Becker et al., 1963), and for studying the influence of various compounds on metabolism (de Meester et al., 1967).

This brief review will be concerned chiefly with the biochemistry of the organism, but the subject of the replication and transcription of nuclear DNA during the growth cycle is not reviewed in depth since a recent review by Cummins (1968), which the reader is urged to consult, has covered this subject admirably. There are more complete general reviews on the myxomycetes by Alexopoulos (1962,

1963, 1966), by von Stosch (1965), and by Gray and Alexopoulos (1968). Although these reviews are concerned chiefly with morphology, they also contain information of biochemical interest. The literature on protoplasmic streaming is far from complete, but a number of comprehensive papers on this subject by Kamiya (1965, 1966) and by various authors in the book by Allen and Kamiya (1964) are available, and many of the most recent papers on the subject have been included in the present chapter.

Physarum has been classed as a slime mold, probably because of the morphological similarity of the fruiting structures to those of fungi. Such structures, of course, are not necessarily characteristic of fungi, since they have also been found in other classes of organisms such as the Myxobacteriales. The myxomycetes resemble fungi in their sporulation pattern, but are probably more closely related to protozoa in other characteristics (Kudo, 1954). For example, the myxamebas and the swarm cells which emerge from germinating spores are very similar to protozoa, the mitochondria of the plasmodia bear a morphological similarity to those in protozoa, and the pattern of the unsaturated fatty acids of *Physarum* is identical to those in the soil amebas *Acanthamoeba* and *Hartmannella,* but not similar to those in fungi or even in *Dictyostelium discoideum* (Korn et al., 1965). An article on the classification of the myxomycetes and related organisms has appeared recently (Cohen, 1967). Attempts to classify these and other organisms on the basis of their tryptophan pathway (Hütter and DeMoss, 1967), their type of cytochrome *c* (Yamanaka et al., 1962; Yamanaka, 1967), and their pattern of unsaturated fatty acids (Korn et al., 1965) have been reported. At all events, it is plausible that the myxomycetes arose from protozoa and became diverted in an evolutionary cul-de-sac to form a separate class with certain features common to both uni- and multicellular forms.

LIFE CYCLE

A diagrammatic representation of the life cycle of *Physarum polycephalum* is shown in Figure 1. The stages are distinct and separate, and pass from one to another in an orderly sequence initiated by changes in the environment. The plasmodium is a yellow, flat mass of multinucleated protoplasm. It is classed as a true slime mold because it is acellular, but since mitosis is synchronous even when the plasmodium is 7 cm in diameter, the plasmodium may be regarded as a giant "cell" biochemically. During this vegetative period the plasmodium continues to grow as long as nutrients are adequate. When food becomes limited the plasmodium has two courses for survival: it may form a sclerotium, which consists of many small spherules, each containing one or more diploid nuclei, or it may sporulate, provided that the plasmodium has been starved for a minimum of 4 days and then exposed to light for a few hours. It is only at the end of sclerotial or sporangial formation that the organism may be considered as being "multicellular"—during the remainder of its life cycle it is either acellular or unicellular. Spores containing single haploid nuclei are formed, which germinate to form amebas (myxamebas) when they come in contact with a suitable moist environment. These amebas grow, have ameboid

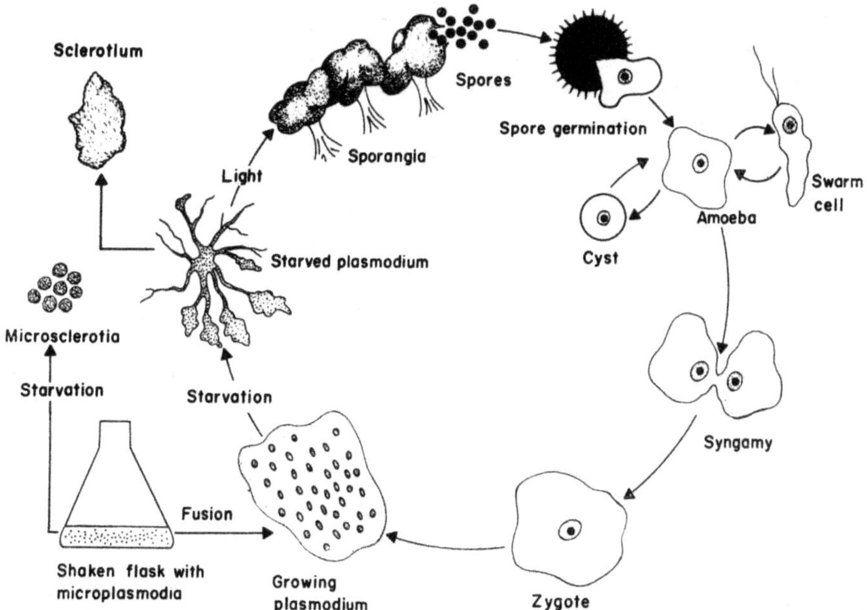

FIG. 1. Diagrammatic representation of the life cycle of *Physarum polycephalum*. The various stages are not drawn to scale; e.g., the plasmodium is much larger than the ameba or zygote, and the spores are relatively much smaller than the size shown. The plasmodium is multinucleated, as depicted by the many small nuclei. Sclerotia can be induced to grow again by placing them in a suitable medium, but arrows to indicate this are not shown to avoid cluttering the figure.

movement on nutrient surfaces, and are similar in size and appearance to the soil amebas.

When the amebas that have been growing on a nutrient surface are placed in a less concentrated medium, they are transformed into motile swarm cells containing one or two flagella. Ross (1967a) reports such transformation when the amebas of *Perichaena vermicularis* are changed from a thin film of medium to a thicker one, and N. S. Kerr (1965) observed a similar change when *Didymium nigripes* is placed in a 0.05M Na-K-phosphate buffer at pH 6.5. Flagellum formation in the latter was inhibited by streptomycin. Aldrich (1968), examining *Physarum flavicomum* with the electron microscope, observed two basal bodies prior to flagellum development and noted that the flagellum in the swarm cells had the typical "nine plus two" arrangement of the microtubules. The species studied by Ross, by N. S. Kerr, and by Aldrich are also myxomycetes, and it is reasonable to assume, therefore, that flagellum formation in the latter is very similar to that observed in the other species. The biochemical changes that accompany the shift from one form of these cells to the other would make an interesting study. The amebas may also survive adverse conditions by becoming encysted; thus *Physarum* has a third pathway to follow in the event of an unfavorable environment.

After a certain period, two amebas or two flagellated cells may fuse to form a zygote, which forms a plasmodium by nuclear division and accompanying growth of protoplasm. If zygotes happen to be close together they may coalesce with one another to form a plasmodium or they may also coalesce directly with plasmodia. In this review, I will follow Ross' (1967c) suggestion and use the term coalescence to describe the nonsexual union between zygotes, between zygotes and plasmodia, and between plasmodia. The term fusion or syngamy will be used to indicate the sexual union of amebas, which normally results in karyogamy. The exact sequence of events involved in syngamy and plasmodial formation is poorly understood for *Physarum.* Most of the information about these events has been obtained from the study of closely related species such as *Didymium nigripes* (N. S. Kerr, 1967) and *Didymium iridis* (Ross, 1967c, and Collins, 1961), and a general discussion of this subject is presented by von Stosch (1965). At all events, Dee (1962, 1966a, 1966b) has demonstrated that two amebas will fuse if they have different mating types. Four mating types (mt_1 to mt_4) have been found. Dee has also selected mutant strains of amebas and has demonstrated recombination between genetic markers. She has shown more recently that the coalescence of plasmodia is controlled by four alleles (f_1 to f_4) of one gene (f), that this locus is without effect on the fusion of amebas, and that the *mt* locus is without effect on the coalescence of plasmodia and is unlinked to f (Carlile and Dee, 1967, Poulter and Dee, 1968). She suggests that the action of the f factors is to inhibit fusion between dissimilar strains rather than to promote fusion between identical strains. In one strain of *P. polycephalum,* von Stosch (1965) observed a variation from the usual sexual cycle. He reports the occurrence of apogamy, in which meiosis in the spores is omitted and the germinating spores remain diploid, yet the amebas undergo nuclear division and form plasmodia.

In summary, there are several distinct and separate stages in the life cycle of *Physarum,* each offering a model system for certain studies concerning specific features of growth and differentiation. There are two vegetative stages (plasmodium and ameba) for investigations on growth in the absence of differentiation; and differentiation may be studied free of growth during the periods of sporulation, sclerotial formation, during the shift between amebas and flagellated cells, or during the formation of plasmodia from amebas. The organism is acellular or unicellular during its entire life cycle except at the very end of sporulation, when it becomes cellular with the formation of uninucleate spores. During the plasmodial stage, nuclear division occurs in the absence of cellular division, and during sclerotial formation, cell cleavage occurs without mitosis. Thus, one can study separately the processes associated with cytokinesis and karyokinesis. The organism survives adverse conditions in three ways: by sporulation, by sclerotial formation, or by the encystment of the amebas.

METHODS OF CULTURE

Daniel and Rusch (1961) first reported the successful axenic culture of the plasmodium of *Physarum polycephalum* on a semidefined medium. Subsequent studies showed that *Physarum* could be grown on a completely defined medium consisting of three amino acids (methionine, glycine, and arginine), two vitamins (biotin and thiamin), glucose, hematin, and a salts mixture (Daniel et al., 1962, 1963). The growth rate is increased considerably, however, when 13 amino acids are used, and the rate is further improved when tryptone (Difco) is used in place of amino acids. For routine culture, we used a semidefined medium, since such medium is not only easier to prepare but supports a more rapid rate of growth. In this medium tryptone (Difco) and yeast extract (Difco) are used in place of amino acids and vitamins, respectively. Cultures are maintained in shaken flasks since it is easier to maintain sterility, to obtain better aerobic conditions, and to make routine transfers. Such flasks are shaken at 170 reciprocations/min and incubated in the dark at 21 to 22°C. At this temperature the length of the growth phase is longer than at higher temperatures, thus increasing the time between routine transfers. Most studies on mitosis are done on surface cultures prepared by permitting microplasmodia from the shaken flask to coalesce on a surface such as a millipore membrane or on filter paper (Daniel and Baldwin, 1964; Guttes and Guttes, 1964a). The surface cultures are incubated at 26°C to decrease the intermitotic phase somewhat and thus hasten the period at which observations of mitosis and various chemical determinations can be made. Under these conditions the mitotic cycle of the plasmodium is about 8 hours. A plasmodium which is about 1 cm in diameter immediately after coalescence attains a size of approximately 8 cm in 24 hours, by which time the plasmodium has completed mitosis III and the S period, and yields about 17 mg of protein, 3.5 mg of RNA, and 300 μg of DNA. Microplasmodia may also be cultured in 30-liter or 50-gallon conventional baffled fermentors yielding 6 to 10 g (dry weight) per liter (Brewer et al., 1964; Brewer, 1965), and surface cultures 14 cm in diameter with nearly synchronous mitosis may also be obtained (Mohberg and Rusch, 1967, 1969). In starved plasmodia, the nuclei undergo synchronous mitoses at intervals of 24 to 39 hours. Addition of growth medium to starved plasmodia at late interphase caused a delay of mitosis, whereas the addition of growth medium during early interphase advanced the onset of mitosis (Guttes and Guttes, 1962).

GROWTH CYCLE—PLASMODIUM

Mitosis

CYTOLOGY. At approximately one hour before metaphase, the centrally located nucleolus starts to swell and begins to move towards the nuclear membrane. Shortly thereafter the chromosomes start to condense and begin to move away from the nuclear membrane. These events continue slowly, and the nucleolus begins to break up about 20 minutes before metaphase and is no longer distinguishable at about 5 minutes before metaphase; the rest of mitosis is completed about 10 minutes after the appearance of metaphase. As soon as the nuclei have divided, many tiny, densely staining, irregular bodies appear in the nucleus, and these gradually fuse over a period of about 2 hours to form the nucleolus. This period of nuclear reconstruction is concurrent with the S period, except that a low level of DNA synthesis continues for another hour. The changes beginning with early prophase and continuing through reconstruction are characteristic, and the experienced investigator can thus estimate the approximate time of mitosis at any time during this 3-hour period (Guttes et al., 1961; Guttes and Guttes, 1964a). For very accurate timing of mitosis, however, it is necessary to make frequent examinations with the phase contrast microscope of pieces of the plasmodium smeared on a glass slide. For determinations of the effect of various treatments on mitosis, the plasmodium may be cut into pieces, some of which are subjected to the experimental changes, with the other pieces serving as the controls.

One of the unusual features of mitosis in the plasmodium, as seen with the light microscope, is that the nuclear membrane does not disappear (Guttes et al., 1961; S. J. Kerr, 1967). Examination of the nuclear membrane with the electron microscope, however, raises some question about the integrity of this membrane during mitosis. Both Kessler (1964) and S. Guttes et al. (1968) observed a partial dissolution in late anaphase or early telophase. However, Goodman and Ritter (in press), who have used a different method of fixation, feel that the membrane remains intact during mitosis. Aldrich (1969) noted a deformation of the nuclear envelope near the poles, although most of the envelope remained intact during mitosis in *Physarum flavicomum*. At all events, if there is disintegration of the nuclear membrane, it is only partial and for a very brief period, and restoration occurs within a matter of minutes. Although there may be some question about the integrity of the nuclear membrane in the plasmodium, Goodman and F. Haugli of our laboratory (unpublished experiments) observed that the nuclear membrane of the myxameba of *Physarum polycephalum* disintegrates during mitosis, and Koevenig (1964), Schuster (1965), S. J. Kerr (1967), Ross (1967a, 1968), and Aldrich (1969) also report such disintegration during mitosis in the myxamebas of closely related species. Aldrich (1969) also comments on the presence of the centriole in the myxameba and its absence in the plasmodium.

Another feature of mitosis noted with the electron microscope is the highly irregular shape of the nuclear membrane during anaphase; this suggests a vigorous movement of the membrane during this stage. It is also of interest that a clump of fibers with a diameter of about 70 Å and some other material are left in the center of the interzone when the daughter chromosomes separate at anaphase, but there are no spindle fibers in the interzone connecting the daughter plates (Guttes et al., 1968). Crawley (1966) noted that some nucleoli in interphase contain particles, about 200 Å in diameter, apparently associated with a coarse filamentous structure. He also stated that the spindle fibers in dividing nuclei were tubules about 220 Å in diameter with walls about 60 Å thick.

CHROMOSOMES. Counts on the number of chromosomes have not been in complete agreement, but the reports by Ross and by Koevenig and Jackson are among the most recent and perhaps the most reliable on this subject. Ross' (1966) observations were made on squashed preparations of nuclei in metaphase. In one strain obtained from the McArdle Laboratory and grown in pure culture for many years, the chromosome count was 50, whereas in another strain obtained from the Carolina Biological Supply House only 58 percent of the plasmodia had a chromosome count of 50. Twenty-five percent of the nuclei had 90 chromosomes, whereas the remainder had counts of 20, 40, 70, and 180. The more variable count of the latter strain may be due to the fact that this "wild" strain had been grown in pure culture for only a short period, whereas the strain from the McArdle Laboratory had been in pure culture for several years and was thus adapted to a more uniform environment. Koevenig and Jackson (1966) reported about 56 chromosomes in a strain also obtained originally from the McArdle Laboratory.

Synthesis of Macromolecules during the Cycle

DNA. The synthesis of DNA in the plasmodium starts immediately after telophase (there is no G1 period), reaches its full rate within 10 minutes, continues at a high rate for about 2 hours, and then decreases rapidly to a very low level by 3 hours after telophase. Autoradiographic experiments with ^{3}H-thymidine confirm the chemical determinations and show that DNA synthesis starts simultaneously in more than 99 percent of the nuclei, but there is some variation in the duration of the synthesis among individual nuclei (Nygaard et al., 1960; Sachsenmaier, 1964; Braun et al., 1965). In late interphase less than 1 percent of the nuclei become labeled, and these are probably a few abnormal nuclei which are heavily labeled (Guttes et al., 1967).

The nuclei maintain the mechanism for the control of DNA synthesis even when isolated from the plasmodium. Nucleotide triphosphates are incorporated into the DNA of nuclei, but only in those isolated during the S period (Brewer and Rusch, 1965).

There is also an ordered temporal sequence in the synthesis of DNA, as shown by double labeling experiments. Tritiated thymidine was supplied to the plasmodium in the second half of one S period, and then ^{14}C-bromodeoxyuridine was added to the same culture before the start of the next S period to make the DNA synthesized in the second S period denser than the DNA replicated during the first S period. The DNA was isolated from the plasmodium at various times during the second S period and centrifuged in a cesium chloride equilibrium gradient. The new heavy DNA contained no tritium when the DNA was extracted one hour or less after the second mitosis, but after one complete cycle of replication the DNA was labeled with both tritium and ^{14}C. These results showed that DNA molecules which were replicated during the second half of one S period were also replicated during the second half of the following S period (Braun et al., 1965). In a similar experiment, Braun and Willi (1969) divided the S period into five parts and measured DNA synthesis during these intervals; the results were in agreement with the earlier observations.

The molecular weight of single-stranded DNA from the plasmodium has been determined by alkaline gradient centrifugation (McGrath and Williams, 1967). The average weight during the S period ($\sim 1.5 \times 10^{7}$ daltons) was less than that during the G2 period ($\sim 4 \times 10^{7}$ daltons). On the basis of a chromosome number of 50 per nucleus and a DNA content of 1 $\mu\mu$g per nucleus, they concluded that, at pH 12, each chromosome dissociates into 300 (single-stranded) pieces of DNA. In addition to the principal DNA, a heavy satellite of DNA has been observed (Braun et al., 1965; Guttes et al., 1967). This satellite has recently been studied in greater detail by Holt and Gurney (1969). It represents about one percent of the total DNA and is not to be confused with the mitochondrial DNA. Its synthesis proceeds throughout the cycle, with the possible exception of the S period.

RNA. The synthesis of RNA in the plasmodium also shows a rhythmical pattern. The incorporation of ^{3}H-uridine during mitosis is very low but increases immediately after telophase (simultaneously with the incorporation of thymidine into DNA), reaches a peak about 2½ hours after mitosis, decreases to a lower level, and then begins to rise again about 2 hours before metaphase (Mittermayer et al., 1964; Braun et al., 1966b). In the experiments by Rusch and associates, 10-minute incubations of the plasmodium in ^{3}H-uridine at various times during the mitotic cycle were used, but the 10-minute pulse period in mitosis exceeded the time during which chromosomes were completely condensed, and it was possible that the incorporation during this period reflected RNA synthesis just prior to or following the condensation of the chromosomes. Indeed this was the case, since Kessler (1967) observed no incorporation into nuclear RNA during metaphase and anaphase after 5-minute incubations in ^{3}H-uridine. Thus, transcription was turned off during this brief period. The rhythmical pattern of incorporation of ^{3}H-uridine into RNA, seen in the whole plasmodium, was also observed in nuclei isolated at various periods of the growth cycle (Mittermayer et al., 1966b).

It might be expected that the amounts of the different classes of RNA would vary during the cycle, and preliminary experiments have verified this expectation. Although the sucrose density gradient patterns of the various kinds of RNA were similar at different times of the growth cycle, the 18S peak was high 1 hour before mitosis, and during mitosis there was a partial shift of the labeling pattern to the lighter side (Braun et al., 1966a). Less than 3 percent of the total plasmodial RNA was localized in the nuclei, but more than 50 percent of pulse-chase labeled RNA was found in nuclei after as long a chase as 30 minutes. The sedimentation profile of this RNA was nearly identical to that of the total plasmodial RNA, which finding indicates that most of the RNA was stored in the nucleus and transferred only later to the cytoplasm (Braun et al., 1966b). Analysis of base ratios of the RNA synthesized after 10 minutes of exposure to ^{32}P-orthophosphate showed a distinct shift in the ratio as the plasmodium passed through mitosis. RNA made early in the cycle was rapidly labeled and contained a little more adenylic acid, whereas that made at the end of the cycle tended to have more guanylic acid (Cummins et al., 1966b). Another indication that different kinds of RNA are transcribed at certain times in the growth cycle was that actinomycin D reduced the incorporation of ^{3}H-uridine into RNA by about one half during the first half of the growth cycle but caused an almost complete inhibition of uridine uptake during the latter half of the cycle (Mittermayer et al., 1964).

The capacity of isolated nuclei to incorporate ^{3}H-uridine into RNA permitted a study of transcription in greater detail. Determinations of the nearest neighbor frequency of the bases in the RNA synthesized by such nuclei showed that dinucleotides beginning with adenine (A) or uridine (U) tended to decrease in relative frequency as the cell cycle progressed (Cummins and Rusch, 1967). Thus, there was a slight shift in DNA transcription from predominantly AT-rich (T = thymidine) regions early in the growth cycle to GC-rich (G = guanine, C = cytosine) regions of nuclear DNA later in the cycle. From these data, Cummins (1968) estimated that the RNA synthesized in nuclei isolated just after mitosis consists of one third ribosomal RNA and two thirds DNA-like RNA, whereas the RNA synthesized in nuclei shortly before mitosis contains two thirds ribosomal RNA and one third DNA-like RNA.

PROTEINS AND POLYSOMES. The total protein content increased at a steady rate and essentially doubled during interphase. In contrast to this steady increase in total protein, the rate of incorporation of ^{3}H-lysine into protein showed a biphasic pattern, with one peak occurring shortly after mitosis and another occurring late in interphase. These peaks paralleled those observed in the synthesis of RNA, except that ^{3}H-lysine incorporation into protein had a distinct lag phase of about 1 hour immediately after mitosis, whereas uridine incorporation into RNA began immediately after telophase (Mittermayer et al., 1966a). The polysome pattern remained stable throughout the cycle except for a brief period, about 30 minutes after mitosis (Mittermayer et al., 1966a), during which time there was an increase of ribosomes and a decrease of polysomes. It is not known whether the brief breakdown of polysomes is related to the lag phase of lysine incorporation. Recently Schiebel et al. (1969) observed that heat shock of 40°C for periods of 10 or 40 minutes reduced the uptake of amino acids into protein and caused a reduction of polyribosomes of more than 50 percent. These results suggest that heat shock interferes with protein synthesis by decreasing the amount of polyribosomes.

ENZYMES AND HISTONES. Only a few enzymes have been examined for their activity during the growth cycle. Sachsenmaier and Ives (1965) reported that the specific activity of glucose-6-phosphate dehydrogenase remained almost constant during the growth cycle, whereas thymidine kinase activity started to increase approximately 50 minutes prior to prophase (about 1¼ hours before metaphase) and reached its maximum at the end of mitosis. Brewer and Rusch (1966) observed that DNA polymerase activity increased about 1 hour before metaphase. This was discovered when spermine and exogenous DNA were added to isolated nuclei. The addition of spermine and DNA not only increased the incorporation of ^{3}H-dATP into acid-insoluble material from isolated nuclei tenfold, but this increase was sufficient to demonstrate the presence of the polymerase prior to the S period.

Histone was extracted from isolated nuclei, and the amount paralleled the amount of DNA during the cycle; this indicates that histone and DNA are synthesized concurrently. The histone was separated into three major and two minor bands in polyacrylamide gel (Mohberg and Rusch, 1964). Methionine metabolism in plasmodia was also investigated, but not in relation to the growth cycle (Daniel and Babcock, 1966).

REGULATORY MECHANISMS

Mitotic synchrony may result from a built-in mechanism within each nucleus that initiates mitosis in every nucleus at the same time, but it does not appear plausible that such a mechanism would function repeatedly with a high degree of precision. It seems more reasonable that synchronous control is mediated through uniformly distributed substances in the cytoplasm, and this view is supported by results obtained when plasmodia in various stages of the growth cycle were coalesced with one another, as summarized in a review by Cummins and Rusch (1968). Recently Chin and Bernstein (1968) reported that the pool of adenosine triphosphate decreased from a high level at prophase to a minimum during mitosis and increased again in the postmitotic period. They suggested that such results reflect the energy requirements for mitosis.

Control of Mitosis

COALESCENCE OF PLASMODIA AND MITOTIC SYNCHRONY. Evidence for the view that mitotic synchrony is mediated through cytoplasmic substances may be obtained by allowing many microplasmodia to coalesce. Plasmodia grown in shaken flasks are present as many tiny microplasmodia, each having synchronous mitosis but with no synchrony among the various pieces. When such microplasmodia are placed on a surface and thus allowed to coalesce, the very first mitosis in the large coalesced plasmodium is nearly synchronous, except for a very few nuclei, and by the second mitosis after coalescence 99 percent of the nuclei in the plasmodium enter metaphase within 5 minutes of one another. This degree of synchrony continues in subsequent mitoses in plasmodia up to 7 cm in diameter (Guttes and Guttes, 1964a). Larger plasmodia show less synchrony (Guttes et al., 1961), and there is a variation of as much as 20 minutes in the time of metaphase in plasmodia that are 14 cm in diameter (Mohberg and Rusch, 1969).

Further experiments demonstrating the control of synchrony in coalesced plasmodia are the following: two plasmodia (A and B) were prepared at such times that mitosis occurred about 2 hours earlier in A than in B; a piece of plasmodium A was then superimposed on a piece of plasmodium B of about the same size, about 2 hours before mitosis in plasmodium A (the plasmodium nearest to mitosis); mitosis occurred in the coalesced pair of plasmodial pieces, not at the time expected for A or even that for B, but midway between the two times, or one hour after uncoalesced A and one hour before uncoalesced B. In the second experiment the sizes of pieces A and B were varied in ratios of A:B as 2:1 and as 1:2. Examination of the coalesced plasmodium showed that the onset of mitosis was accelerated in the nuclei of plasmodium B when the ratio A:B was 2:1, and was retarded in plasmodium A when the ratio was 1:2. In the last experiment two plasmodia were used which were timed similarly to A and B, as described above, and overlapped at one edge. When this was done, a gradient in the time of mitosis from the point of coalescence was observed. Mitosis was accelerated on the B side of the coalescence ridge and was retarded on the A side of the line of coalescence (Rusch et al., 1966).

In a somewhat related experiment, Guttes and Guttes (1963a) demonstrated that nearly all nuclei from microplasmodia which had coalesced into a large plasmodium underwent synchronous mitosis in the first postcoalescence mitosis, except for a very few that had only recently divided just prior to coalescence. It had been shown previously that plasmodia coalesce best during periods of starvation and that the addition of nutrients to coalescing plasmodia delays their coalescence. These investigators found that the vast majority of starved microplasmodia had coalesced after 1½ hours, and the coalescence of the small percentage which had not yet done so was further delayed by adding nutrients to the culture 1½ hours after the microplasmodia were first placed on filter paper. During the next few hours the remaining single microplasmodia gradually coalesced into the main plasmodium. Eventually the nuclei in the large plasmodium entered prophase, except for a small percentage of the nuclei (1.2 percent) that had just coalesced and were in the period of late reconstruction. The presence of from two to four small nucleoli in these few nuclei indicated that they had undergone mitosis within the previous 45 minutes, just prior to coalescence. This experiment shows that all nuclei can be induced to undergo mitosis after the period of nuclear reconstruction has been completed.

Recently Guttes et al. (1969) have reported on further experiments on mitotic synchrony. In these experiments, surface plasmodia in various stages of the cycle were fragmented into smaller plasmodia by placing them in a fluid medium in flasks which were shaken vigorously for 15 minutes. The fragmented plasmodia from different stages were then allowed to coalesce. Nuclei from plasmodia that were in prophase at the time of fragmentation proceeded through mitosis without appreciable delay following coalescence, but mitosis was delayed by about 4 hours in nuclei from plasmodia fragmented more than 1 hour before metaphase. When microplasmodia in a stage approximately 1 hour prior to mitosis were coalesced with microplasmodia whose nuclei were in the postmitotic stage, mitosis in the composite microplasmodium began at a time closer to that of the premitotic controls.

The above results prove that mitosis is not dependent on a previously and irrevocably determined trigger, since the timing can be altered. Therefore, the best explanation for the presence of synchronous mitosis is that substances are formed in the cytoplasm, increase in amount during the G2 period, reach a maximum just before mitosis, and are transferred to the nucleus shortly before mitosis. These experiments also show that the nucleus does not require a certain "latent" period before it can undergo mitosis but is ready to divide after nuclear reconstruction has been completed. From such experiments it appears that a G2 period is not necessary per se for mitosis, but we will indicate below some of the events that transpire during the G2 period and which are essential for mitosis. Two questions immediately arise as the result of these experiments: Is complete DNA replication required before mitosis can occur? What is the purpose of the G2 period if nuclei obtained from the very beginning of the G2 period can divide?

ROLE OF DNA. The results of the experiment on coalescing tiny plasmodia containing nuclei in the late S period with a large plasmodium just ready to enter mitosis indicate that only those nuclei from the late S period would undergo mitosis. The results suggested, but did not prove, that DNA replication was complete in such nuclei. To gain further information on this point, DNA synthesis was inhibited with FUDR (2×10^{-5}M); it was found that the onset of mitosis was greatly delayed and did not occur until some time after DNA had been completely replicated.

When the inhibiting effect of FUDR was reversed with thymidine ($4 \times 10^{-4}M$), the onset of mitosis occurred approximately 6 hours after the replication of DNA was completed. Thus, DNA replication was essential, and a G2 period of about 6 hours was required before the onset of mitosis (Rusch et al., 1964; Sachsenmaier and Rusch, 1964; Sachsenmaier, 1966).

The importance of DNA synthesis to mitosis was also shown by exposing *Physarum* to ultraviolet light. An ultraviolet lamp that emitted about 90 percent of its energy at 2,537 Å was used at a dose rate of 96.5 ergs/mm^2/sec. The total dose in most experiments was 14,500 ergs/mm^2, but some experiments were done with higher or lower total doses. The ultraviolet light delayed mitosis, and the effect was most pronounced when the exposure occurred during the first 30 percent of the intermitotic period. Since ultraviolet light also depressed DNA synthesis, it appears that one reason for the delay in mitosis was the interference with the replication of DNA (Devi et al., 1968). Somewhat similar experiments were also reported by Sachsenmaier (1966).

It appears likely from these experiments that DNA must be completely replicated before mitosis can occur. Once such replication is complete, the nucleus can enter mitosis, but under normal conditions in the growth cycle an additional period is required for the orderly transcription and synthesis of substances needed, not only for mitosis, but also for the plasmodium to increase its size. These events occur during the G2 period.

ROLE OF RNA. As one might expect, certain messenger RNA's are associated with mitosis. Actinomycin D (150 to 250 μg/ml), added any time during the greater portion of interphase, delayed the onset of mitosis by more than one hour but had little or no effect on mitosis when added during the 90-minute period prior to metaphase (Mittermayer et al., 1965). Sachsenmaier (1966) confirmed these findings and also showed that a smaller amount of actinomycin D (100 μg/ml) partially inhibited RNA synthesis without affecting the onset of mitosis. He suggested that the RNA molecules required for mitosis are less sensitive to the inhibitor than are certain other RNA molecules. More recently Sachsenmaier et al. (1967) observed that actinomycin C was a better inhibitor of RNA synthesis in *Physarum* than was actinomycin D. Actinomycin C (250 μg/ml) had a rapid effect on inhibiting the incorporation of ^{3}H-uridine, but ^{3}H-leucine incorporation decreased slowly over several hours, and the results suggested an average half-life of about 3 hours for the bulk of mRNA molecules. In this case, the inhibitor delayed mitosis when added as late as 35 minutes before telophase. However, actinomycin D did not cause a comparable inhibition of mitosis when added this late in the cycle (Sachsenmaier and Becker, 1965; Mittermayer et al., 1965). These investigators also found that actinomycin C did not inhibit thymidine kinase activity when added later than 90 minutes prior to telophase. These results demonstrate that certain mRNA's required for mitosis or for DNA replication are transcribed at different times; those needed for mitosis are completed 35 minutes before telophase, whereas the mRNA necessary for the synthesis of thymidine kinase is transcribed at least 90 minutes prior to telophase.

The large amounts of actinomycin D required to inhibit RNA synthesis resulted either from the relative impermeability of the cell membrane or from inactivation of the inhibitor within the cytoplasm. At all events, only tiny amounts (1 μg/ml) are required to inhibit the incorporation of ^{3}H-uridine into the RNA in isolated nuclei (Mittermayer et al., 1966b). This level is comparable to the amounts required for similar inhibition in certain bacteria or in mammalian cells in culture.

Role of Protein. Of the various substances required for the initiation and regulation of mitosis, proteins must play an essential role in the actual initiation of mitosis. Actidione (cycloheximide) at levels of 5 to 20 μg/ml inhibited the incorporation of labeled amino acids into proteins without drastically inhibiting the incorporation of nucleic acid precursors into RNA. This inhibitor completely blocked the ensuing mitosis when added at any time prior to about 15 minutes before metaphase; when it was added between 7 and 15 minutes before metaphase, the duration of mitosis was prolonged from the usual 15 minutes to about an hour, but mitosis was eventually completed even though nuclear amino acid incorporation was drastically reduced during this period. The results suggest that the last proteins necessary for the initiation of mitosis are completed by 15 minutes before metaphase but that some proteins, probably those concerned with energy requirements, are synthesized during mitosis (Cummins et al., 1965; Cummins et al., 1966a). Sachsenmaier et al. (1967) reported that the increase in thymidine kinase activity was prevented immediately upon the addition of actidione (50 μg/ml). This enzyme starts to increase in activity 50 minutes before prophase and reaches its maximum by the end of mitosis, but it is not necessary for mitosis per se.

As mentioned above, actidione added to the culture medium between 7 and 15 minutes prior to metaphase causes a prolongation of mitosis; an elevated temperature applied to the plasmodium during approximately this same period also produces an unusual, but different, effect on mitosis. The optimal temperature for growth of the plasmodium is about 26°C; raising the temperature to 37° or 40°C for periods of 10 to 30 minutes delays the onset of mitosis. There is no delay when plasmodia are subjected to elevated temperature during the S period; the greatest delay is noted when the treatment is applied about two hours before mitosis. The delayed mitosis, once it occurs, is entirely normal. However, if a 10-minute period of the same heat shock is started 15 to 20 minutes before metaphase, mitosis does not occur, although fibers in the nucleus, resembling mitotic spindles, are arranged irregularly. After a time, these spindles disappear, the nuclei undergo reconstruction without a true mitosis having occurred, and DNA synthesis follows, with the result that twice the normal level of DNA is formed. It would appear that the effect of the heat shock 15 to 20 minutes before mitosis is to disrupt the normal arrangement of spindles, resulting in a pseudomitosis during which the biochemical events required for DNA replication proceed normally (Brewer and Rusch, 1968).

Irradiation. It is of interest that irradiation with ultraviolet light (14,500 ergs/mm^2) during early prophase and as late as 13 minutes before metaphase has an effect similar to that of heat shock: it prevents mitosis and causes a reversion to interphase. Although the experiments are not directly comparable, heat shock and ultraviolet irradiation have another similarity in that the mitotic cycles following the first post–heat shock or post–irradiation mitosis were shortened (Devi et al., 1968; Brewer and Rusch, 1968). Nygaard and Guttes (1962) subjected the plasmodium to 9,000 r of x-radiation at various times in the cycle, and the results were similar in some respects to the effect of elevated temperature shocks. They found that x-radiation in late prophase delayed the onset of mitosis, yet permitted DNA replication. The effect of x-radiation on protein synthesis was not reported. The various products needed for mitosis are apparently not equally affected by heat shock, ultraviolet light, or x-radiation. It is possible that these agents may block some substances that prevent mitosis while other products are unaffected and persist into the next cycle, thereby reducing the time required for their synthesis in that cycle. This implies some type of feedback control.

McGrath et al. (1964) noted a 40 percent increase in specific activity of DNA in *Physarum* after irradiation with 25 kr of x-radiation, but the DNA per nucleus was unchanged, and it appears, therefore, that the irradiation had not prevented mitosis. These authors suggested that their results may have been caused by a change in the precursor pool or by a replacement of the damaged sites on DNA.

Control of DNA Replication

ROLE OF PROTEIN. The role of protein synthesis in initiating and maintaining DNA replication during the S period was studied by Cummins and Rusch (1966), who found that actidione (cycloheximide) at 10 μg/ml inhibited protein synthesis instantly, completely, yet reversibly. The addition of the inhibitor during late prophase or in metaphase permitted both mitosis and nuclear reconstruction, and about 20 percent of the usual amount of DNA was synthesized even though protein synthesis was blocked. The addition of the inhibitor during the S period permitted a similar partial replication of nuclear DNA, and the proportion of total nuclear DNA replicated was about the same regardless of when, during the S period, the inhibitor was added. Such results suggest that completion of an early unit of DNA replication is necessary to trigger the synthesis of proteins that act to initiate later replication. It is estimated that there may be a considerable number of units of replication during the S period in *Physarum*. Proteins synthesized during the G2 period and during early prophase induce an early replicating part of nuclear DNA. These have been called "initiators" by Cummins (1968). When this early nuclear DNA replication has been completed, the synthesis of "initiators" for later replicating DNA is triggered. This sequence continues for several units and is responsible for maintaining the temporal order of replication.

ROLE OF NUCLEUS. There also appears to be a control of DNA replication within the nucleus of *Physarum*. It was previously mentioned that only nuclei isolated from plasmodia during the S period could synthesize DNA (Brewer and Rusch, 1965). Recent experiments by Guttes and Guttes (1968) have shown that nuclei from plasmodia in the S period will continue to synthesize DNA when transferred to plasmodia in the G2 period, but nuclei from plasmodia in the G2 period transplanted into S-period plasmodia do not synthesize DNA. These results are not in agreement with a similar experiment done with *Ameba proteus* (Prescott and Goldstein, 1967). This discrepancy may be due to differences in the permeability of the nuclear membranes in the two organisms.

The information concerning the biochemical regulation of mitosis in *Physarum* is still very incomplete, but a few brief conclusions may be summarized: (1) DNA synthesis occurs immediately after mitosis and lasts for about 3 hours (there is no G1 period); DNA molecules replicated at one point in one S period are replicated during a similar temporal segment of the next S period; nuclei from S-period plasmodia transplanted into G2 plasmodia continue DNA synthesis, but nuclei from plasmodia in the G2 period do not synthesize DNA when placed in S-period plasmodia; (2) although mitosis can be experimentally induced in nuclei that have just completed DNA replication, the G2 period serves for the transcription and synthesis of substances essential for mitosis and for the enlargement of the plasmodium; (3) mitosis, or a pseudomitosis, is necessary to activate "initiators" for the replicating parts of the genome, and it appears to provide a mechanism for realigning the transcription machinery at the starting point in the temporal cycle; (4) the kinds of RNA change during the growth cycle; RNA made early in the cycle is transcribed from relatively more AT-rich molecular regions of nuclear DNA than that synthesized later in the cycle; (5) the last essential mRNA for mitosis is completed about a half hour before metaphase, whereas the last essential protein for this process is synthesized only 15 minutes before metaphase; and (6) mitosis does not occur until after DNA has been replicated.

MITOCHONDRIA

When Nygaard et al. (1960) first reported that DNA synthesis in the plasmodium occurred during a 3-hour period immediately after mitosis, they also observed a low level of incorporation of ^{14}C-labeled orotic acid into DNA thymine during the G2 period. At the time it was not known whether this incorporation resulted from an imperfect nuclear synchrony or was an artifact. Sachsenmaier (1964) suggested that such incorporation might represent extranuclear synthesis of DNA, and Guttes and Guttes (1964b) were the first to demonstrate that this was indeed the case. They presented autoradiographic data showing the incorporation of tritiated thymidine into the mitochondria in a form that was insoluble in acid and insensitive to ribonuclease but unstable against deoxyribonuclease, and they concluded that this represented DNA synthesis by the mitochondria. Recently Kessler and his associates (Seavey et al., 1967) confirmed this observation and also showed that mitochondrial DNA can be stained with auramine O, a fluorescent dye that is visible with the light microscope (Grunfeld and Kessler, 1967).

The next advance was made by Evans (1966), who isolated DNA that had been labeled with ^{3}H-thymidine from isolated nuclei and from the whole plasmodium and subjected the extracts to density gradient centrifugation in cesium chloride for 68 hours. He found a satellite DNA band having a density of 1.686, which he assumed was from the mitochondria since it was present in the whole plasmodium but not in the material obtained from the isolated nuclei. The density of the main DNA was 1.700. Guttes et al. (1967) and Holt and Gurney (1969) investigated this problem in greater detail. Guttes et al. (1967) used both autoradiographic and cesium chloride gradient equilibrium sedimentation techniques and found that mitochondrial DNA incorporated thymidine at all times of the mitotic cycle, and this was enhanced during a short period immediately preceding mitosis. Starvation of the plasmodium did not produce a periodicity in DNA synthesis. The densities of the principal and the mitochondrial DNA components were 1.697 and 1.686, respectively, corresponding to guanine-cytosine contents of 38 and 23 percent, respectively. The results of Holt and Gurney (1969) are in excellent agreement with these results. The mitochondrial DNA represented from 5 to 10 percent of the total DNA. Guttes et al. (1967) concluded that the mitochondrial population is not synchronized with regard to DNA replication, but this does not exclude the possibility that a periodicity of DNA replication exists for single mitochondria. Evidence for mitochondrial division was first suggested by Guttes et al. (1966). Further support for this suggestion was obtained by adding actinomycin C to the plasmodial medium, which then inhibited the division of the nucleoid and the mitochondrion (Guttes et al., 1969b). Mitochondria prelabeled with ^{3}H-thymidine may be identified through at least one mitotic cycle when transplanted from one plasmodium to another (Guttes and Guttes, 1967).

Brewer et al. (1967) found that isolated mitochondria incorporated ^{3}H-dATP into an acid-insoluble product having the same buoyant density as that of *Physarum* mitochondrial DNA. The reaction required all four deoxyribonucleoside triphosphates and Mg^{2+}, was unaffected by exogenous DNA, deoxyribonuclease, or ATP, and was inhibited by Ca^{2+} and spermine. Again, there was no evidence of a periodicity, since the DNA polymerase activity was observed in mitochondria isolated at all times of the growth cycle. This lack of periodicity differs from the DNA polymerase activity in the nucleus, which activity was detected only in those nuclei isolated from plasmodia in the S period (Brewer and Rusch, 1965).

Another study of the mitochondria compared the nearest neighbor frequencies of RNA synthesized *in vitro* on purified DNA from nuclei and from mitochondria. The frequency of CpG in nuclear DNA was not a random arrangement of nucleotides and was similar to that for higher organisms; whereas the frequency of CpG in mitochondrial DNA was near the random value, a relationship commonly observed in bacterial DNA. These results suggest that mitochondrial DNA, in terms of evolution, may be more closely related to bacterial DNA than it is to the chromosomal DNA of higher organisms (Cummins et al., 1967).

PROTOPLASMIC MOVEMENT

One of the striking features of the plasmodium that has attracted attention for a number of years is the protoplasmic streaming and movement. It is this feature that stimulated Seifriz (1936, 1937, 1942), Kamiya (1960, 1961, 1966, 1968), Loewy (1952), and others to study the physical and biochemical properties associated with this movement. Streaming protoplasm can be readily seen with the aid of a dissecting microscope in any plasmodium that is not too thick; the protoplasm first streams rapidly in one direction, then slows its rate, reverses the direction, and thus continues in a constant to and fro streaming. The time consumed for each direction is approximately 1 minute if the plasmodium is relatively stationary, but if the plasmodium begins to migrate the period of the flow increases in the direction of the migration at the expense of the time spent in streaming in the reverse direction. Migration ceases during mitosis, but the to and fro streaming of the protoplasm continues during this time (Guttes and Guttes, 1963c). It is of interest that the plasmodium remains compact and in one place as long as the nutrients are adequate. When nutrients become deficient, the plasmodium loses its compactness, begins to migrate about the surface, and forms a net-like appearance consisting of many anastomosing strands. When such plasmodium encounters food, locomotion stops until the food is depleted. Besides protoplasmic streaming, the plasmodium exhibits an active pinocytosis and represents suitable material for studies on this subject (Guttes and Guttes, 1960).

A number of theories have been proposed to explain protoplasmic streaming. Park and Robinson (1967) suggested that the formation of vacuoles in the plasmodium attracts water from the ground plasm and thereby increases the viscosity of the cytoplasm; this in turn creates a tension in the fibrils which is responsible for protoplasmic streaming. They found a chloroform extract of *Fusarium oxysporum* which contained a vacuolation factor that temporarily checked streaming when added to the anterior lobes of a migrating plasmodium. Anderson (1964) has studied the ion concentration in migrating plasmodia and has shown that potassium, but not sodium, has an essential function in migration. Sodium and potassium are exchanged at the advancing front, and potassium is retained preferentially over sodium. Plasmodia whose direction of migration has been oriented by passing an electric current through the supporting agar lose about ten times more potassium (Anderson, 1962). Fast potential changes of about +1 to +5 mV were recorded only from an advancing front or a site of injury. Slow potentials, recorded simultaneously from different regions, were not in phase with one another (Miller et al., 1968). Stewart and Stewart (1959) and Stewart (1964) proposed a theory that streaming results from the response of the fluid protoplasm to the pattern of pressure gradients. Waves of local relaxation and contraction are caused by temporary, reversible, and local weakenings of the gel-like meshwork and spread through the plasmodium.

MYXOMYOSIN AND FIBRILLAR STRUCTURES. Another explanation of the motivating force is based on the presence of an ATP-sensitive actomyosin-like "contractile" protein in the plasmodium. This was first described by Loewy (1952) and later named myxomyosin by Ts'o et al. (1957). Recent experiments have shown that this protein consists of myosin B-like protein which is made up of a myosin A-like protein and a F-actin. Plasmodial F-actin, a fibrous polymer, consists of polymerized plasmodial G-actin, and the latter can be copolymerized with rabbit striated G-actin on the addition of monovalent cations or divalent cations. Myosin B has an ATP sensitivity and an ATPase activity as high as those of myosin B from the striated muscle of the rabbit. Electron micrographs of myosin B reveal a filament with a diameter of about 100 Å with globular particles of myosin A attached to its surface. When ATP is added to the filament, F-actin with a diameter of 75 Å appears. Myosin A-like protein (plasmodium myosin A) can be isolated from myosin B in the presence of ATP and Mg^{2+} by ultracentrifugation. Plasmodial myosin A can combine with either plasmodial F-actin or with muscle F-actin to form actomyosin-like complexes (Hatano and Oosawa, 1966b; Hatano et al., 1967; and Hatano and Tazawa, 1968). The fine structure of the fibrils in glycerinated plasmodia has been described by Nagai and Kamiya (1968) and is in general agreement with the findings of Hatano and associates. Hatano and Oosawa (1966a) also reported that the amino acid compositions of plasmodial and muscle actin are very similar. Nakajima (1964) observed that the enzymatic properties of plasmodial and muscle myosin are very similar: 2,4-dinitrophenol stimulates the enzymatic activity at both high and low K^+ concentrations; *p*-chloromercuribenzoic acid inhibits the ATPase activity, and this inhibition is partially reversed by cysteine; monoiodoacetate has no effect on the ATPase activity at either high or low K^+ concentration, and EDTA (ethylenediamine tetraacetate) inhibits ATPase with both high and low K^+ concentrations. It is apparent, therefore, that the actomyosin-like protein, which we will refer to as myxomyosin, has many physiological and enzymatic similarities to the properties of muscle actomyosin. It is also known that ATP placed on the plasmodia or injected into them causes a strong increase in protoplasmic streaming (Nakajima, 1964). All of these facts point to myxomyosin as the motivating force in protoplasmic movement, but the exact mechanism by which this occurs is not known.

One suggestion concerning the mechanism is that the myxomyosin is a chief constituent of fibrillar structures that may play a role as the motivating force. These structures were first described by Wohlfarth-Bottermann (1962, 1964), and his observations have been confirmed by others (Porter et al., 1965; Rhea, 1966; Crawley, 1966; and Nagai and Kamiya, 1966). These fibrillar structures consist of many filaments laid parallel to one another and bearing a close resemblance to a smooth muscle fiber; these filaments are approximately 70 Å in diameter. According to Wohlfarth-Bottermann, these structures form a coherent network made up of branching, anastomosing fibrils that extend for considerable distances, and in some places they appear to join the outermost plasma membrane. The fibrils are arranged longitudinally and circularly to the streams of protoplasm and are positioned in a way that could cause streaming. Most of the evidence indicates that the fibrillar structures consist chiefly, if not entirely, of myxomyosin. Nakajima and Allen (1965) examined the fibrils with a polarizing microscope and found a complex and changing pattern of birefringence associated with the streaming protoplasm. Their studies indicate that the changing birefringence can be explained in terms of coordinated contractions of fibrils. Furthermore, these fibrillar structures remain after extraction with glycerol, and such fibers still contract in the presence of ATP (Nakajima, 1964; Kamiya and Kuroda, 1965; Nagai and Kamiya, 1966), which response is similar to that of muscle fibers following similar treatment

(Szent-Györgyi, 1953). Although the fibrillar structures can be made to contract in the presence of ATP, their exact participation in the motile force is not yet known, and direct evidence for their contraction in the living plasmodium has yet to be demonstrated. Nakajima and Allen (1965) believe that they appear and disappear regularly, following the change in phase of streaming. The suggestion has also been made that they produce movement by some kind of sliding on one another.

Kamiya and associates have also checked on the effect of increased air pressure on plasmodial streaming. They placed a small plasmodium in a device that had two, three, or four chambers that were interconnected with small openings; the plasmodia in these chambers were also connected with one another by a narrow plasmodial strand, and an increased pressure of from 10 to 40 cm of water in one chamber forced the endoplasm out of one plasmodium into its adjoining "twin," leaving behind a plasmodial ectoplasmic gel that was deficient in endoplasm. Release of pressure permitted the endoplasm to flow back and reestablish a normal equilibrium of the streaming protoplasm between the "twin" plasmodia. The experiments showed that the fibrils were found mainly in the endoplasm-rich "twin." The experiments also demonstrated that the return of the endoplasm to the depleted "twin" did not result from a simple elastic type of rebound but was caused by something which is not yet understood (Kamiya and Yoneda, 1967; Kamiya and Takata, 1967; Takata et al., 1967). Certain physical factors, such as intercapillary velocity distribution and the torsion of the plasmodial strands, have also been studied (Kamiya, 1965).

Although no nerve-like structures have been detected in the plasmodium, it may possess a very primitive system for the control of the motivating force. At all events, a homogenate of the plasmodium hydrolyzes acetylcholine, β-methylcholine, and benzoylcholine, and the activity towards acetylcholine is inhibited by eserine and caffeine (Nakajima and Hatano, 1962). These authors concluded that the plasmodium contains either specific acetylcholine or nonspecific cholinesterase, and it is possible that these may be involved in altering the concentration of ions as observed by Anderson (1962, 1964). These results suggest that the plasmodium possesses a functional primitive "neuromotor system" at the biochemical level (Nakajima and Hatano, 1962). It has also been demonstrated that coordinated ciliary action of *Tetrahymena* is dependent on acetylcholine esterase activity (Seaman and Houlihan, 1951). Hoitink and Van Dijk (1967) also report evidence for the probable occurrence of acetylcholine esterase and acetylcholine in the plasmodium of *Physarella oblonga,* a species closely related to *P. polycephalum.* They also found that adrenalin and noradrenalin caused a shortening, whereas acetylcholine caused a lengthening, in the duration of protoplasmic streaming.

In summary, the motivating force for protoplasmic streaming in *Physarum* appears to be dependent on myxomyosin, which is a major constituent of the fibrillar structures. Although these structures may be less complex morphologically than muscle fibers, less is known about their mechanism of action in protoplasmic streaming than is known about the physiology of muscle contraction. There also appears to be a very primitive "neuromotor system" in the plasmodium that may control the streaming, but again our knowledge concerning the manner in which this system may exert its control is less well understood than how the more complicated nervous system in higher organisms exerts its influence.

SPORULATION

One of the pathways the plasmodium may follow when nutrients become limiting is to sporulate. The sporulating plasmodium provides especially good material for the study of the biochemical events associated with differentiation, since there is no growth during this process. Much of the information concerning the biochemical events associated with sporulation has been summarized by Daniel (1966). It should be remembered that sporulation is one mechanism for survival, is the form in which this organism is dispersed, and may result in variation by recombination. Thus, it is not surprising that a depletion of nutrients is one of the conditions necessary to initiate the events leading to sporulation. In the laboratory, sporulation is induced by starving the plasmodium for a minimum of 4 days on a surface moistened with a salt mixture and niacin, and then exposing it to visible light for at least 4 hours (Gray, 1953; Daniel and Rusch, 1962a). During the period of starvation, the plasmodium decreases in size, the amounts of DNA, RNA, protein, and glycogen all decrease, but in spite of this, mitosis and DNA synthesis still occur although there is less synchrony and the periods between mitoses are greatly increased. During this period, the plasmodium is actively migrating in "search" of food and appears to cannibalize itself in order to survive. The gross morphological changes that characterize sporulation are completed in about 16 to 18 hours after the period of illumination. Although sporulation does not occur without niacin, the exact function of this vitamin is unknown; it is probably involved in the increase of pyridine nucleotide during the period of illumination (Daniel and Rusch, 1962b). Starvation causes a reduction in glycogen, and such depletion is essential for sporulation; furthermore, glucose (100 μg/ml) inhibits sporulation. Aerobic conditions are also necessary for sporulation (Daniel, 1966).

During the starvation of the plasmodium the intermitotic period is increased to 24 to 36 hours (Guttes et al., 1961). It is possible that a G1 period occurs during starvation, but this point has not been established. If a G1 period did exist, one might expect that the synthesis of macromolecules and other substances required for sporulation would be initiated during this period. As mentioned previously, a G1 period does not occur in the growing plasmodium, and this is consistent with the idea that all the products made during growth are used only for growth. It appears that the G1 period is that time when specialized substances needed for special functions, but not for growth per se, are synthesized, and this period is found only in those cells that specialize to some degree.

Plasmodia cultured vigorously for a year or more with repeated transfers in shaken flasks may not sporulate as regularly as cultures that have recently been started from sclerotia. The exact reason for this is unknown but may be due to a selection during the growth period for genetic factors favoring growth rather than sporulation. A number of problems still remain in achieving good sporulation in all cultures, and it is necessary to pay close attention to many small details to obtain a high percentage of sporulating cultures.

A period of illumination is essential for inducing sporulation in the pigmented, but not in the nonpigmented, species of the myxomycetes (Gray, 1938). The illumination causes a decrease, then an increase, and then a second drop in ATP. Light also causes a rapid decrease in respiration, inhibits glucose uptake, increases the amount of nonheme ferrous iron, and increases the pH of the starvation medium, perhaps by altering the cell membrane. In addition, it induces changes which lead to the melanization of the sporangium (Daniel, 1966). The precise chemical nature of the yellow pigment in *P. polycephalum* is not known, although suggestions have been made that the pigment may be pteridines (Wolf, 1959), peptides (Dresden, 1959), riboflavin (Gray, 1953, 1955), flavone (Seifriz and Zetzmann, 1935), or some type of phenolic compound (Nair and Zabka, 1966). Brewer (1965) isolated three pigments: pigment A, a water-soluble, nonaromatic hydrochloride compound containing a conjugated hexane chromophore and with strongly basic nitrogen functions; pigment B, an amphoteric compound containing a polyene chromophore; and pigment C, a pigment very similar to pigment B. Daniel (1966) made a fairly extensive study of the effect of light on the pigments.

Studies on certain enzymes that accompanied sporulation were made by Ward and his associates. Ward (1958) found three times as much cytochrome oxidase activity in spores as in plasmodia, and about six times as much ascorbic acid oxidase activity in the plasmodia as in spores. In addition, acrylamide electrophoresis was used to fractionate various proteins from both the growing plasmodia and the presporangial stages. There was a 60 percent decrease in α-amylase activity after the plasmodium was committed to sporulation, and a further decrease to undetectable levels as visible morphological changes occurred. Polysaccharidase B activity also decreased in the presporangical stage (Zeldin and Ward, 1963a, 1963b).

Questions concerning the genetic control of sporulation are being studied, and it appears, from preliminary data, that the mRNA's necessary for sporulation are probably induced by starvation and light. Such RNA's are completed about 2 hours after the period of illumination, since actinomycin D applied any time after this period has no effect on sporulation. There is a critical turning point in the cycle of the organism that occurs about 3 to 3½ hours after the end of illumination. Prior to this time the starved and illuminated plasmodia will start to grow again when returned to the growth medium, but after this time they will go on to sporulate even when placed in the growth medium. Approximately 12 to 13 hours after the end of illumination mitosis occurs, and almost immediately thereafter multiple cleavages in the sporangia occur, separating the nuclei into mononucleate spores. Soon after cleavage is completed, melanization of the spore wall occurs and is completed approximately 16 hours after the end of illumination (H. Sauer and K. Babcock, unpublished results). Aldrich (1967) reported that meiosis I starts in the nuclei of spores about 20 hours past the cleavage stage in *P. flavicomum*, but the exact time of meiosis I is still a matter of contention. Von Stosch (1965) also reported that both meiotic divisions occur in the maturing spore within 2 days after sporangial mitosis. Some preliminary evidence indicates that meiosis II starts about 18 hours after meiosis I. The observations concerning meiosis have been made with several species of myxomycetes cultured under various conditions, and it is understandable, therefore, that considerable variations were found (Therrien, 1966; Koevenig, 1964; Ross, 1961, 1967b; von Stosch, 1935; Wilson and Ross, 1955). A discussion of this problem is presented by Ross (1967b), who postulated that the nuclear cycle is not necessarily in complete synchrony with the morphological cycle, and that the mechanisms which trigger the developmental pathway leading to sporulation may also trigger an independent sequence of nuclear events leading to meiosis. In some cases the morphological changes leading to sporulation may proceed faster than those leading to meiosis, and in other cases the reverse situation may occur.

SCLEROTIUM FORMATION

In addition to sporulation, there is a second pathway for survival which the plasmodium may follow in the event of an adverse environment—it may form a sclerotium. The sclerotium consists of clusters of spherules, each containing one or more diploid nuclei. Walls are laid down between the spherules, and the whole mass appears cellular. Camp (1936) was the first to outline precisely a procedure for the development of large sclerotia. He placed large plasmodia in a moist, dark chamber and then slowly desiccated them under a bell jar. Hodapp (1942) and Luyet and Gehenio (1944, 1945) continued and expanded the studies started by Camp, and all obtained sclerotia that were very resistant to adverse conditions, such as 10-minute exposure to liquid air or to 2 weeks at 50°C. In addition to the use of heat and desiccation, Jump (1954) induced the formation of sclerotia by low temperature, high osmotic pressure, sublethal concentrations of heavy metals, and occasionally by starvation. Stewart and Stewart (1961) observed that sclerotia could also be induced in the liquid media of shaken flasks that were employed for culturing the plasmodia. Daniel and Rusch (1961), Hemphill (1962), and Guttes and Guttes (1963b) have reported in detail the technique for obtaining sclerotia in shaken flasks containing a nonnutrient salts medium. Now sclerotia are prepared routinely in shaken flasks at regular intervals in our laboratory, and such sclerotia are dried on sterile filter paper, stored, and used to maintain stocks for long periods of time. When a piece of filter paper containing a sclerotium is placed in a shaken flask containing nutrients, growing microplasmodia are obtained after a few days (Daniel and Baldwin, 1964). There have been very few reports on the biochemical changes that accompany sclerotial formation, but Sullivan (1953) reports decreases in bound water, glycogen, reducing sugars, and mucoprotein and an increase in lipids. It was suggested that there is a conversion of carbohydrates to fats.

Sclerotial formation and sporulation have several things in common. They both result from deprivation of nutrients, but they differ in that the induction of sporulation requires a longer period of starvation, the presence of niacin, plus exposure to light. In both cases the nuclei are separated into spores, or spore-like clusters, by a series of multiple cleavages of the protoplasm. This cleavage may be similar to that which accompanies cell division and is probably not temporally associated with nuclear division, but this latter point is not yet established.

Sclerotial formation occurs uniformly throughout the plasmodia in the absence of growth. This stage provides ideal material for the study of cell division and membrane formation. The process of sclerotization and subsequent plasmodial reconstruction in *Physarum* as seen in the light microscope was described in some detail by Jump (1954). More recently Stewart and Stewart (1961) examined the formation of membranes during sclerotization with the aid of the electron microscope. They found that preexisting vesicles lined up and coalesced to form new membranes for the spherules as well as for the sclerotial walls. Their data support the concept that biological membranes are interconvertible at certain stages in their formation.

CONCLUSIONS

Physarum polycephalum is an outstanding organism for the study of a number
of growth. This stage provides ideal material for the study of cell division and
chemical events associated with growth, mitosis, differentiation, cytokinesis, and
detail by Jump (1954). More recently Stewart and Stewart (1961) examined the
on physical and biochemical properties of a primitive motile protoplasm and of a possible primitive "neuromotor" system.

On the basis of studies to date we may conclude the following:

1. DNA molecules replicated at one point in one S period are replicated during a similar temporal segment of the next S period; nuclei that have just completed DNA synthesis may be induced to undergo mitosis immediately, without an intervening G2 period.
2. The G2 period is for the transcription and the synthesis of substances essential for mitosis and for the enlargement of the protoplasm.
3. Mitosis, or pseudomitosis, is necessary to activate "initiators" for the replicating parts of the genome and appears to provide a mechanism for realigning the transcription machinery at the starting point in the temporal cycle.
4. Mitosis does not occur until after DNA has been replicated.
5. The kinds of RNA made change during the growth cycle, and RNA made early in the cell cycle is transcribed from relatively more AT-rich molecular regions of nuclear DNA than that synthesized later in the cell cycle.
6. The last essential mRNA for mitosis is completed about a half hour before metaphase, whereas the last essential protein for this proces is synthesized only 15 minutes before metaphase.
7. Starvation in the presence of niacin for a minimum of 4 days followed by a minimum exposure to light of 4 hours is required to induce sporulation.
8. The plasmodium has reached a point of no return and is irreversibly committed to sporulate about 3 hours after exposure to light.
9. The motivating force for protoplasmic streaming is very similar to muscle actomyosin.

10. There appears to be a primitive "neuromotor" system that controls streaming.
11. The plasmodium begins to migrate only when nutrients become inadequate.
12. The plasmodium contains a factor that controls coalescence.
13. The myxamebas have mating types.

ACKNOWLEDGMENTS

I am especially grateful to Drs. Joseph E. Cummins, Edmund Guttes, and Sophie Guttes for permitting me to see their recent papers prior to publication, and to Dr. Ilse Riegel for assistance with the manuscript. Valuable suggestions during the preparation of this review were made by my present associates: Drs. Eugene Goodman, Justin McCormick, Joyce Mohberg, Helmut Sauer, and Winfried Schiebel, as well as Karlee Babcock, Judith Blomquist, Alice DeVries, and Loralee Sauer.

This chapter is dedicated to those who have worked with me on the *Physarum* problem, including the persons mentioned above and my former associates, Drs. Eugene Brewer, Richard Braun, John W. Daniel, Christian Mittermayer, Oddvar F. Nygaard, and Wilhelm Sachsenmaier.

Notes added in proof

To p. 305: Hiramaru et al. (1969. J. Biochem., 65:693–700, 701–708) have reported the presence of four ribonucleases and two nucleases, and Braun and Behrens (1969. Biochim. Biophys. Acta, in press) have described one ribonuclease in the plasmodium.

To p. 308: The addition of FUDR and uridine during the G2 period did not affect RNA synthesis, but these two compounds, when added during the S period, immediately reduced RNA synthesis. These results suggest that part of the newly synthesized DNA becomes immediately functional in the synthesis of RNA (Rao and Gontcharoff, 1969. Exp. Cell Res., 56:269–274).

REFERENCES

Aldrich, H. C. 1967. The ultrastructure of meiosis in three species of Physarum. Mycologia, 49:127–148.

——— 1968. The development of flagella in swarm cells of the myxomycete *Physarum flavicomum*. J. Gen. Microbiol., 50:217–222.

——— 1969. The ultrastructure of mitosis in myxamoebae and plasmodia in *Physarum flavicomum*. Amer. J. Bot., 56:290–299.

Alexopoulos, C. J. 1962. Introductory Mycology, 2nd ed., New York, John Wiley and Sons, Inc.

——— 1963. The myxomycetes II. Bot. Rev., 29:1–78.

——— 1966. Morphogenesis in myxomycetes. *In* The Fungi, an Advanced Treatise, Ainsworth, G. C., and Sussman, A. S., eds., New York, Academic Press, Inc., Vol. 2, pp. 211–234.

Allen, R. D., and N. Kamiya, eds. 1964. Primitive Motile Systems in Cell Biology. New York, Academic Press, Inc.

Anderson, J. D. 1962. Potassium loss during galvanotaxis of slime mold. J. Gen. Physiol., 45:567–574.

——— 1964. Regional differences in ion concentration in migrating plasmodia. *In* Primitive Motile Systems in Cell Biology, Allen, R. D., and Kamiya, N., eds., New York, Academic Press, Inc., pp. 125–136.

Becker, J., J. W. Daniel, and H. P. Rusch. 1963. Growth inhibition of *Physarum polycephalum* for the evaluation of chemotherapeutic agents. Cancer Res. (Suppl.), 23:1910–1929.

Braun, R., C. Mittermayer, and H. P. Rusch. 1965. Sequential temporal replication of DNA in *Physarum polycephalum*. Proc. Nat. Acad. Sci. U.S.A., 53:924–931.

——— C. Mittermayer, and H. P. Rusch. 1966a. Sedimentation patterns of pulse-labeled RNA in the mitotic cycle of *Physarum polycephalum*. Biochim. Biophys. Acta, 114: 27–35.

——— C. Mittermayer, and H. P. Rusch. 1966b. Ribonucleic acid synthesis *in vivo* in the synchronously dividing *Physarum polycephalum* studied by cell fractionation. Biochim. Biophys. Acta, 114:527–535.

——— and H. Willi. 1969. Time sequence of DNA replication in Physarum. Biochim. Biophys. Acta, 174:246–252.

Brewer, E. N. 1965. Culture and chemical composition of the slime mold, *Physarum polycephalum*. Ph.D. dissertation, Univ. of Wisconsin, Madison, Wisconsin.

——— A. DeVries, and H. P. Rusch. 1967. DNA synthesis by isolated mitochondria of *Physarum polycephalum*. Biochim. Biophys. Acta, 145:686–692.

——— S. Kuraishi, J. C. Garver, and F. M. Strong. 1964. Mass culture of a slime mold, *Physarum polycephalum*. Appl. Microbiol., 12:161–164.

——— and H. P. Rusch. 1965. DNA synthesis by isolated nuclei of *Physarum polycephalum*. Biochem. Biophys. Res. Commun., 21:235–241.

——— and H. P. Rusch. 1966. Control of DNA replication: effect of spermine on DNA polymerase activity in nuclei isolated from *Physarum polycephalum*. Biochem. Biophys. Res. Commun., 25:579–584.

——— and H. P. Rusch. 1968. Effect of elevated temperature shocks on mitosis and on the initiation of DNA replication in *Physarum polycephalum*. Exp. Cell Res., 49:79–86.

Camp, W. G. 1936. A method of cultivating myxomycete plasmodia. Bull. Torrey Bot. Club, 63:205–210.

Carlile, M. J., and J. Dee. 1967. Plasmodial fusion and lethal interaction between strains in a myxomycete. Nature (London), 215:832–834.

Chin, B., and I. A. Bernstein. 1968. Adenosine triphosphate and synchronous mitosis in *Physarum polycephalum*. J. Bact., 96:330–337.

Cohen, A. L. 1967. The "late–formers"—changing concepts of the opsimorphs (*Myxomycetes, Acrasieae, Myxobacteria*). Arch. Mikrobiol., 59:59–71.

Collins, O. R. 1961. Heterothallism and homothallism in two myxomycetes. Amer. J. Bot., 48:674–683.

Crawley, J. C. W. 1966. Fine structure and cytoplasmic streaming in *Physarum polycephalum*. J. Roy. Micr. Soc., 85:313–322.

Cummins, J. E. 1968. Nuclear DNA replication and transcription during the cell cycle of Physarum. *In* The Cell Cycle: Gene-Enzyme Interactions, Padilla, G., Whitson, G., and Cameron, I., eds., New York, Academic Press, Inc.

——— J. C. Blomquist, and H. P. Rusch. 1966a. Anaphase delay after inhibition of protein synthesis between late prophase and prometaphase. Science, 154:1343–1344.

——— E. N. Brewer, and H. P. Rusch. 1965. The effect of actidione on mitosis in the slime mold *Physarum polycephalum*. J. Cell Biol., 27:337–341.

——— and H. P. Rusch. 1966. Limited DNA synthesis in the absence of protein synthesis in *Physarum polycephalum*. J. Cell Biol., 31:577–583.

——— and H. P. Rusch. 1967. Transcription of nuclear DNA in nuclei isolated from plasmodia at different stages of the cell cycle of *Physarum polycephalum*. Biochim. Biophys. Acta, 138:124–132.

——— and H. P. Rusch. 1968. Natural synchrony in a slime mold. Endeavour, 27: 124–129.

——— H. P. Rusch, and T. E. Evans. 1967. Nearest neighbor frequencies and the phylogenetic origin of mitochondrial DNA in *Physarum polycephalum*. J. Molec. Biol., 23:281–284.

——— G. E. Weisfeld, and H. P. Rusch. 1966b. Fluctuation of ^{32}P distribution in rapidly labeled RNA during the cell cycle of *Physarum polycephalum*. Biochim. Biophys. Acta, 129:240–248.

Daniel, J. W. 1966. Light–induced synchronous sporulation of a myxomycete—the relation of initial metabolic changes to the establishment of a new cell state. *In* Cell Synchrony, Cameron, I. L., and Padilla, G. M., eds., New York, Academic Press, Inc., pp. 117–152.

——— and K. Babcock. 1966. Methionine metabolism of the myxomycete *Physarum polycephalum*. J. Bact., 92:1028–1035.

——— K. Babcock, A. H. Sievert, and H. P. Rusch. 1963. Organic requirements and synthetic media for growth of the myxomycete *Physarum polycephalum*. J. Bact., 86:324–331.

——— and H. H. Baldwin. 1964. Methods of culture for plasmodial myxomycetes. *In* Methods in Cell Physiology, Prescott, D. M., ed., New York, Academic Press, Inc., Vol. 1, pp. 9–41.

——— J. Kelley, and H. P. Rusch. 1962. Hematin-requiring plasmodial myxomycete. J. Bact., 84:1104–1110.

——— and H. P. Rusch. 1961. The pure culture of *Physarum polycephalum* on a partially defined soluble medium. J. Gen. Microbiol., 25:47–59.

——— and H. P. Rusch. 1962a. Method for inducing sporulation of pure cultures of the myxomycete *Physarum polycephalum*. J. Bact., 83:234–240.

——— and H. P. Rusch. 1962b. Niacin requirement for sporulation of *Physarum polycephalum*. J. Bact., 83:1244–1250.

Dee, J. 1962. Recombination in a myxomycete, *Physarum polycephalum* Schw. Genet. Res., 3:11–23.

——— 1966a. Genetic analysis of actidione–resistant mutants in the myxomycete *Physarum polycephalum* Schw. Genet. Res., 8:101–110.

——— 1966b. Multiple alleles and other factors affecting plasmodium formation in the true slime mold *Physarum polycephalum* Schw. J. Protozool., 13:610–616.

de Meester, C., R. Lambert, and A. Wiaux. 1967. Effect of some dithiocarbamates on respiratory activity of a myxomycete. Physiol. Plantarum, 20:697–701.

Devi, V. R., E. Guttes, and S. Guttes. 1968. Effects of ultraviolet light on mitosis in *Physarum polycephalum*. Exp. Cell Res., 50:589–598.

Dresden, C. F. 1959. Pigments of *Physarum polycephalum*. Dissertation Abstr., 20:869–870. Ann Arbor, Michigan, University Microfilms, Inc.

Evans, T. E. 1966. Synthesis of a cytoplasmic DNA during the G_2 interphase of *Physarum polycephalum*. Biochem. Biophys. Res. Commun., 22:678–683.

Goodman, E. M., and H. Ritter. Plasmodial mitosis in *Physarum polycephalum*: a phase contrast and electron microscopic study. Arch. Protistenk. (in press).

Gray, W. D. 1938. The effect of light on the fruiting of myxomycetes. Amer. J. Bot., 25:511–522.

——— 1953. Further studies on the fruiting of *Physarum polycephalum*. Mycologia, 45:817–824.

——— 1955. Riboflavin synthesis in cultures of *Physarum polycephalum*. Ohio J. Sci., 55:212–214.

——— and C. J. Alexopoulos. 1968. Biology of the Myxomycetes. New York. The Ronald Press Co.

Grunfeld, C., and D. Kessler. 1967. Detection of mitochondrial DNA in the slime mold *Physarum polycephalum* by a fluorescent Feulgen staining method. J. Cell Biol., 35:168A.

Guttes, E., R. V. Devi, and S. Guttes. 1969a. Synchronization of mitosis in *Physarum polycephalum* by coalescence of postmitotic and premitotic plasmodial fragments. Experientia (in press).

——— and S. Guttes. 1960. Pinocytosis in the myxomycete *Physarum polycephalum*. Exp. Cell Res., 20:239–241.

——— and S. Guttes. 1962. Cell growth and mitosis in *Physarum polycephalum*. I. The effect of an increased nutrient supply upon the onset of mitosis. Exp. Cell Res., 26: 205–209.

——— and S. Guttes. 1963a. Initiation of mitosis in post-mitotic nuclei of *Physarum polycephalum*. Experientia, 19:13–15.

——— and S. Guttes. 1963b. Starvation and cell wall formation in the myxomycete *Physarum polycephalum*. Ann. Bot., N. S., 27:49–53.

——— and S. Guttes. 1963c. Arrest of plasmodial motility during mitosis in *Physarum polycephalum*. Exp. Cell Res., 30:242–244.

——— and S. Guttes. 1964a. Mitotic synchrony in the plasmodia of *Physarum polycephalum* and mitotic synchronization by coalescence of microplasmodia. *In* Methods in Cell Physiology, Prescott, D. M., ed., New York, Academic Press, Inc., Vol. 1, pp. 43–54.

——— and S. Guttes. 1964b. Thymidine incorporation by mitochondria in *Physarum polycephalum*. Science, 145:1057–1058.

——— and S. Guttes. 1967. Transplantation of nuclei and mitochondria of *Physarum polycephalum* by plasmodial coalescence. Experientia, 23:713–718.

——— S. Guttes, and R. V. Devi. 1969b. Division stages of the mitochondria in normal and actinomycin-treated plasmodia of *Physarum polycephalum*. Experientia, 25:66–67.

——— S. Guttes, and H. P. Rusch. 1961. Morphological observations on growth and differentiation of *Physarum polycephalum* grown in pure culture. Develop. Biol., 3:588–614.

——— P. C. Hanawalt, and S. Guttes. 1967. Mitochondrial DNA synthesis and the mitotic cycle in *Physarum polycephalum*. Biochim. Biophys. Acta, 142:181–194.

Guttes, S., and E. Guttes. 1968. Regulation of DNA replication in the nuclei of the slime mold, *Physarum polycephalum*: transplantation of nuclei by plasmodial coalescence. J. Cell Biol., 37:761–772.

——— E. Guttes, and R. A. Ellis. 1968. Electron microscope study of mitosis in *Physarum polycephalum*. J. Ultrastruct. Res., 22:508–529.

——— E. Guttes, and R. Hadek. 1966. Occurrence and morphology of a fibrous body in the mitochondria of the slime mold *Physarum polycephalum*. Experientia, 22:452–454.

Hatano, S., and F. Oosawa. 1966a. Isolation and characterization of plasmodium actin. Biochim. Biophys. Acta, 127:488–498.

——— and F. Oosawa. 1966b. Extraction of an actin–like protein from the plasmodium of a myxomycete and its interaction with myosin A from rabbit striated muscle. J. Cell. Physiol., 68:197–202.

——— and M. Tazawa. 1968. Isolation, purification and characterization of myosin B from myxomycete plasmodium. Biochim. Biophys. Acta, 154:507–519.

——— T. Totsuka, and F. Oosawa. 1967. Polymerization of plasmodial actin. Biochim. Biophys. Acta, 140:109–122.

Hemphill, M. D. 1962. Studies on a resting phase of *Physarum polycephalum* in axenic liquid cultures. M. S. dissertation. Madison, Wisconsin, University of Wisconsin.

Hodapp, E. L. 1942. Some factors influencing sclerotization in Mycetozoa. Biodynamica, 4:33–46.

Hoitink, A. W. J. H., and G. Van Dijk. 1967. The influence of neurohumoral transmitter substances on protoplasmic streaming in the myxomycete *Physarella oblonga*. J. Cell. Physiol., 67:133–140.

Holt, C. E., and E. G. Gurney. 1969. Minor components of the DNA of *Physarum polycephalum*: cellular location and metabolism. J. Cell Biol., 40:484–496.

Hütter, R., and J. A. DeMoss. 1967. Organization of the tryptophan pathway: a phylogenetic study of the fungi. J. Bact., 94:1896–1907.

Jump, J. A. 1954. Studies on sclerotization in *Physarum polycephalum*. Amer. J. Bot., 41:561–567.

Kamiya, N. 1960. Physics and chemistry of protoplasmic streaming. Ann. Rev. Plant Physiol., 11:323–340.

——— 1961. Protoplasmaströmung. Protoplasma, 53:600–614.

——— 1965. Rheology of cytoplasmic streaming. *In* Proc. IV International Congress on Rheology, Part I, New York, John Wiley & Sons, pp. 105–121.

——— 1966. Motilität des Plasmas der lebenden Zelle. Naturwiss. Rundsch., 19:270–282.

——— 1968. The mechanism of cytoplasmic movement in a myxomycete plasmodium. *In* 22nd Symposium of the Society for Experimental Biology: Aspects of Cell Motility, New York, Academic Press, Inc., pp. 199–214.

——— and K. Kuroda. 1965. Movement of the myxomycete plasmodium. I. A study of glycerinated models. Proc. Jap. Acad., 41:837–841.

——— and T. Takata. 1967. Movement of the myxomycete plasmodium. V. The motive force of endoplasm-rich and endoplasm-poor plasmodia. Proc. Jap. Acad., 43:537–540.

——— and M. Yoneda. 1967. Movement of the myxomycete plasmodium. IV. Dislocation of endoplasm and its effect on motive force production. Proc. Jap. Acad., 43:531–536.

Kerr, N. S. 1965. Inhibition by streptomycin of flagella formation in a true slime mold. J. Protozool., 12:276–278.

——— 1967. Plasmodium formation by a minute mutant of the true slime mold *Didymium nigripes*. Exp. Cell Res., 45:646–655.

Kerr, S. J. 1967. A comparative study of mitosis in amoebae and plasmodia of the true slime mold *Didymium nigripes*. J. Protozool., 14:439–445.

Kessler, D. 1964. An autoradiographic study of rapidly synthesized ribonucleic acid with reference to the mitotic cycle in the slime mold, *Physarum polycephalum*. Ph.D. dissertation, Part II, Univ. of Wisconsin, Madison, Wisconsin.

——— 1967. Nucleic acid synthesis during and after mitosis in the slime mold, *Physarum polycephalum*. Exp. Cell Res., 45:676–680.

Koevenig, J. L. 1964. Studies on the life cycle of *Physarum gyrosum* and other myxomycetes. Mycologia, 56:170–184.

——— and R. C. Jackson. 1966. Plasmodial mitoses and polyploidy in the myxomycete *Physarum polycephalum*. Mycologia, 58:662–667.

Korn, E. D., C. L. Greenblatt, and A. M. Lees. 1965. Synthesis of unsaturated fatty acids in the slime mold *Physarum polycephalum* and the zooflagellates *Leishmania tarentolae, Trypanosoma lewisi,* and *Crithidia sp.*: a comparative study. J. Lipid Res., 6:43–50.

Kudo, R. R. 1954. Protozoology, 4th ed., Springfield, Illinois, Charles C Thomas, Publisher, p. 427.

Loewy, A. G. 1952. An actomyosin-like substance from the plasmodium of a myxomycete. J. Cell. Comp. Physiol., 40:127–156.

Luyet, B. J., and P. M. Gehenio. 1944. The lethal action of desiccation on the sclerotia of Mycetozoa. Biodynamica, 4:369–375.

——— and P. M. Gehenio. 1945. The role of water in the lethal action of heat on dormant protoplasm. Biodynamica, 5:339–352.

McGrath, R. A., and R. W. Williams. 1967. Interruptions in single strands of the DNA in slime mold and other organisms. Biophys. J., 7:309–317.

——— R. W. Williams, and R. B. Setlow. 1964. Increased ^{3}H–thymidine incorporation into DNA of irradiated slime mould. Int. J. Radiat. Biol., 8:373–380.

Miller, D. M., J. D. Anderson, and B. C. Abbott. 1968. Potentials and ionic exchange in slime mold plasmodia. Comp. Biochem. Physiol., 27:633–646.

Mittermayer, C., R. Braun, T. G. Chayka, and H. P. Rusch. 1966a. Polysome patterns and protein synthesis during the mitotic cycle of *Physarum polycephalum*. Nature (London), 210:1133–1137.

——— R. Braun, and H. P. Rusch. 1964. RNA synthesis in the mitotic cycle of *Physarum polycephalum*. Biochim. Biophys. Acta, 91:399–405.

——— R. Braun, and H. P. Rusch. 1965. The effect of actinomycin D on the timing of mitosis in *Physarum polycephalum*. Exp. Cell Res., 38:33–41.

——— R. Braun, and H. P. Rusch. 1966b. Ribonucleic acid synthesis *in vitro* in nuclei isolated from the synchronously dividing *Physarum polycephalum*. Biochim. Biophys. Acta, 114:536–546.

Mohberg, J., and H. P. Rusch. 1964. Isolation of nuclei and histones from the plasmodium of *Physarum polycephalum*. J. Cell Biol., 23:61A.

——— and H. P. Rusch. 1967. Large-scale production of *physarum polycephalum* plasmodia. J. Cell Biol., 35:96A.

——— and H. P. Rusch. 1969. Growth of large plasmodia of the myxomycete *Physarum polycephalum*. J. Bact., 97:1411–1418.

Nagai, R., and N. Kamiya. 1966. Movement of the myxomycete plasmodium. II. Electron microscopic studies on fibrillar structures in the plasmodium. Proc. Jap. Acad., 42:934–939.

——— and N. Kamiya. 1968. Movement of the myxomycete plasmodium. VI. Fibrillar structures in the glycerinated plasmodium. Proc. Jap. Acad., 44:1044–1047.

Nair, P., and G. G. Zabka. 1966. Pigmentation and sporulation in selected myxomycetes. Amer. J. Bot., 53:887–892.

Nakajima, H. 1964. The mechanochemical system behind streaming in *Physarum*. *In* Allen, R. D., and Kamiya, N., eds., Primitive Motile Systems in Cell Biology, New York, Academic Press, Inc., pp. 111–123.

——— and R. D. Allen. 1965. The changing pattern of birefringence in plasmodia of the slime mold, *Physarum polycephalum*. J. Cell Biol., 25:365–374.

——— and S. Hatano. 1962. Acetylcholinesterase in the plasmodium of the myxomycete, *Physarum polycephalum*. J. Cell. Physiol., 59:259–263.

Nygaard, O. F., and S. Guttes. 1962. Effects of ionizing radiation on a slime mould with synchronous mitosis. Int. J. Radiat. Biol., 5:33–44.

——— S. Guttes, and H. P. Rusch. 1960. Nucleic acid metabolism in a slime mold with synchronous mitosis. Biochim. Biophys. Acta, 38:298–306.

Park, D., and P. M. Robinson. 1967. Internal water distribution and cytoplasmic streaming in *Physarum polycephalum*. Ann. Bot. (London) N. S., 31:731–738.

Porter, K. R., N. Kawakami, and M. C. Ledbetter. 1965. Structural basis of streaming in *Physarum polycephalum*. J. Cell Biol., 27:78A.

Poulter, R. T. M., and J. Dee. 1968. Segregation of factors controlling fusion between plasmodia of the true slime mould *Physarum polycephalum*. Genet. Res., 12:71–79.

Prescott, D. M., and L. Goldstein. 1967. Nuclear–cytoplasmic interaction in DNA synthesis. Science, 155:469–470.

Rhea, R. P. 1966. Electron microscopic observations on the slime mold *Physarum polycephalum* with specific reference to fibrillar structures. J. Ultrastruct. Res., 15:349–379.

Ross, I. K. 1961. Further studies on meiosis in the myxomycetes. Amer. J. Bot., 48:244–248.

——— 1966. Chromosome numbers in pure and gross cultures of myxomycetes. Amer. J. Bot., 53:712–718.

——— 1967a. Growth and development of the myxomycete *Perichaena vermicularis*. I. Cultivation and vegetative nuclear divisions. Amer. J. Bot., 54:617–625.

——— 1967b. Growth and development of the myxomycete *Perichaena vermicularis*. II. Chromosome numbers and nuclear cycles. Amer. J. Bot. 54:1231–1236.

——— 1967c. Syngamy and plasmodium formation in the myxomycete *Didymium iridis*. Protoplasma, 64:104–119.

——— 1968. Nuclear membrane behavior during mitosis in normal and heteroploid myxomycetes. Protoplasma, 66:173–184.

Rusch, H. P. 1962. The use of *Physarum polycephalum* for studies on growth and differentiation. *In* Biological Interactions in Normal and Neoplastic Growth (Henry Ford Hospital International Symposium), Brennan, M. J., and Simpson, W. L., eds., Boston, Little, Brown & Co., Inc., pp. 21–24.

——— R. Braun, J. W. Daniel, C. Mittermayer, and W. Sachsenmaier. 1964. The role of DNA and RNA in mitosis and differentiation in *Physarum polycephalum. In* Cellular Control Mechanisms and Cancer, Emmelot, P., and Mühlbock, O., eds., Amsterdam, Elsevier Publishing Company, pp. 80–85.

——— W. Sachsenmaier, K. Behrens, and V. Gruter. 1966. Synchronization of mitosis by the fusion of the plasmodia of *Physarum polycephalum.* J. Cell Biol., 31:204–209.

Sachsenmaier, W. 1964. Zur DNS– und RNS–Synthese im Teilungscyclus synchroner Plasmodien von *Physarum polycephalum.* Biochem. Z. 340:541–547.

——— 1966. Analyse des Zellcyclus durch Eingriffe in die Makromolekül-Biosynthese. *In* Probleme der biologischen Reduplikation, Sitte, P., ed., New York, Springer-Verlag New York Inc., pp. 139–160.

——— and J. E. Becker. 1965. Wirkung von Actinomycin D auf die RNS-Synthese und die synchrone Mitosetätigkeit in *Physarum polycephalum.* Mschr. Chem., 96: 754–765.

——— D. v. Fournier and K. F. Gürtler. 1967. Periodic thymidine kinase production in synchronous plasmodia of *Physarum polycephalum:* inhibition by actinomycin and actidion. Biochem. Biophys. Res. Commun., 27:655–660.

——— and D. H. Ives. 1965. Periodische Änderungen der Thymidinkinase–Aktivität im synchronen Mitosecyclus von *Physarum polycephalum.* Biochem. Z., 343:399–406.

——— and H. P. Rusch. 1964. The effect of 5-fluoro-2′-deoxyuridine on synchronous mitosis in *Physarum polycephalum.* Exp. Cell Res., 36:124–133.

Schiebel, W., T. A. Chayka, A. DeVries, and H. P. Rusch. 1969. Decrease of protein synthesis and breakdown of polyribosomes by elevated temperature in *Physarum polycephalum.* Biochem. Biophys. Res. Commun., 35:338–345.

Schuster, F. L. 1965. Ultrastructure and morphogenesis of solitary stages of the true slime molds. Protistologica, 1:49–72.

Seaman, G. R., and R. K. Houlihan. 1951. Enzyme systems in *Tetrahymena geleii* S. II. Acetylcholinesterase activity. Its relation to motility of the organism and to coordinated ciliary action in general. J. Cell. Comp. Physiol., 37:309–321.

Seavey, D., P. Goldmark, and D. Kessler. 1967. Mitochondrial DNA synthesis in the slime mold *Physarum polycephalum.* J. Cell Biol., 35:187A.

Seifriz, W. E. 1936. Protoplasm, New York, McGraw-Hill Book Co.

——— 1937. A theory of protoplasmic streaming. Science, 86:397–398.

——— 1942. Some physical properties of protoplasm and their bearing on structure. *In* A Symposium on the Structure of Protoplasm (Monograph Amer. Soc. Plant Physiol.), Seifriz, W., ed., Ames, Iowa, The Iowa State College Press, pp. 245–264.

——— and M. Zetzmann. 1935. A slime mould pigment as indicator of acidity. Protoplasma, 23:175–179.

Stewart, P. A. 1964. The organization of movement in slime mold plasmodia. *In* Primitive Motile Systems in Cell Biology, Allen, R. D., and Kamiya, N., eds., New York, Academic Press, Inc., pp. 69–78.

——— and B. T. Stewart. 1959. Protoplasmic movement in slime mold plasmodia: the diffusion drag force hypothesis. Exp. Cell Res., 17:44–58.

——— and B. T. Stewart. 1961. Membrane formation during sclerotization of *Physarum polycephalum* plasmodia. Exp. Cell Res., 23:471–478.

Sullivan, A. J. 1953. Some aspects of the biochemistry of dormancy in the myxomycete *Physarum polycephalum.* Physiol. Plantarum, 6:804–815.

Szent-Györgyi, A. 1953. Chemical Physiology of Contraction in Body and Heart Muscle, New York, Academic Press, Inc., p. 43.

Takata, T., R. Nagai, and N. Kamiya. 1967. Movement of the myxomycete plasmodium. III. Artificial polarization in endoplasm distribution in a plasmodium and its bearing on protoplasmic streaming. Proc. Jap. Acad., 43:45–50.

Therrien, C. D. 1966. Microspectrophotometric measurement of nuclear deoxyribonucleic acid content in two myxomycetes. Canad. J. Bot., 44:1667–1675.

Ts'o, P. O. P., L. Eggman, and J. Vinograd. 1957. Physical and chemical studies of myxomyosin, an ATP-sensitive protein in cytoplasm. Biochim. Biophys. Acta, 25:532–542.

von Stosch, H. A. 1935. Untersuchungen über die Entwicklungsgeschichte der Myxomyceten. Sexualität und Apogamie bei Didymiaceen. Planta, 23:623–656.

——— 1965. Wachstums- und Entwicklungsphysiologie der Myxomyceten. *In* Handbuch der Pflanzenphysiologie, Ruhland, F., ed., New York, Springer-Verlag New York Inc., Vol. 15, pp. 641–679.

Ward, J. M. 1958. Shift of oxidases with morphogenesis in the slime mold, *Physarum polycephalum*. Science, 127:596.

Wilson, C. M., and I. K. Ross. 1955. Meiosis in the myxomycetes. Amer. J. Bot., 42:743–749.

Wohlfarth-Bottermann, K. E. 1962. Weitreichende, fibrilläre Protoplasmadifferenzierungen und ihre Bedeutung für die Protoplasmaströmung. I. Elektronenmikroskopischer Nachweis und Feinstruktur. Protoplasma, 54:514–539.

——— 1964. Differentiations of the ground cytoplasm and their significance for the generation of the motive force of ameboid movement. *In* Primitive Motile Systems in Cell Biology, Allen, R. D., and Kamiya, N., eds., New York, Academic Press, Inc., pp. 79–109.

Wolf, F. T. 1959. Chemical nature of the photoreceptor pigment inducing fruiting of plasmodia of *Physarum polycephalum*. *In* Photoperiodism and Related Phenomena in Plants and Animals, Withrow, R. B., ed., Washington, D. C., Amer. Ass. Advance. Sci., Publication No. 55, pp. 321–326.

Yamanaka, T. 1967. Cytochrome *c* and evolution. Nature (London), 213:1183–1186.

——— H. Nakajima, and K. Okunuki. 1962. Purification and some properties of a *c*-type cytochrome from a slime mould, *Physarum polycephalum*. Biochim. Biophys. Acta, 63:510–512.

Zeldin, M. H., and J. M. Ward. 1963a. Acrylamide electrophoresis and protein pattern during morphogenesis in a slime mould. Nature (London), 198:389–390.

——— and J. M. Ward. 1963b. Protein changes during photo-induced morphogenesis in *Physarum polycephalum*. Bact. Proc., p. 114.

Index

GPSR Compliance
The European Union's (EU) General Product Safety Regulation (GPSR) is a set of rules that requires consumer products to be safe and our obligations to ensure this.

If you have any concerns about our products, you can contact us on

ProductSafety@springernature.com

In case Publisher is established outside the EU, the EU authorized representative is:

Springer Nature Customer Service Center GmbH
Europaplatz 3
69115 Heidelberg, Germany

www.ingramcontent.com/pod-product-compliance
Lightning Source LLC
LaVergne TN
LVHW080845170826
845678LV00006B/1718

9781468484816